DELIVERY STRATEGIES *for* ANTISENSE OLIGONUCLEOTIDE THERAPEUTICS

DELIVERY STRATEGIES
for
ANTISENSE OLIGONUCLEOTIDE THERAPEUTICS

Edited by
Saghir Akhtar

CRC Press
Taylor & Francis Group
Boca Raton London New York

CRC Press is an imprint of the
Taylor & Francis Group, an **informa** business

PREFACE

Antisense nucleic acids (oligonucleotides) can act sequence specifically to modulate gene expression in living cells and thus are being considered for therapeutic application in diseases where the genetic target and its sequence have been identified. Indeed, the concept of antisense oligonucleotide therapeutics has rapidly developed from being a purely cell culture phenomenon to one that is now undergoing clinical trial evaluations for the treatment of leukemias and viral infections such as HIV, herpes simplex, and human papilloma virus.

A major issue in the development of antisense nucleic acids as potential therapeutic agents has been their effective delivery to target cells. The large molecular weights and often polar nature of oligonucleotides coupled with their sensitivity to nuclease degradation have posed new challenges for biopharmaceutical scientists in developing suitable drug delivery systems. With contributions from leading experts in the antisense field, it is the aim of this volume to highlight the major hurdles to effective delivery and then to discuss the state-of-the-art strategies currently being developed to improve the delivery and targeting of oligonucleotides to the desired sites of action in the body.

An overview of the concepts involved in antisense oligonucleotide therapeutics (which includes the antisense, antigene, and ribozyme strategies) can be acquired from the introductory chapters by A. Gewirtz and colleagues, N. Chaudhary et al., and by D. Elkins and J. Rossi in Section 1. The *in vivo* pharmacokinetic behavior of oligonucleotides will be an important consideration in the design of a suitable formulation or delivery system and this area is therefore covered in Section 2 with contributions from the groups of L. Neckers and V. Vlassov.

The important issue of improving the biological stability of nucleic acids with a view to improving delivery to target sites is covered in Section 3. This section includes a review by E. Wickstrom on the general strategies for modifying nucleic acid structure to improve nuclease resistance followed by chapters reviewing important developments in the use of self-stabilized oligonucleotides (Agrawal et al.), circular oligonucleotides (E. Kool), peptide nucleic acids (J. Hanvey and L. Babbiss), and stabilized RNA for antisense and ribozyme applications (F. Eckstein and colleagues).

Section 4 addresses the major issues of improving membrane transport, manipulating subcellular distribution, and improving delivery by cell-specific targeting of antisense oligonucleotide therapeutics. This section begins with chapters by A. Krieg and by R. Juliano's group reviewing the mechanisms by which oligonucleotide analogs enter living cells. A general opinion emerging from these (and other) studies is that uptake of oligonucleotides is reduced by some form of endocytosis with subsequent accumulation of these macromolecules into acidic endosomal/lysosomal compartments. This is followed by the eventual escape of oligonucleotides into the cytosol and/or efflux out of the cell by exocytosis. Strategies which may bypass endocytosis and thus improve the intracellular bioavailability of oligonucleotides include the use of cationic and other liposome formulations, and these are covered in chapters by F. Bennet and by A. Thierry and G.B. Takle. Alternative strategies aimed at enhancing endosomal exit of oligonucleotides that have entered cells by endocytosis include the use of pH-sensitive liposomes (C. Malvy and colleagues) and pH-sensitive viral fusion peptides, a strategy being developed for the delivery of DNA in gene therapy, but which has obvious implications in oligonucleotide therapy (E. Wagner and colleagues). The final two chapters highlight the progress being made in achieving cell-specific delivery of oligonucleotides by either forming antibody-oligonucleotide conjugates (C. Gooden and A. Epenetos) or by conjugating oligonucleotides to cell-specific ligands to promote receptor-mediated endocytosis (E. Carmichael et al.).

I hope that this volume, the first to focus solely on the delivery of these interesting macromolecules, will serve both to provide an overview of the current strategies being pursued in the effective delivery of antisense oligonucleotide therapeutics, and to stimulate new contributions to this exciting and rapidly developing field.

Saghir Akhtar
Birmingham, U.K.

CONTRIBUTORS

Sudhir Agrawal, Ph.D.
Senior Vice President
Department of Discovery
Hybridon, Inc.
Worcester, Massachusetts

Saghir Akhtar, Ph.D.
Lecturer
Pharmaceutical and Biological Sciences
Aston University
Birmingham, UK

Helle Aurup, Ph.D.
Professor
Max-Planck-Institute Für
    Experimental Medizin
Göttingen, Germany

Anna Avroutskaya
Research Technician
Department of Radiation Oncology
University of North Carolina at Chapel
    Hill
Chapel Hill, North Carolina

Lee E. Babiss
Department of Cell Biology
GLAXO
Research Triangle Park, North Carolina

C. Frank Bennett, Ph.D.
Director
Department of Molecular Pharmacology
ISIS Pharmaceuticals
Carlsbad, California

Jeffery S. Bishop
Department of Biology
Triplex Pharmaceutical Corporation
The Woodlands, Texas

Ellen P. Carmichael, Ph.D.
Staff Scientist
The Immune Response Corporation
Carlsbad, California

Nilabh Chaudhary, Ph.D.
Senior Research Scientist
Department of Biology
Triplex Pharmaceutical Corporation
The Woodlands, Texas

Christine Chavany, Ph.D.
Visiting Fellow
Clinical Pharmacology Branch
NIH
Bethesda, Maryland

Henry C. Chiou, Ph.D.
Staff Scientist
Department of Gene Therapy
The Immune Response Corporation
Carlsbad, California

P. Dan Cook
ISIS Pharmaceuticals, Inc.
Carlsbad, California

Patrick Couvreur, Ph.D.
Full Professor
URA CNRS 1218
University of Paris XI
Chatenay-Malabry, France

Fritz Eckstein, Ph.D.
Professor
Max-Planck-Institute Für
    Experimental Medizin
Göttingen, Germany

David A. Elkins, B.S.
Graduate Student
Center for Molecular Biology and Gene
    Therapy
Loma Linda University School of
    Medicine
Loma Linda, California

Agamemnon A. Epenetos, Ph.D.
Professor
Department of Clinical Oncology
Hammersmith Hospital
London, UK

Mark A. Findeis, Ph.D.
Staff Scientist
Department of Chemistry
Pharmaceutical Peptides, Inc.
Cambridge, Massachusetts

Daniel A. Geselowitz, Ph.D.
Chemist
Department of Nuclear Medicine
NIH
Bethesda, Maryland

Alan M. Gewirtz, M.D.
Associate Professor
Department of Pathology and Internal
  Medicine
University of Pennsylvania School of
  Medicine
Philadelphia, Pennsylvania

Calvin Stephen Roy Gooden, B.Sc.
Imperial Cancer Research Fund
  Oncology Unit
Hammersmith Hospital
London, UK

Charles J. Guinosso
ISIS Pharmaceuticals, Inc.
Carlsbad, California

Judith K. Guy-Caffey, Ph.D.
Research Scientist
Department of Biology
Triplex Pharmaceutical Corporation
The Woodlands, Texas

Jeffery C. Hanvey
Department of Cell Biology
GLAXO
Research Triangle Park, North Carolina

Olaf Heidenreich, Ph.D.
Professor
Max-Planck-Institute Für
  Experimental Medizin
Göttingen, Germany

Jeffery Hughes, Ph.D.
Assistant Professor
Department of Pharmaceutics
University of Florida
Gainesville, Florida

Krishna Jayaraman, Ph.D.
Director of Oligonucleotide Chemistry
Department of Biology
Triplex Pharmaceutical Corporation
The Woodlands, Texas

R.L. Juliano, Ph.D.
Professor and Chair
Department of Pharmacology
University of North Carolina at Chapel
  Hill
Chapel Hill, North Carolina

Valery Karamyshev, Ph.D.
Researcher
Institute of Bioorganic Chemistry
Novosibirsk, Russia

Eric T. Kool, Ph.D.
Assistant Professor
Department of Chemistry
University of Rochester
Rochester, New York

David Kregenow, B.S.
Medical Student
University of Pennsylvania School of
  Medicine
Philadelphia, Pennsylvania

Arthur M. Krieg, M.D.
Assistant Professor
Department of Internal Medicine
University of Iowa
Iowa City, Iowa
  and
Staff Physician
Department of Internal Medicine
Veterans Administration Medical Center
Iowa City, Iowa

Claude Malvy, Ph.D.
Director of Research
URA 147
CNRS
Villejuif, France

June Ray Merwin, Ph.D.
Department of Cell Biology
The West Company
Lionville, Pennsylvania

Marina Nechaeva
Researcher
Institute of Bioorganic Chemistry
Novosibirsk, Russia

Leonard M. Neckers, Ph.D.
Head, Tumor Cell Section
Clinical Pharmacology Branch
NIH
Bethesda, Maryland

Ludmila Pautova, Ph.D.
Senior Researcher
Institute of Bioorganic Chemistry
Novosibirsk, Russia

Mariusz Z. Ratajezak, Ph.D.
Senior Research Scientist
Department of Pathology
University of Pennsylvania School of
  Medicine
Philadelphia, Pennsylvania

Catherine Ropert, Ph.D.
Research Scientist
Department of Pharmaceutical Science
Rhone Roulene Rorer
Vitry Cedex, France

John J. Rossi, Ph.D.
Associate Director
Center for Molecular Biology and Gene
  Therapy
Loma Linda University School of
  Medicine
Loma Linda, California

Elena Rykova
Researcher
Institute of Bioorganic Chemistry
Novosibirsk, Russia

Henri M. Sasmor
ISIS Pharmaceuticals, Inc.
Carlsbad, California

George L. Spitalny, Ph.D.
Chief Scientific Officer
The Immune Response Corporation
Carlsbad, California

Garry B. Takle, Ph.D.
Senior Scientist
Innovir Labs, Inc.
New York, New York
  and
Adjunct Faculty
Department of Molecular Parasitology
Rockefeller University
New York, New York

Jinyan Tang, Ph.D.
Director
Department of Process Chemistry
Hybridon, Inc.
Worcester, Massachusetts

Jamal Temsamani, Ph.D.
Associate Director
Department of Discovery
Hybridon, Inc.
Worcester, Massachusetts

Alain R. Thierry, Ph.D.
Visiting Scientist
Lab Tumor Cell Biology
NCI/NIH
Bethesda, Maryland
  and
Adjunct Assistant Professor
Department of Radiation Medicine
Georgetown University
Washington, D.C.

Valentin V. Vlassov, Ph.D.
Professor
Institute of Bioorganic Chemistry
Novosibirsk, Russia

Ernst Wagner, Ph.D.
Group Leader
Institute of Molecular Pathology
Vienna, Austria
  and
Senior Scientist
Department of Therapeutic Vaccines
Bender and Co.
Vienna, Austria

Eric Wickstrom, Ph.D.
Professor
Department of Pharmacology
Thomas Jefferson University
Philadelphia, Pennsylvania

L.A.Yakubov, Ph.D.
Senior Researcher
Institute of Bioorganic Chemistry
Novosibirsk, Russia

TABLE OF CONTENTS

Chapter 1

Disrupting the Flow of Genetic Information with Antisense Oligodeoxynucleotides: Research and Therapeutic Applications

David Kregenow, Mariusz Z. Ratajczak, and Alan M. Gewirtz

CONTENTS

I. INTRODUCTION

The ability to efficiently disrupt gene expression has become an important experimental approach for studying the biological function of a gene at the single-cell level and in intact organisms.[1-3] This capability also possesses promising pharmaceutical applications because of the potential utility of downregulating the expression of genes which contribute to disease states. The downregulation of over- or aberrantly expressed proto-oncogenes or oncogenes in malignant cells, for example, could inhibit cell proliferation, force cells to undergo apoptosis, or lead to reversion of the malignant phenotype. Inhibition of intimal or smooth muscle cell proliferation might be of use in the treatment of cardiovascular disease.[4-6] Finally, inhibiting viral and parasitic genes responsible for replication may be useful in the treatment of many infectious diseases.[7-12] No doubt many other applications await elucidation.

Relevant technologies for disrupting gene expression include homologous recombination, which, in taking advantage of natural crossover events that occur during cell division actually destroys the targeted gene[2,3] and the use of reverse complementary DNA[7,13-15] or RNA[14,15] to interfere with utilization of the mRNA of the target gene. These are the so-called "antisense strategies". Each of these approaches for manipulating gene expression relies on specific nucleotide base pairing for targeting the gene of interest, but only the antisense approaches are applicable in fully developed organisms. Accordingly, it is the antisense strategies which have received the most attention and development with regard to therapeutic application.

The antisense mRNA strategy relies on the transfection and subsequent expression of a plasmid carrying the cDNA of the gene of interest subcloned into the vector in an antisense orientation.[14,15] After transfection into the cell, the plasmid expresses the antisense mRNA which, because of its complementarity, is capable of hybridizing

0-8493-4778-5/95/$0.00+$.50

1

exclusively with the mRNA of the gene of interest. The mechanism of downregulation of sense mRNA by antisense mRNA mimics some physiological regulatory mechanisms observed in primitive organisms, where both strands of genomic DNA are transcribed, producing sense and antisense mRNA molecules.[16] The sense-antisense mRNA hybrids are excluded from intracellular metabolism, and the translation of the mRNA is disturbed. Thus, gene expression is disrupted at the mRNA level. It is also possible to engineer into the antisense sequence a catalytic mRNA sequence which often increases the efficiency of mRNA destruction. Antisense RNA containing hammerhead or other catalytic sequences are called ribozymes.[17]

The antisense oligodeoxynucleotide (ODN) strategy relies on the introduction of short sequences of synthetic ODN into cells. Such ODN may be microinjected into cells or internalized by cells from the extracellular milieu. These ODN are able to hybridize within the cell to the appropriate DNA or mRNA sense sequence of the gene of interest.[1,7,13-15] Thus, DNA ODN can disrupt gene expression at the levels of transcription and translation. In addition, DNA ODN which show binding affinity to important intracellular proteins can be used as a means of selectively excluding these proteins from intracellular metabolism, thus downregulating gene expression at the protein level.[13] The relative ease of synthesis and use of antisense ODN predict that they would be the preferred form of molecule for both research and therapeutic purposes. Accordingly, these molecules are the subject of intense scrutiny and experimentation. This review will focus on recent developments in this rapidly expanding area.

II. MOLECULAR INTERACTIONS OF ANTISENSE OLIGOMERS

In order to more fully understand how antisense ODN may perturb the flow of genetic information and thereby disrupt gene expression, it is necessary to review the process through which genes are expressed in living cells. Briefly, genetic information is encoded by the nucleotide sequence of chromosomal DNA. When a gene is to be expressed, it is transcribed from DNA into the corresponding nucleotide sequence of a mRNA molecule within the nucleus. Subsequently, the mRNA is processed, leaves the nucleus, and enters the cytoplasm. Here it is translated on ribosomes into the amino acid sequence of a protein. It is this protein which is responsible for producing the biological effect of the gene. Synthetic ODN can influence gene expression at each level — from DNA to protein (Figure 1).

First, under certain conditions, ODN can bind to sense sequences in the chromosomal DNA and create a triple helix.[13,19] This directly inhibits the transcription of the genetic information encoded by the DNA into mRNA.[21,22] The formation of a triple helix structure requires, however, that the targeted sense sequence within the DNA molecule contain a pyrimidine- or purine-rich motif.[13,20] The binding of ODN to such a motif follows the Hogsten's binding principle where thymidine binds to adenine-thymidine pairs, creating a triplet (TAT) and protonated cytosine recognizes guanine-cytosine pairs, creating another triplet (C⁺GC). According to this principle, a purine-rich sequence on one strand of DNA such as AGAAAGGAGAAAAAGGGG will bind along the major grove of the double helix with synthetic ODN containing the sequence TC⁺TTTC⁺C⁺TC⁺TTTTTC⁺C⁺C⁺C⁺. The effectiveness of the antisense strategy to directly inhibit transcription is limited by the paucity of sufficiently long runs of pyrimidine- or purine-rich motifs in the genomic sequence of interest to allow the formation of stable triple helices. In spite of this limitation, the formation of a triple helix has already been shown to inhibit the expression of several genes, and a number of novel approaches to overcome this technical constraint have also been used with success. For example, the antisense ODN can be designed with molecular links which hybridize to separate motifs on opposite strands.[13] In addition, ODN conjugated with EDTA-Fe are able to generate

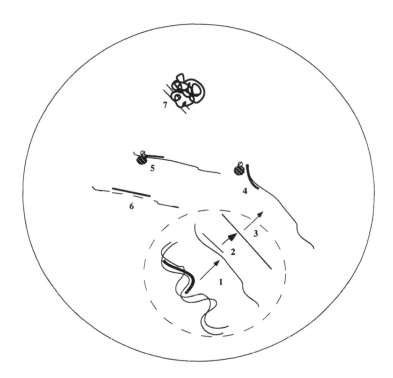

Figure 1 Potential sites at which ODN may disrupt gene expression. (1) Triple helix formation with chromosomal DNA disrupts transcription. (2) Hybridization to nuclear RNA disrupts RNA processing and/or (3) translocation to the cytoplasm. (4) Hybridization to mRNA disrupts the binding of ribosomes and/or (5) translation into protein. (6) Hybridization to mRNA provides a substrate for RNase H. (7) ODN interactions with proteins disrupt protein function.

free radicals and destroy the structure of a gene at the site of hybridization.[19] This strategy has been employed successfully to cleave a part of the human fourth chromosome.[23]

The second potential interaction of ODN relies on the fact that antisense ODN can hybridize according to Watson-Crick base pairing inside the cell to the corresponding sense sequence in mRNA molecules and prevent their translation into protein. Synthetically created antisense ODN are designed against specific sense sequences in the mRNA molecules of the gene of interest.[24-28] Antisense ODN may impair translation after hybridizing to its corresponding mRNA sense sequence through a number of mechanisms: (1) hybridization to pre-mRNA in the nucleus may impair subsequent splicing and processing to mRNA; (2) hybridization can impair the translocation of mRNA from the nucleus to the cytoplasm; (3) hybridization can prevent the binding of ribosomes to the mRNA molecule; (4) hybridization can impair the translation process itself; and (5) the hybridized antisense oligomer-mRNA duplex is a substrate for intracellular digestion by RNase H.

The third potential point of oligomer interference with gene expression, also shown in Figure 1, relies on the ability of ODN to bind to proteins important in cellular homeostasis and thereby exclude them from metabolism.[13] For example, this strategy has been shown to completely eliminate the actions of DNA regulatory proteins known as transcription factors.[29] The DNA regulatory proteins bind to short ODN containing the specific DNA sequence they recognize, leaving an insufficient number of protein molecules available to bind to the regulatory sequences in the gene promoter. The techniques for synthesizing ODN capable of binding to other biologically important proteins are still being refined, but many such ODN have already been synthesized. For instance, short DNA molecules, known in the literature as aptamers, have been created which show binding affinity to other intracellular proteins.[13] In general, the antisense strategy has been applied successfully to disrupt the expression of genes coding proteins such as growth factors, cellular receptors, adhesion molecules, second messengers, nuclear factors, enzymes, and structural proteins.[1]

III. BIOCHEMISTRY OF ANTISENSE OLIGOMERS

Antisense ODN, as indicated above, are short stretches of synthetic DNA. They are typically 10 to 30 bases long. Sequences of this length satisfy the requirement for sequence uniqueness,[31] usually assured if the ODN is at least 15 bases in length, and are likely to be taken up by cells. Shorter sequences are of course easier and less costly to synthesize, but lack sequence specificity, while longer sequences are more expensive to produce, may not be taken up by cells as easily, and also suffer from specificity breakdown due to the looping out of nonhybridizing sequence.

From a pharmaceutical point of view, the ideal antisense drug should be stable in the intracellular and extracellular melieu, be able to cross cell membranes, demonstrate hybridization specificity, and have little in the way of nonsequence-related toxicity.[1,30] The last two of these properties are a function of the antisense ODN sequences themselves and their ability to interact specifically with intracellular molecules. The first two properties are functions of the biochemical properties of the components of the ODN and thus subject to biochemical modification.

The development of phosphoramidite chemistry and its subsequent automation has enabled the large-scale synthesis of ODN with several advantageous properties. It is well known that unmodified phosphodiester ODN are readily digested by 5′ or 3′ exonucleases as well as endonucleases.[1,30,31] The presence of these degrading enzymes in biological fluids such as cytoplasmic and nuclear extracts restricts the application of unmodified ODN, especially for use *in vivo*. There are multiple strategies in use to modify ODN in order to increase their resistance to degradation by nucleases.[1,30,31] The first generation of modified ODN include modification of the phosphodiester bridge between nucleotides by creating methylphosphonate or phosphorothioate derivatives (Figure 2), changing the glycosidic linkage from the β to the α anomeric form, and capping or altering the 3′ and 5′ ends of the oligomer molecule. There have also been attempts to use phosphotriester, phosphorodithioate, and phosphoroselenite derivatives. Second-generation modifications include the synthesis of phosphorothioate-phosphodiester co-polymers which possess some of the properties of both unmodified and modified ODN.[32] Such derivatives are, for example, more stable in biological fluids than unmodified phosphorodiesters and hybridize to target sequences better than modified phosphorothioates. Another interesting modification is the peptide-nucleic acid analog (PNA) in which the entire deoxyribose-phosphate backbone has been exchanged with a chemically different, but structurally homomorphous, polyamide (peptide) backbone composed of (2-aminoethyl) glycine units.[33,34] Such derivatives can bind to single-stranded genomic DNA by Watson-Crick base pairing after displacing the other DNA strand, and to double-stranded genomic DNA by forming a triple helix in the major groove following the rules of Hogsten base pairing. Unfortunately, these compounds do not have the ability, at least as presently synthesized, to enter living cells. The creation of oligomer derivatives resistant to nucleases enables their utilization in *in vivo* trials. These modifications, however, change some of the biochemical properties of ODN. Some of the properties of unmodified phosphodiester and modified methylphosphonate and phosphorothioate ODN are shown in Table 1.

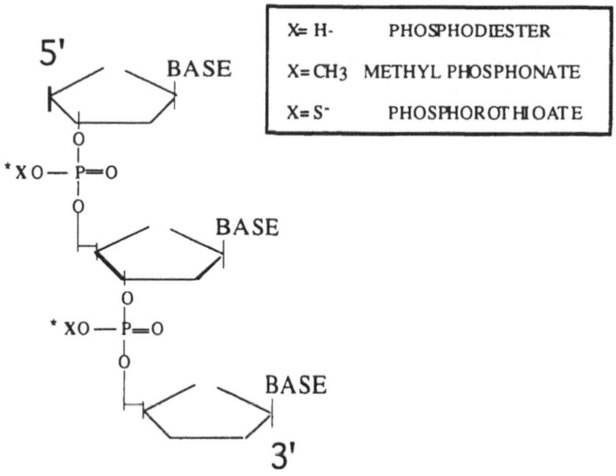

5'

BASE

X= H-	PHOSPHODIESTER
X= CH3	METHYL PHOSPHONATE
X= S-	PHOSPHOROTHIOATE

Figure 2 Commonly employed first-generation modifications of the phosphodiester backbone designed to promote stability of the molecule without sacrificing the ability to enter cells and hybridized with message. Note that methylphosphonates are quite adept at crossing cell membranes because they are hydrophobic. Nevertheless, they are poorly soluble and are not permissive of RNase H binding. Phosphorothioates do not cross membranes easily, but are soluble and do allow the binding of RNAse H.

Table 1 **Properties of normal and modified oligodeoxynucleotides**

Property	Phosphodiester	Methylphosphonate	Phosphorothioate
Biological stability	–	+++	++
Cellular uptake	+	++	+/–
Hybridization efficacy	+++	+	++
RNase H activation	+++	–	+++

Note: – Poor; + Fair; ++ Good; +++ Excellent

ODN, with the exception of methylphosphonate derivatives, are polyanionic. For this reason they do not easily cross cell membranes.[1,30,31] In this regard they are similar to most antineoplastic agents which are of low molecular weight, hydrophilic, and incapable of passively diffusing across cell membranes. The uptake of ODN into cells was found to be dependent upon time, concentration, energy, and temperature. These are all character-istics of an active process.[37] These findings led to the discovery that ODN are taken up by cells by two active processes: fluid phase endocytosis and receptor-mediated endocy-tosis. Several receptor proteins responsible for uptake of ODN have been described. Once across the membrane, their intracellular distribution is also somewhat controversial. Fluorescent-labeled ODN microinjected into the cytoplasm of cells are found to rapidly accumulate in the nucleus.[35] However, when fluorescent-labeled ODN are placed in tissue culture media, the labeled molecules accumulate in vacuoles, presumably endosomes and lysosomes, within the cytoplasm forming a punctate perinuclear pattern.[36] In contrast to ODN localization when the compounds are introduced by microinjection, there is no visible fluorescence in the nucleus itself, indicating that the release of ODN from these vacuoles is an inefficient process. Methylphosphonate derivatives, on the other hand, are uncharged molecules that have been reported to enter cells via passive diffusion.[39]

Since ODN uptake (save for methylphosphonates) is inefficient, several different strategies have been developed to augment this process on the theory that increased uptake will translate into increased efficacy. ODN have been conjugated with synthetic polypeptides such as poly-L-lysine, or with acridine, cholesterol, poly-rA, and transferrin.[37,40-43] Liposomes have also been utilized as oligomer delivery systems.[40] Liposomes containing ODN can additionally be targeted to reach cell surface determinants with appropriate monoclonal antibodies.[43] The poly-L-lysine modification may serve to mask the negative charge of the ODN and are hypothesized to destabilize endosomal membranes, thus permitting their escape from the endosomal compartment. Conjugation with cholesterol permits the binding of ODN to apolipoprotein E and to low-density lipoprotein (LDL) which allows internalization of the ODN via the LDL receptor.[44] It has also been found that synthetic cationic lipids such as N-[1(-2,3-dioleyloxy)propyl]-n_1n_1n-trimethylammonium chloride (DOTMA) (Lipofectin®) and some antifungal drugs (amphotericin B) increase oligomer uptake by cells.[45] Lipofectin® not only increases the cellular uptake of ODN, but also increases the accumulation of ODN in the nucleus. The results of oligomer uptake in the presence of Lipofectin® resemble those of microinjection. The list of compounds which increase oligomer uptake is still expanding, and many laboratories are continuing to search for new delivery systems.

It seems certain that efficacy of antisense ODN will depend upon many factors besides their ability to enter cells. Accordingly, there have been many attempts to increase the effectiveness of ODN once inside the cell. For instance, it is widely believed that the efficacy of an antisense ODN can be dependent upon the secondary and tertiary structures of the targeted mRNA molecule.[46] If the target sequence is buried within the mRNA molecule, it may not be accessible for binding to the antisense ODN. Thus it is generally believed that open loop regions in mRNA, rather than sequences likely to be involved in intramolecular secondary structure, will be more amenable to targeting by the antisense strategy.[48] One of the more interesting strategies used to increase the efficacy of antisense ODN involves the synthesis of ODN containing RNA degrading enzymes known as ribozymes.[17,31] In this case, antisense ODN bind the targeted sense sequence in a mRNA molecule and the ribozyme subsequently cleaves it. Another factor affecting antisense oligomer efficacy is the base sequence of the antisense oligomer itself. An oligomer may have a substantial amount of either external or internal self-complementarity. This must be avoided in order to prevent bimolecular self-association or intramolecular hairpin formation, respectively.[47] In addition, the position of the optimal target sequence in the mRNA molecule may vary from one gene to another. Encouraging results have been obtained by targeting cap sites, translation initiation sites, intron-exon splice junction sites, and even sequences in the 3′ untranslated region of the mRNA molecule. The strategy may differ depending on the gene of interest and often requires targeting of different sequences in order to find the most effective antisense oligomer. Nevertheless, the complexity of the events surrounding the ability of the ODN to find and then interact with its target mRNA are enormous, as illustrated in Figure 3.

IV. PHARMACODYNAMICS OF ANTISENSE ODN

The first pharmacokinetic studies concerning the distribution of ODN were performed in mice. After intravenous injection of phosphorothioated ODN it was found that ODN of different compositions and lengths ranging from 20 to 27 bases exhibited similar plasma kinetics. They demonstrated half-lives of 40 to 72 h, and little metabolic degradation was observed in the ODN excreted in the urine.[49] The ODN accumulated in the kidney and liver at concentrations which exceeded plasma levels. In mice, the brain apparently is a privileged site into which very little ODN accumulates.[49]

In monkeys, the plasma half-life of ODN is similar to that of rodents. The organs showing the highest accumulation of ODN were kidney, liver, thymus, bone marrow, lymph nodes, salivary glands, lungs, and pancreas. Moderate accumulation was found in muscle, gastrointestinal tract, and trachea. The lowest accumulations were found in brain, spinal cord, cartilage, skin, and prostate.[49,50]

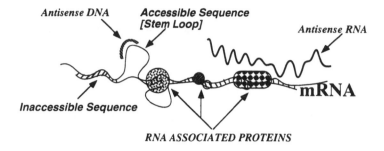

Figure 3 Interactions of antisense DNA and RNA with target mRNA. The ability of an antisense molecule to exert its biological effect is governed by its ability to interact with the targeted mRNA sequence. This is turn is a function of accessibility of the target sequence which is determined (at least) by the secondary structure and associated proteins of the mRNA. Only "uncovered", single-stranded mRNA is accessible. Since antisense RNA molecules are much longer than oligodeoxynucleotides, the ability of these RNA molecules to interact with their targets is theoretically even more constrained. RNA unwindases, which relax mRNA secondary structure, may be of help in this regard.

As with any new pharmaceutical agent, the potential side effects of antisense ODN are important to both consider and determine. ODN, depending on their target, may disturb the expression of genes not only in neoplastic cells, but also in normal, healthy cells. This raises the issue of differential sensitivity between normal and diseased cells. It is imperative that abnormal cells be significantly more sensitive to the disruption of the function of a gene compared to normal cells so that it is feasible to select a dose at which abnormal cells cannot function properly, but normal cells can tolerate. We might also expect some nonspecific side effects of infused ODN which are not related to their ability to interact specifically with nucleotides or proteins. For example, one potential side effect of treatment with phosphorothioated ODN may be thrombocytopenia. Mice receiving high doses of phosphorothioated ODN display a decrease in platelet counts[57]. It is hypothesized that the thrombocytopenia is related to the polyanionic nature of the ODN. Again, it will be necessary to determine if there is a dose range which achieves sufficient disruption of a targeted gene in abnormal cells without undue systemic toxicity.

In our experience, mice treated with phosphorothioated human c-myb antisense ODN (which are unable to target murine c-myb mRNA because of sequence specificity) at doses ten times higher than those found to be effective in the inhibition of *in vivo* human leukemia cell growth do not show any morphological changes in liver or kidney.[51] Nevertheless, we expect that increasing the dose of ODN increases the likelihood of organ damage. Therefore, every oligomer needs careful evaluation *in vivo* in an animal model before clinical use. In the first reported case where phosphorothioated ODN were infused into a patient with ALL, only slight increases in γ-Glutamyl transpeptidase (γGTP) values, transient metallic taste, nausea, and vomiting were described. This patient was treated with a continuous infusion of p53 phosphorothioated ODN at a dose 0.05 mg/kg/h (total dose 700 mg).[50]

V. THE TROUBLE WITH ANTISENSE

Though seemingly simple in theory, antisense experiments have been notoriously difficult to carry out in some cases, and have at times also been difficult to reproduce.[7,30] The reasons why these difficulties arise have been alluded to above and may be placed into three major categories. First, problems with reproducibility can be explained by many factors, including differences in oligomer synthesis, handling, storage, and purification.

Differences in experimental conditions and sensitivity between subtypes of cell lines may also be factors. An additional important variable, as with any technique, is the experience of the research team using this strategy. Second, problems related to the physical ODN-target mRNA interactions are also difficult to predict. Cellular mRNAs are thought to have highly ordered secondary and tertiary structures. As mentioned above (and shown in Figure 3), if the targeted sequence is buried within the core of a structurally complicated mRNA molecule, or rendered inaccessible by an associated protein, then hybridization of the ODN and mRNA may not be possible.[46] Generally, it is difficult to predict the secondary structure of mRNA molecules, and accessible regions must be determined empirically by synthesizing ODN to multiple sites and selecting those which are most effective.

An additional important problem for which the antisense strategy is criticized focuses on effects created by nonspecific interactions of the ODN. Recent papers indicate that some of the observed activities of ODN in tissue cultures may be produced predominantly through nonantisense mechanisms. For example, it was reported that four consecutive guanines ("G-Box") within a larger phosphorothioated oligomer have an antiproliferative effect which is not due to an antisense mechanism.[52] The products of oligomer degradation such as 2'-deoxyadenosine or high concentrations of thymidine are toxic to cells and inhibit cell proliferation.[53,54] Micromolar concentrations of dAMP or dGMP can have pronounced cytotoxic or cytostatic effects on cells, particularly those derived from hematopoietic tissues. In addition, phosphorothioated analogs may bind to ribosomes, resulting in nonspecific inhibition of protein synthesis.[55] Hybridization through partially matched sequences has also been raised as a problem.[56] The latter seems a less important consideration, since it has been shown that as few as two base mismatches result in loss of antisense effectiveness.[57] Further, the experimental system employed to demonstrate this effect was the frog oocyte, which is typically kept at a lower, more permissive temperature than mammalian cells in culture.

Finally, it is important to note that nonspecific interactions may prove to be advantageous. There are some data that phosphorothioated antisense ODN designed to act against viral mRNAs in order to disrupt viral proliferation may also act nonspecifically as viral polymerase inhibitors or may nonspecifically inhibit viral absorption onto cell membranes. These actions are thought to be the result of interactions between the phosphorothioated backbone and cell surface or viral proteins.[8]

VI. POTENTIAL APPLICATIONS OF THE ANTISENSE STRATEGY IN CLINICAL MEDICINE

There are many potential clinical situations where antisense ODN may be useful therapeutically. The most advanced work accumulated on the use of these compounds has been performed in the areas of clinical oncology, viral and parasitic diseases, and cardiovascular disease.

The targets for antisense oligomer treatment were first identified in leukemias.[58] Antisense ODN may ultimately be applied in both *ex vivo* bone marrow purging[25-27] and *in vivo* treatments.[28] They may be able to be used alone or in combination with cytostatic agents, radiotherapy, or other antileukemic therapies such as cytotoxic monoclonal antibodies, activated Natural Killer (NK) cells, or positive selection of healthy hematopoietic stem cells.[28,59]

There have been several attempts to apply the antisense strategy to the treatment of viral diseases.[7-12] There are encouraging data concerning the use of these compounds to inhibit the replication of herpes simplex virus type I, human immunodeficiency virus, vesicular stomatitis virus, influenza virus and papilloma virus. Antisense ODN have also been used as an experimental treatment of some parasitic disorders, such as malaria.

It has also been reported that antisense ODN against c-myb and proliferative cell nuclear antigen (PCNA) mRNA inhibits smooth muscle proliferation both *in vitro* and *in vivo*.[4,5] This strategy may prove useful in the prevention of restenosis after coronary angioplasty.

The list of diseases in which the antisense strategy may be applied clinically is growing rapidly along with further developments of the strategy itself. There is much optimism as the first trials using these compounds therapeutically in animal models is currently in progress.[4,28,60,61,62]

VII. PERSPECTIVES ON THE THERAPEUTIC APPLICATIONS OF ANTISENSE DNA OLIGOMERS IN THE TREATMENT OF HUMAN LEUKEMIAS — AN EXAMPLE OF THE UTILIZATION OF THIS STRATEGY IN CLINICAL MEDICINE

Disrupting the expression of genes thought to play a role in initiating and/or maintaining a malignant phenotype is an exciting and challenging problem. As noted above, however, there is now a large body of evidence to suggest that this will be feasible. Antisense ODN may first find practical application as therapeutic agents in the treatment of hematological disorders.[25-28,58,59] Thus, these diseases have been actively scrutinized for potential antisense gene targets. Save for neogenes produced directly by chromosomal translocations (bcr-abl, E2A/PBX, DEC/CAN, PML/RARα), all proto-oncogenes thought to play a role in malignant cell transformation are also expressed to some degree in normal cells. Demonstrating differential sensitivity to loss of the function of the target gene therefore becomes critical.

In our laboratory we have successfully disrupted the expression of two important proto-oncogenes involved in the proliferation of hematopoietic cells. The first is the c-kit proto-oncogene which encodes a tyrosine kinase receptor.[27] The second proto-oncogene is c-myb, which encodes a DNA-binding transcription regulatory protein.[24-26] We were interested in determining what role each proto-oncogene plays in the proliferation of normal and malignant cells and, more importantly, whether or not there is differential sensitivity to the abrogation of expression of these proto-oncogenes between normal and malignant progenitors.

First, normal human bone marrow progenitors were exposed to antisense ODN targeted against c-myb or c-kit proto-oncogenes and the proliferative capacity was assayed in *in vitro* cultures. The results indicated that the expression of c-myb is crucial for the proliferation of normal human hematopoietic progenitors.[58] The downregulation of c-myb with antisense ODN inhibited the proliferation of hematopoietic progenitors from all cell lineages: granulocyto-monocytic, erythroid, and megakaryocytic.[63] In addition, further studies indicated that c-myb is also crucial for proliferation of human lymphocytes at the G1/S transition of the cell cycle.[24]

In contrast, the effect of downregulating c-kit was more lineage restricted.[27] We found that in cultures of human hematopoietic bone marrow progenitors, the erythroid colony-forming cells (both colony forming unit-erythroid [CFU-E] and burst forming unit-erythroid [BFU-E]) and colony-forming cells of mixed lineages (CFU-Mix) were more sensitive to disruption of c-kit expression than granulocyto-monocytic and megakaryocytic progenitors. We have observed the inhibition of the CFU-Mix, BFU-E, and CFU-E colony formation even when exogenous c-kit receptor ligand, known as kit ligand (KL) or stem cell factor (SCF), was not included in the culture medium. In contrast, the inhibition of CFU-granular macrophage (GM) and CFU-Megakaryocyte (MG) colony formation was observed only when the bone marrow progenitors were stimulated with media containing multiple growth factors including KL.

Next, we turned our attention towards understanding the role both proto-oncogenes play in the proliferation of malignant human hematopoietic progenitors.[25-27,58] As mentioned, we were particularly interested in examining the existence of a differential sensitivity to the toxic effects of c-myb and c-kit antisense ODN between normal and malignant progenitors. If this were the case, these ODN could have therapeutic applications as bone marrow purging agents or as systemic anticancer drugs.

First we demonstrated that c-myb antisense ODN at doses which fail to inhibit the growth of normal human hematopoietic progenitors were able to inhibit the growth of malignant hematopoietic progenitors. We found this to be true in progenitors of both acute and chronic myelogenous leukemias.[25,26] The latter disease has been of particular utility for study because the cells of the maligant clone express the tumor-specific bcr/abl gene. Using this marker, it has been learned that normal progenitor cells exist in CML marrow for some time during the chronic and, likely, accelerated stages of disease. For example, the reappearance of normal hematopoiesis has occurred in CML patients after autologous bone marrow transplants, intensive chemotherapy treatment, and interferonα therapy. Such reappearance has also been reported in the long-term bone marrow cultures of CML bone marrow cells. The advantage in survival rates of malignant CML progenitors, as opposed to the survival rates of nonmalignant CML progenitors, might depend on the bcr-abl gene product which inhibits the process of apoptosis in the malignant cells. The c-myc and c-myb proto-oncogenes might play a direct role in the regulation of bcr-abl gene expression. Therefore, influencing bcr-abl gene expression and apoptotis might provide a theoretical explanation of the success of disrupting expression of the c-myb gene.

As shown below, we exposed CML blood or marrow cells derived from 11 patients to c-myb sense, antisense, and scrambled ODN.[26] We were able to inhibit CFU-GM colony formation in 8 out of the 11 patients. This inhibition was antisense sequence specific, dose dependent, and ranged between 58 and 93%. The inhibition was statistically significant in seven of the eight cases. Of particular importance is the fact that four out of seven patients with chronic phase disease reacted positively to this treatment, suggesting that the antisense DNA might be effective in controlling the disease in such patients. To determine whether CML CFU had been reduced or even eliminated after exposure to the antisense ODN, we examined cells in the residual colonies for bcr-abl mRNA expression. In each case where antisense myb inhibited growth, bcr-abl expression (as detected by reverse transcription-polymerase chain reaction [RT-PCR]) was either greatly decreased or nondetectable. Moreover, the DNA extracted from the CML cells that had been exposed to c-myb antisense ODN showed laddering characteristic of apoptosis.

We have also performed experiments using c-kit ODN to deplete the c-kit receptor in pathological hematopoietic cells. As is shown in Table 2, c-kit ODN were able to inhibit CFU-GM colony formation in two out of seven acute leukemia patients, five out of ten chronic leukemia patients, and 100% of the polycythemia vera (PV) cases tested. In normal hematopoietic cell cultures not stimulated with exogenous KL, only erythroid colony formation, as opposed to granulocyto-monocytic colony formation, was inhibited by the c-kit antisense ODN.

The data presented above suggest that unmodified antisense DNA ODN may prove useful in *ex vivo* bone marrow purging.[58] Whether appropriately modified antisense ODN would have *in vivo* use as a systemic antileukemia treatment is also clearly of interest.[28] To address this issue we created an animal model of human leukemias — we transplanted a bcr-abl human CML leukemia line, K562, into severe combined immune deficient (SCID) mice ("SCID mice biological model"). These cells have been shown to express the c-myb proto-oncogene and are sensitive to c-myb antisense ODN in *in vitro* studies. After transplantation, the mice were monitored for the presence of circulating blast cells in their peripheral blood. These circulating blast cells were usually demonstrable within 5 to 6 weeks after transplantation. Neurologic abnormalities were a prominent part of the

Table 2   Effect of c-kit antisense ODN on CFU-GM colony formation in leukemia patients

Disease type	No. pts. studied	No. pts. with decrease in colonies	% Decrease in colony number
Acute lymphocytic leukemia (ALL)	3	1	68%
Acute myelogenous leukemia (AML)	4	1	63%
Chronic myelogenous leukemia (CML)	10	5	65%
Polycythemia vera (PV)	4	4	75%
Total	21	11	Avg = 68%

clinical picture. The involvement of the central nervous system with meningeal infiltration and packing of the subarachnoidal space by hematopoietic human blasts was found in 100% of the transplanted mice. In addition to the brain, a second organ, the ovary, was consistently involved by the leukemic process in this model.

We employed phosphorothioated, nuclease-resistant ODN to examine the *in vivo* efficacy of antisense ODN. When blast cells appeared in the peripheral blood, we treated the mice with phosphorothioated ODN specific for codons two through nine of the human c-myb proto-oncogene sequence. The ODN were delivered by continuous infusion for 7 or 14 d using subcutaneously implanted minipumps. After 7 d a statistically significant survival advantage was noted in mice receiving antisense, but not sense, ODN. Repeating these experiments with more mice and a longer infusion confirmed the efficacy of the antisense ODN for prolonging survival of the mice. In fact, on day 16, when all of the various control mice had died, seven out of ten antisense-treated mice were still alive. In aggregate, mice treated with antisense ODN survived 3.5 to 8 times longer than those that were untreated or those receiving control ODN. In addition, autopsies on animals receiving c-myb antisense ODN showed significantly less disease in the central nervous system and ovaries.

VIII.   CONCLUSIONS

It is apparent from the work presented here that much optimization will be needed before antisense DNA can be used as an effective, single-agent therapy for treating human cancers. In this regard, our studies with c-myb and c-kit serve as a paradigm for such treatment strategies. We freely acknowledge, however, that other molecular targets, depending on the disease in question, may be of greater utility. Optimization of the approach will also require greater knowledge of the antisense DNA's mechanism of cell killing. Increasing DNA uptake may increase the cytostatic and cytocidal efficiency of these compounds. If uptake is primarily by pinocytosis, then increasing extracellular DNA concentration and exposure time may be all that is required. Our results with repeated DNA infusions[51] support this notion and our toxicity data suggest that this will be feasible. Nevertheless, antisense DNA-mediated cytotoxicity may also be dependent on the trafficking of the compounds once they enter the cell. Translocating phosphorothioate ODN from pinocytotic vesicles to the endoplasmic reticulum where they presumably associate with ribosomes may be a crucial problem to solve. Finally, cellular "defense" mechanism, such as increasing transcription of the targeted message, may also be factors to consider in planning effective treatment strategies with these agents. Accordingly, while this area remains in its scientific infancy, these *in vivo* studies and those of our colleagues[4,5,28,49,50,56,60-62] convince us that modulation of gene expression with antisense DNA is a therapeutic strategy worth pursuing.

REFERENCES

1. Stein, C.A., Cheng, Y.C. (1993). Antisense oligodeoxynucleotides as therapeutic agents — is the bullet really magical? *Science* 261: 1004.
2. Yamamura, K., Wakasugi, S. (1991). Manipulating the mouse genome: new approaches for the dissection of mouse development. Dev. Growth Differ. 32: 93.
3. Capecchi, M. (1989). The new mouse genetics: altering the genome by gene targeting. *Trends Genet.* 5: 70.
4. Simons, M., Edelman, E.R., DeKeyser, J.L., Langer, R., Rosenberg, R.D. (1992). Antisense c-myb oligonucleotides inhibit intimal arterial smooth muscle cell accumulation *in vivo*. *Nature* 359: 67.
5. Speir, E., Epstein, S.E. (1992). Inhibition of smooth muscle cell proliferation by an antisense oligodeoxynucleotide targeting the messenger RNA encoding proliferating cell nuclear antigen. *Circulation* 86: 538.
6. Morishita, R., Gibbsons, G.H., Ellison, K.E., Nakajima, M., Zhang, L., Kaneda, Y., Ogihara, T., Dzau, V.J. (1993). Single intraluminal delivery of antisense cdc2 kinase and proliferating cell nuclear antigen oligonucleotides results in chronic inhibition of neointimal hyperplasia. *Proc. Natl. Acad. Sci. U.S.A.* 90: 8474.
7. Milligan, J.F., Matteucci, M.D., Martin, J.C. (1993). Current concepts in antisense drug design. *J. Med. Chem.* 36: 1923.
8. Gao, W.Y., Hanes, R.N., Vazquez-Padua, M.A., Stein, C.A. (1990). Inhibition of herpes simplex virus type 2 growth by phosphorothioate oligodeoxynucleotides. *Antimicrob. Agents Chemother.* 34: 806.
9. Smith, C.C., Aurelian, L., Reddy, M., Miller, P.S., Howley, P.M. (1986). Antiviral effect of an oligo (nucleotide methylophosphonate) complementary to the splice-junction of herpes simplex virus type 1 immediate early pre-mRNAs 4 and 5. *Biochemistry* 83: 2787.
10. Agrawal, S., Tang, J.Y. (1992). GEM91 — an antisense oligodeoxynucleotide phosphorothioate as a specific therapeutic agent for AIDS. *Antisense Res. Dev.* 2: 261.
11. Agris, C.H., Blake, K.R., Miller, P.S., Reddy, P.M., Ts'o POP (1986). Inhibition of vesicular stomatitis virus protein synthesis and infection by sequence-specific oligodeoxyribonucleoside methylophosphonates. *Biochemistry* 25: 6268.
12. Cowsert, L.M., Fox, M.C., Zon, G., Mirabelli, C.K. (1993). *In vitro* evaluation of phosphorothioate oligonucleotides targeted to the E2 mRNA of papillomavirus: potential treatment for genital warts. *Antimicrob. Agents Chemother.* 37: 171.
13. Riordan, M.L., Martin, J.C. (1991). Oligonucleotide-based therapeutics. *Nature* 350: 442.
14. Colman, A. (1990). Antisense strategies in cell and developmental biology. *J. Cell Sci.* 97: 399.
15. van der Krol, A.R., Mol, J.N.M., Stuitje, A.R. (1988). Modulation of eukaryotic gene expression by complementary RNA or DNA sequences. *Biotechniques* 6: 958.
16. Green, P.J., Pines, O., Inouye, M. (1986). The regulation of antisense RNA in gene regulation. *Annu. Rev. Biochem.* 55: 569.
17. Snyder, D.S., Wu, Y., Wang, J.L., Rossi, J.J., Swiderski, P., Kaplan, B.E., Froman, S.J. (1993). Ribozyme mediated inhibition of bcr-abl gene expression in a Philadelphia chromosome-positive cell line. *Blood* 82: 600.
18. Hijiya, N., Gewirtz, A. (1992). Oncogenes, molecular medicine, and bone marrow transplantation. *J. Hematother.* 1: 369.
19. Moser, H.E., Dervan, P.B. (1987). Sequence-specific cleavage of double helical DNA by triple helix formation. *Science* 234: 645.
20. Maher, J.L., Wold, B., Dervan, P.B. (1989). Inhibition of DNA binding proteins by oligonucleotide-directed triple helix formation. *Science* 245: 725.

21. Cooney, M., Czernuszewicz, G., Postel, E.H., Flint, S.J., Hogan, M.E. (1988). Site-specific oligonucleotide binding represses transcription of the human c-myc gene *in vitro*. *Science* 241: 456.

22. Moffat, A.S. (1991). Triplex DNA finally comes of age. *Science* 252: 1374.

23. Strobel, S.A., Doucette-Stamm, L.A., Riba, L., Housman, D.E., Dervan, P.B. (1991). Site-specific cleavage of human chromosome 4 mediated by triple-helix formation. *Science* 254: 1639.

24. Gewirtz, A.M., Anfossi, G., Venturelli, D., Valpreda, S., Sims, R., Calabretta, B. (1989). G_1/S transition in normal human T-lymphocytes requires the nuclear protein encoded by c-myb. *Science* 245: 180.

25. Calabretta, B., Sims, R.B., Valtieri, M., Caracciolo, D., Szczylik, C., Venturelli, D., Ratajczak, M., Beran, M., Gewirtz, A.M. (1991). Normal and leukemic hematopoietic cells manifest differential sensitivity to inhibitory effects of c-myb antisense oligodeoxynucleotides: an *in vitro* study relevant to bone marrow purging. *Proc. Natl. Acad. Sci. U.S.A.* 88: 2351.

26. Ratajczak, M.Z., Hijiya, N., Catani, L., DeRiel, K., Luger, S.M., McGlave, P., Gewirtz, A.M. (1992). Acute- and chronic-phase chronic myelogenous leukemia colony-forming units are highly sensitive to the growth inhibitory effects of c-myb antisense oligodeoxynucleotides. *Blood* 79: 1956.

27. Ratajczak, M.Z., Luger, S.M., DeRiel, S.M., Abrahm, J., Calabretta, B., Gewirtz, A.M. (1992). Role of the KIT protooncogene in normal and malignant human hematopoiesis. *Proc. Natl. Acad. Sci. U.S.A.* 89: 1710.

28. Ratajczak, M.Z., Kant, J.A., Luger, S.M., Hijiya, N., Zhang, J., Zon, G., Gewirtz, A.M. (1993). In vivo treatment of human leukemia in a scid mouse model with c-myb antisense oligodeoxynucleotides. *Proc. Natl. Acad. Sci. U.S.A.* 89: 11823.

29. Bielinska, A., Shivdasani, R.A., Zhang, L., Nabel, G.J. (1990). Regulation of gene expression with double-stranded phosphorothioate oligodeoxynucleotides. *Science* 250: 997.

30. Tidd, D.M. (1990). A potential role for antisense oligodeoxynucleotide analogues in the development of oncogene targeted cancer chemotherapy. *Anticancer Res.* 10: 1169.

31. Uhlmann, E., Peyman, A. (1990). Antisense oligonucleotides: a new therapeutic principle. *Chem. Rev.* 90: 543.

32. Ghosh, M.K., Ghosh, K., Cohen, J.S. (1993). Phosphorotioate-phosphodiester oligonucleotide co-polymers: assessment for antisense application. *Anti-Cancer Drug Des.* 8: 15.

33. Frank-Kamenetskij, M. (1991). A change of backbone. *Nature* 354: 505.

34. Nielsen, P.E., Egholm, M., Berg, R.H., Buchargt, O. (1993). Peptide nucleic acids (PNAs): potential anti-sense and anti-gene agents. *Anti-Cancer Drug Des.* 8: 53.

35. Leonetti, J.P., Mechti, N., Degols, G., Gagnor, C., Lebleu, B. (1991). Intracellular distribution of microinjected antisense-oligonucleotides. *Proc. Natl. Acad. Sci. U.S.A.* 88: 2702.

36. Cerruzi, M., Draper, K., Schwartz, J. (1990). Natural and phosphorothioate-modified oligodeoxyribonucleotides exhibit a non-random cellular distribution. *Nucleosides Nucelotides* 9: 679.

37. Crooke, R.M. (1991). *In vitro* toxicology and pharmacokinetics of antisense oligonucleotides. *Anti-Cancer Drug Des.* 6: 609.

38. Yakubov, L.A., Deeva, E.A., Zarytova, V.F., Ivanova, E.M., Ryte, A.S., Yurchenko, L.V. (1989). Mechanism of oligonucleotide uptake by cells: Involvment of specific receptors? *Proc. Natl Acad. Sci U.S.A.* 86: 6454.

39. Miller, P.S. (1991). Oligonucleotide methylophosphonates as antisense reagents. *Biotechnology* 9: 358.

40. Clarenc, J.P., Degols, G., Leonetti, J.P., Milhaud, P., Lebleu, B. (1993). Delivery of antisense oligonucleotides by poly-(L-lysine) conjugation and liposome encapsulation. *Anti-Cancer Drug Des.* 8: 81.

41. Citro, G., Perrotti, D., Cucco, C., D'Agnano, I., Sacchi, A., Zupi, G., Calabretta, B. (1992). Inhibition of leukemia cell proliferation by receptor-mediated uptake of c-myb antisense oligodeoxynucleotides. *Proc. Natl. Acad. Sci. U.S.A.* 89: 7031.

42. Juliano, R.L., Akhtar, S. (1992). Liposomes as a drug delivery system for antisense oligonucleotides. *Antisense Res. Dev.* 2: 165.

43. Leonetti, J.P., Degols, G., Clarenc, J.P., Mechti, N., Lebleu, B. (1993). Cell delivery and mechanisms of action of antisense oligonucleotides. *Prog. Nucleic Acids Res. Mol. Biol.* 44: 143.

44. Krieg, A.M., Tonkinson, J., Matson, S., Zhao, Q., Saxon, M., Zhang, L.M., Bhanja, U., Yabubov, L., Stein, C.A. (1993). Modification of antisense phosphodiester oligodeoxynucleotides by a 5' cholesteryl moiety increases cellular association and improves efficacy. *Proc. Natl. Acad. Sci. U.S.A.* 90: 1048.

45. Bennett, C.F., Chiang, M.Y., Chan, H., Shoemaker, J.E.E., Mirabelli, C.K. (1993). Cationic lipids enhance cellular uptake and activity of phosphorothioated antisense oligonucleotides. *Mol. Pharmacol.* 41: 1023.

46. Mayford, M., Weisblum, B. (1989). Conformational alteration in the ermC transcript in vivo during induction. *EMBO J.* 8: 4307.

47. Zon, G. (1988). Oligonucleotide analogues as potential chemotherapeutic agents. *Pharm. Res.* 5: 539.

48. Wickstrom, E.L., Bacon, T.A., Gonzalez, A., Freeman, D.L., Lyman, G.H., Wickstrom, E. (1988). Human promyelocytic leukemia HL-60 cell proliferation and c-myc protein expression are inhibited by an antisense pentadecadeoxynucleotide targeted against c-myc RNA. *Proc. Natl. Acad. Sci. U.S.A.* 85: 1028.

49. Iversen, P. (1993). In vivo studies with phosphorothioate oligonucleotides: rationale for systemic therapy, in *Antisense Research and Applications,* Crooke ST, Lebleu B, Eds. CRC Press, Boca Raton, p. 461.

50. Bayever, E., Iversen, P., Smith, L., Spinolo, J., Zon, G. (1992). Guest editorial: systemic human antisense therapy begins. *Antisense Res. Dev.* 2: 109.

51. Hijiya, N., Zhang, J., Ratajczak, M.Z., DeRiel, K., Herlyn, M., Gewirtz, A.M. (1993). The biologic and therapeutic significance of c-Myb expression in human melanoma. *Proc. Natl. Acad. Sci. U.S.A.* 91: 1199, 1994.

52. Yaswen, P., Stampfer, M.R., Gosh, K., Cohen, J. (1993). Effects of sequence of thioated oligonucleotides on cultured human mammary epithelial cells. *Antisense Res. Dev.* 3: 67.

53. Doida, Y., Okada, S. (1967). Synchronization of L5178Y cells by successive treatment with excess thymidine and colcemid. *Exp. Cell Res.* 48: 540.

54. Scharenberg, J.G.M., Rijkers, G.T., Toebes, E.A.H., Stall, G.E.J., Zegers, B.J.M. (1988). Expression of deoxyadenosine and deoxyguanosine toxicity at different stages of lymphocyte activation. *Scand. J. Immunol.* 28: 87.

55. Stein, C.A., Cohen, J.S. (1989). Phosphorotioate oligodeoxynucleotide analogues, in *Oligodeoxynucleotides. Antisense Inhibitors of Gene Expression,* Cohen JS, Ed., Macmillan, Basingstoke, pp. 97–117.

56. Woolf, T.M., Melton, D.A., Jennings, C. (1992). Specificity of antisense oligonucleotides *in vivo. Proc. Natl. Acad. Sci. U.S.A.* 89: 7305.

57. Anfossi, G., Gewirtz, A.M., Calabretta, B. (1989). An oligomer complementary to c-myb-encoded mRNA inhibits proliferation of human myeloid leukemia cell lines. *Proc. Natl. Acad. Sci. U.S.A.* 86: 3379.

58. Gewirtz, A.M. (1992). Therapeutic applications of antisense DNA in the treatment of human leukemia. *Ann. N.Y. Acad. Sci.* 660: 178.

59. Skorski, T., Nieborowska-Skorska, M., Barietta, C., Malaguarnera, L., Szczylik, C., Chen, S.T., Lange, B., Calabretta, B. (1993). Highly efficient elimination of Philadelphia leukemic cells by exposure to bcr/abl antisense oligodeoxynucleotides combined with mafosfamide. *J. Clin. Invest.* 92: 194.
60. Kitajima, I., Shinohara, T., Bilakovics, J., Brown, D.A., Xu, X., Nerenberg, M. (1992). Ablation of transplanted HTLV-I Tax-transformed tumors in mice by antisense inhibition of NF-κB. *Science* 258: 1792.
61. Chiasson, B.J., Hooper, M.L., Murphy, P.R., Robertson, H.A. (1992). Antisense oligonucleotide eliminates *in vivo* expression of c-fos in mammalian brain. *Eur. J. Pharmacol. Mol. Pharmacol.* 227: 451.
62. Wickstrom, E., Bacon, T.A., Wickstrom, E.L. (1992). Down-regulation of c-Myc antigen expression in lymphocytes of Eμ-c-myc transgenic mice treated with anti-c-myc DNA methylophosphonates. *Cancer Res.* 52: 6741.
63. Gewirtz, A.M., Calabretta, B. (1988). A c-myb antisense oligodeoxynucleotide inhibits normal human hematopoiesis *in vitro*. *Science* 245: 1303.

Chapter 2

Cellular Delivery of Ribozymes

David A. Elkins and John J. Rossi

CONTENTS

I. INTRODUCTION

The discovery and characterization of self-cleaving RNAs (see reviews by Cech and Bass[1] and Symons[2]) prompted the adaptation of several of these motifs (most of which are naturally *cis* cleaving) to *trans*-acting and potentially catalytic molecules. Two characteristics of "ribozymes" have made them attractive tools for the inhibition of gene expression: (1) their ability to cleave target RNA molecules and thereby functionally inactivate them and (2) their potential to act on multiple target molecules (catalytic function). In contrast, the mechanism(s) by which antisense RNA or DNA molecules inactivate mRNA are not well understood and are known to vary. For example, the contribution of RNase H activity to oligonucleotide-mediated antisense inhibition can vary widely between cell types. Also, there are instances where standard antisense molecules (both RNA and DNA) have failed (e.g., Kerr et al.[3]). Although ribozyme application has likewise met with obstacles, the cleavage activity and potential catalysis offered by ribozymes (in particular, the hammerhead and the hairpin motifs) render them increasingly popular agents for the inhibition of gene expression for both scientific and medical purposes.

A primary challenge for application of ribozyme technology to an investigative or clinical situation is delivery of the ribozyme to the appropriate cells, as well as localization to the appropriate subcellular compartment. The modalities used for ribozyme delivery may be broadly classified as either (1) delivery of exogenously synthesized ribozyme or (2) "gene therapy": some type of vector is used to insert a ribozyme-encoding gene into the desired cells, to be maintained in either an episomal or chromosomally integrated state. (Many investigators utilize transient expression of ribozyme by nonstable transfection methods; such studies are more rapid, generally, than establishment of stable cell lines, and can serve as the endpoint of investigation or as a stage in the development of therapeutic ribozymes, e.g., in the selection of promoter sequences optimal for the desired target cells.)

0-8493-4778-5/95/$0.00+$.50
© 1995 by CRC Press Inc.

The purpose of this review is to consider the lessons learned — and the questions raised — by intracellular ribozyme experiments to date, with special attention to the aspects of ribozyme delivery. We will also examine various strategies for enhancing interaction of ribozyme and substrate in an intracellular environment. We will consider both investigative and therapeutic applications of ribozymes. Since ribozymes currently represent a minority of the therapeutics being developed for gene therapy applications, much can be learned from studies of protein-coding *trans*-genes and traditional antisense constructs.

Other reviews have focused on details of the various ribozyme motifs (Cech and Bass,[1] Symons,[2] Castanotto et al.[4]); we will therefore not examine them in detail. Most experimental adaptations have exploited the smaller size and greater targeting flexibility of the hammerhead ribozyme (see Figure 1). All current applications utilize the design of Haseloff and Gerlach[5] in which the catalytic core nucleotides reside entirely in the ribozyme RNA. The hairpin ribozyme (see Figure 2) has also been adapted to trans function, and its relatively efficient ligation activity makes it amenable to rapid *in vitro* optimization (or "evolution"; see Burke and Berzal-Herranz[6]) for a given target molecule, as has been performed for the Group I intron of Tetrahymena (Beaudry and Joyce[7]).

II. SYNTHESIS AND DELIVERY OF EXOGENOUS RIBOZYME

Once a target RNA has been chosen and a ribozyme motif designed against it, the first step in exogenous delivery is synthesis of the catalytic RNA. Methods for the chemical synthesis of RNA have advanced rapidly in the past few years (for review, see Usman and Cedergren[8]). The standard solid-phase chemistry uses the same protecting groups employed in conventional DNA synthesis, which confers the dual benefit of convenience and synthetic flexibility (e.g., one can synthesize chimeric RNA-DNA molecules on a standard oligonucleotide synthesis machine). We will briefly explore the various synthetic options afforded by modern chemical RNA synthesis. Note that another method of obtaining ribozyme RNA (for exogenous delivery or *in vitro* analysis) is enzymatic synthesis by purified bacteriophage RNA polymerases, such as T7 or SP6. *In vitro* transcription is, of course, limited to synthesis of "ordinary" RNA, incorporating only the "natural" ribonucleotides recognized by the phage enzymes. Many investigators utilize *in vitro* transcription for initial studies of ribozymes designed against a given target molecule; for example, one might compare the *in vitro* cleavage efficiencies of ribozymes with similar hybridization lengths, but which hybridize to different sites within the target RNA.

The RNA modifications possible during chemical synthesis include base modifications, sugar modifications, or modifications of the phosphodiester backbone. Ribozymes with DNA bases in the flanking arms (the nonenzymatic portions of the molecule) have been shown to resist ribonuclease degradation and, in some cases, to display enhanced catalytic turnover and even an accelerated rate of cleavage, when compared to corresponding all-RNA molecules. Initial studies of DNA-RNA chimeric ribozymes were aimed at identifying the specific 2'-OH groups necessary for the catalytic function of various ribozyme motifs (e.g., Perreault et al.[9]). However, the enhanced nuclease resistance of the chimeric ribozymes prompted functional comparison with all-RNA molecules. Taylor et al.[10] reported that a hammerhead ribozyme with DNA bases in the arms flanking the catalytic core displayed a sixfold increase in turnover, as well as enhanced intracellular stability, compared to the all-RNA counterpart. Hendry et al.[11] analyzed a similar hammerhead chimera and found that its rate constant for substrate cleavage was increased threefold. Shimayama et al.[12] studied chimeric hammerhead ribozymes whose resistance to serum nucleases was tenfold greater than all-RNA ribozymes. Further modifications yielded ribozymes more than a hundred times more stable in serum than the all-RNA molecules.

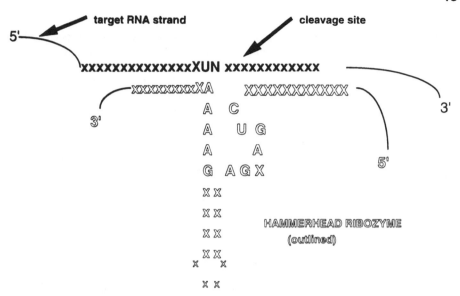

Figure 1 The hammerhead ribozyme. N = any base but G; X = any base. Paired regions are shown parallel to one another.

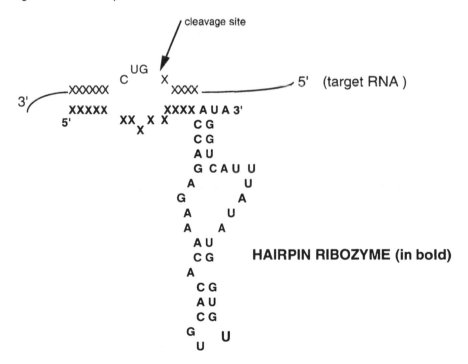

Figure 2 The hairpin ribozyme. X = any base; paired regions are shown parallel to one another. (From Joseph, S., Berzal-Herranz, A., Chowrira, B.M., Butcher, S.E., and Burke, J.M., *Genes Dev.*, 7, 120, 1993. With permission.)

One of the backbone modifications possible with automated chemical synthesis is incorporation of phosphorothioate linkages between the nucleosides. Such linkages render both RNA and DNA oligonucleotides more resistant to serum and intracellular nucleases (Pieken et al.,[13] Shimayama et al.[12]). This modification has been applied to therapeutic antisense oligodeoxynucleotides (e.g., Agrawal et al.[14]). A nonspecific effect of phosphorothioate oligos has been observed against HIV, possibly due to inhibition of viral entry (Stein et al.[15]). Phosphorothioate linkages have been employed in ribozyme and substrate RNAs to delineate various parameters of the cleavage mechanism (e.g., for the hammerhead ribozyme; see Slim and Gait[16] and references therein). Shimayama et al.[12] synthesized chimeric DNA/RNA hammerhead molecules with DNA in the pairing flanks and RNA in the catalytic core; the DNA residues were linked either with standard phosphodiester bonds or with phosphorothioate bonds. As mentioned above, they observed that these chimeric ribozymes were about tenfold more resistant to serum nucleases. They also discovered that the thio-DNA/RNA ribozymes displayed a sevenfold enhancement of the chemical cleavage step; they suggest that this enhancement may be attributed to altered structure at the catalytic site.

Several investigators have examined the activity and nuclease stability of ribozymes with modified sugar residues. Pieken et al.[13] reported that 2'-fluoro and 2'-amino substitutions in a hammerhead molecule increased resistance to serum nucleases; unfortunately, most of the modified ribozymes were much less catalytically active. Heidenreich and Eckstein[17] showed that a more selective substitution protocol, combined with a partial phosphorothioate backbone, resulted in less inhibition of ribozyme activity while retaining the considerable advantage of nuclease resistance. They also demonstrated that 2'-fluoro-substituted molecules are more active than comparable 2'-deoxy analogs. While the cost of RNA synthesis is currently a serious obstacle to the medical application of chemically synthesized ribozymes, the advantages of such modifications (especially in combination, as demonstrated by Heidenreich and Eckstein[17]) may justify their further development as human therapeutic agents.

Another modification reported recently is the use of 2'-O-allyl and 2'-O-methyl groups. Paolella et al.[18] showed that synthetic ribozymes retaining only six 2'-hydroxyl residues were catalytically active and quite nuclease resistant. Keller and Haner[19] described the synthesis of oligonucleotides with 2' substitutions including lipophilic and intercalating groups and tertiary amino acids. They observed that such molecules remain sequence specific in their hybridization properties; some of them display properties potentially useful in therapeutic antisense molecules (DNA or RNA), especially enhanced catalysis for ribozymes.

Base modifications within ribozyme molecules have proven useful in functional delineation of the catalytic center, or active site, of ribozymes. By substituting bases lacking certain functional groups, such as exocyclic amino groups, it has been possible to identify the role of these groups in catalysis. For example, a recent study by Fu et al.[20] demonstrates the critical role played in cleavage by the N^2-amino group of the G_{12} residue within the hammerhead sequence: loss of this group reduces cleavage efficiency 450-fold. Loss of amino groups at other residues impacts cleavage efficiency much less dramatically. Tuschl et al.[21] performed similar replacement studies which help to define the pairing interactions and metal-binding roles played by several guanosine residues within the hammerhead core. Such studies may lead to ribozymes more active and more stable in the intracellular environment.

The chemical stability of ribozyme RNAs is a prime consideration for exogenous delivery. The intracellular stability of ribozymes (or any other RNA molecule) may be further enhanced by incorporation of a secondary structure at the 3' or 5' termini of the molecules. An example is the work of Sioud et al;[22] in which they describe the stabilizing effect of a stem-loop at the 3' end of a hammerhead ribozyme, delivered to human cells via liposomal encapsulation. This type of modification pertains also to ribozymes synthesized from a *trans*-gene, particularly when the ribozyme sequence will be relatively short.

In order to examine the inhibitory activity of a ribozyme on a cellular RNA target, one is faced with the task of delivering the presynthesized ribozyme RNA into the cells under investigation. Naked RNA is quickly destroyed by nucleases present in serum. Moreover, the cellular uptake of free RNA is rather poor. While a cellular mechanism has been described for the uptake of deoxy-oligonucleotides (Loke et al.,[23] Yakubov et al.[24]), none has yet been described for ribonucleotides. Various methods for accomplishing the intracellular delivery of exogenous ribozymes are under consideration or have been tested in cell culture. While most of these methods have been applied primarily to delivery of DNA, they are likely to function as well for RNA, since they do not rely on receptor-mediated uptake of naked nucleic acids.

To date, the most popular vehicle for delivering presynthesized ribozyme RNA to target cells is liposome encapsulation (for extensive review, see Juliano and Akhtar[25]). Liposomes are microscopically small spheres of phospholipid bilayer which can either encapsulate nucleic acids within the aqueous center or form lipid-nucleic acid complexes which protect DNA and RNA from nuclease attack. Various liposome formulations have been used for the delivery of a number of therapeutic compounds, including antibiotics, anticancer agents, and hormones. Broadly speaking, liposomes may be classified as either cationic or noncationic, based on the type of lipids used in the formulation. (Note that cationic lipids do not encapsulate nucleic acids, but instead form complexes with them, based on the negatively charged phosphate groups in the DNA or RNA and the positively charged lipid molecules.) A further modification of liposome delivery is incorporation of monoclonal antibodies, coupled to the vesicles, for targeting of a specific cell population. Leonetti et al.[26] reported the use of antibody-targeted liposomes (encapsulating antisense DNA oligonucleotides) to inhibit vesicular stomatitis virus replication in murine cell culture.

Most of the nucleic acid applications for liposomes have been antisense deoxyribo-nucleotides, for which a sizable body of literature exists. Only a few groups have reported the use of liposomes for delivery of ribozymes. Sioud et al.[22] studied the effects of an anti-TNF-α ribozyme delivered to human cells via cationic liposomes. They found that liposomes mediated delivery of between 2 and 4% of the complexed ribozyme to the target cells, and that this level of intracellular ribozyme was stable and effective in reducing both target mRNA and protein levels by 85 to 90%. As mentioned earlier, they also observed that a 3′-terminal stem-loop — a T7 transcriptional terminator — enhanced intracellular stability of the ribozyme delivered in this fashion.

Advantages of liposome delivery of ribozymes include protection from extracellular nucleases and depending on the lipid formulation used, localization of the delivered ribozyme to the appropriate cellular compartment. Liposome delivery of nucleic acids also ensures that they will not suffer wholesale degradation by lysosomal enzymes, a common problem with delivery of naked DNA oligonucleotides.

There are a number of nonliposomal methods for enhancing stability and cellular uptake of nucleic acids; again, most applications have been with DNA (and will therefore be mentioned with regard to gene therapy approaches). Some of these methods share with cationic liposomes the temporary disruption of the cell membrane, e.g., electroporation and microprojectile transfection. However, gentler methods have been developed. Wagner et al.[27] have modified the transferrin protein with cationic domains to render it capable of binding nucleic acids — primarily DNA, but there is hope that this avenue may prove amenable to RNA delivery as well (Cotten[28]). Wu and Wu[29,30] describe the delivery of foreign DNA to cultured hepatocytes via a polylysine-glyco-protein conjugate which enters cells by a receptor-mediated endocytosis. Gao et al.[31] reported the use of inactivated adenoviral particles to enhance the *in vivo* transforma-tion efficiency of DNA-polylysine conjugates, primarily through endosomal disrup-tion. It is conceivable that these and other nonlipid biopolymers might be adapted to exogenous ribozyme delivery, but no published reports of such applications have yet appeared.

III. ENDOGENOUS EXPRESSION OF RIBOZYME-ENCODING GENES

The concept of gene therapy has progressed from dream to reality within the last 20 years. Its basis is the introduction of a functional gene, encoding a protein (or RNA) beneficial to the patient, into cells requiring the beneficial function. Ultimately, the goal is one-time introduction of the gene and permanent expression of the desired gene product. Obviously, the choice of vector, cell type, and regulatory sequences (transcriptional and post-transcriptional) are key factors in the success of a gene therapy endeavor. As the clinical potential of gene therapy has grown, so have the number and quality of vectors available for delivering genes to various cell types — for example, consult recent reviews by Morgan and Anderson,[32] Mulligan,[33] Tolstoshev,[34] Friedman,[35] and Miller.[36]

The full scope of gene therapy is beyond the limits of this review; however, it is practical to briefly consider the types of vectors available, their strengths and limitations, and the issues involved in adapting them to particular investigative or therapeutic situations. Also, we will examine a number of "unconventional" methods for optimizing the effectiveness of antisense and ribozyme constructs within cells.

Perhaps the most popular vectors for gene delivery to date are amphotropic retroviral vectors (Figure 3) (see reviews by Majors[37] and To and Neiman[38]). Retroviruses were the first vectors developed for gene therapy applications and are therefore the best-characterized group. Their strong points include high transduction efficiency (approaching 100% in many cases) and stable integration into the host cell genome, ensuring that the transduced gene will be passed to all daughter cells. However, transduction by standard retroviral vectors is primarily limited to dividing cells (Miller et al.[39]), and retrovirus insertion poses the potential hazards of oncogene activation or insertional gene disruption. The majority of retroviral vectors developed to date are based on "simple" retroviruses (which lack complex gene regulatory circuits) such as Moloney murine leukemia virus (MoMLV) and avian sarcoma/leukosis virus (ASLV).

Most current retroviral vectors are replication defective: "packaging" cell lines provide the viral structural proteins in *trans*, complementing those genes deleted from the vector. The virions so produced are capable of only a single round of infection. An early obstacle was the occurrence of recombinant wild-type virus; however, this problem has been largely reduced by either eliminating most of the noncoding viral sequences (such as the packaging signal) from the helper virus (Miller and Buttimore[40]) or expressing the gag-pol and the env proteins from separate plasmids (Markowitz et al.[41]). However, the presence of replication-competent virions is an inherent hazard of retroviral delivery (Donahue et al.[42]), and preparations must be screened carefully to reduce the risk of contamination.

Wild-type retroviruses are generally quite restricted in the cell types they infect, which is a disadvantage for gene therapy. One remedy for this limitation is the use of amphotropic packaging cell lines, which express env proteins from a murine retrovirus with an unusually large host range (Miller et al.[39]). Another approach is "pseudotyping": Emi et al.[43] demonstrated the feasibility of constructing packaging cell lines which produce retroviral "pseudotypes," which carry in their envelopes the vesicular stomatitis virus (VSV) G protein, as well as the normal complement of MoMLV proteins. These pseudotype viruses displayed the broad host cell range of VSV, opening the possibility of developing gene therapy vectors with the efficiency and stable integration of a retrovirus, but lacking the severe cell- and tissue-type restrictions. A recent report by Burns et al.[44] illustrates that these pseudotypes can be produced at titers of up to 10^9 infectious units per milliliter — much greater than the 10^2 to 10^3 IU/ml reported earlier — by utilizing transient transfections instead of stable cell lines.

A recent addition to the array of retroviral gene therapy vectors is HIV-1, the causative agent of AIDS and a "complex" retrovirus encoding several small proteins that regulate the expression of viral genes. Poznansky et al.[45] demonstrated that an HIV provirus

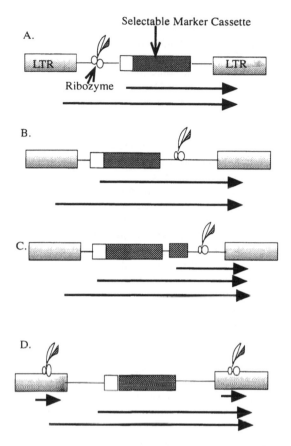

Figure 3 Expression possibilities for ribozymes using a retroviral vector for delivery. Various expression possibilities for a ribozyme gene encoded within a retroviral vector are depicted. (A) The ribozyme is transcribed from the retroviral LTR and is part of a long transcript which includes the selectable marker. (B) The ribozyme is transcribed as part of the selectable marker transcription unit, or as part of the longer LTR-driven transcript. (C) The ribozyme is transcribed from its own promoter or as part of a transcriptional unit from the selectable marker or the viral vector LTR. (D) The ribozyme is cloned into the 3′ LTR and is duplicated in the 5′ LTR during proviral DNA formation. In this setting, the ribozyme is usually included as part of a pol III transcription unit which will not interfere with the LTR promoter.

lacking 19 nucleotides responsible for packaging of the genomic RNA was able to provide in *trans* the functions required to generate replication-defective HIV-1 vector virions. A more recent report from the same group[45a] suggests that HIV sequences upstream and downstream of the 19 nt packaging signal also play a role in packaging.

The use of HIV-1 as a vector might be especially attractive for treatment of HIV-infected persons, since the target population of the replication-defective therapeutic vector is the same as that of the pathogen; if HIV gene regulation signals are maintained in the vector, it might be possible to express a therapeutic, anti-HIV ribozyme only when an HIV provirus is stimulated to express its own genes. Buchschacher and Panganiban[46] reported the construction of HIV-1 vectors which express either the gene for hygromycin resistance or an influenza virus hemagglutinin gene, in a Tat-dependent manner. They propose that such vectors might mediate expression of other foreign genes specifically in HIV-infected cells.

HIV vectors would be even more useful if they prove capable of transferring genes to stem cells. Disadvantages include cytopathicity and restricted cell targeting, although pseudotyping (described above) has been demonstrated for HIV-1 (Lusso et al.),[47] raising the possibility of expanded host cell range for HIV-1 vectors.

One might devise a similar strategy for the selective induction of another anti-HIV gene, including a ribozyme gene, from a non-HIV vector. An example is the work of Weerasinghe et al.,[48] described in detail later. They obtained markedly enhanced expression — and enhanced antiviral effect — of an anti-HIV ribozyme linked in *cis* to a TAR sequence, enabling transactivation by Tat. This strategy might be applied to other pathogenic situations; the basic requirement is to link a promoter element, specific to viral expression, to a ribozyme-expressing gene.

Adenoviral vectors are attractive for their ability to transduce nondividing cells; also, vaccination of U.S. military personnel has demonstrated that adenovirus types 4 and 7 are safe for human use. Construction of adenovirus vectors entails the deletion of the adenoviral E1 or E3 genes, or both, and placement of the therapeutic gene within an appropriate transcriptional cassette. The gene transferred by adenoviral vector remains in the episomal state and is eventually lost in a dividing cell population. A replication-deficient adenovirus vector has been successfully used to establish expression of human α-1-antitrypsin gene in cultured human lung epithelial cells and primary endothelial cells (Rosenfeld et al.,[49] Lemarchand et al.[50]). These studies indicate the potential of adenoviral vectors for ribozyme expression.

Other viral vectors used for cellular delivery of foreign genes include adeno-associated virus (AAV), herpes simplex virus (HSV), and hepatitis delta virus (HDV). AAV is a nonpathogenic parvovirus which integrates specifically and with high efficiency into a region of human chromosome 19 — a feature which obviates many safety concerns and which ensures the permanence of gene insertion (see Chatterjee and Wong[51] for review). A further advantage of AAV vectors is their broad range of cell tropism. AAV vectors to date are limited to about 5 kB of foreign DNA, and packaging of recombinant vectors requires coinfection of the packaging line with adenovirus or herpes virus, raising the possibility of contamination. Chatterjee[52] have used AAV to deliver TAR antisense RNA to human T-lymphocyte and to human embryonic cell lines, blocking HIV-1 replication upon viral challenge of the transduced cells. Given this foundation, AAV could easily be adapted to delivery of ribozyme genes to a broad range of cell types.

A unique advantage of herpesvirus (HSV-1) as a therapeutic vector is its neurotropism. It is also capable of accommodating very large stretches of inserted DNA. HSV-1 vectors have been successfully used to transduce mammalian neurons (Fink et al.[53]). A significant disadvantage, however, is the innate cytopathogenicity of herpesviruses, which dictates that dramatic modifications of the wild-type virus will be necessary prior to clinical use. Toward this end, HSV deletion mutants are being developed which retain the ability to infect neurons and grow to high titer in cell culture, but are attenuated in brain tissue and do not spread in that biological context.[53] Regions of HSV which mediate transcription during the viral latency stage have also been identified; such promoters would be ideal for expression of therapeutic genes following an obligate latent neural infection with a tailored HSV vector.

HDV is a replication-defective virus often found as a satellite of hepatitis B virus (HBV). HDV requires expression of the HBV envelope antigen (in *trans*) in order to establish a productive infection. J. Taylor and colleagues, availing themselves of the hepatotropism of this virus, have constructed modified HDV genomes intended specifically for treatment of chronic human hepatitis B infection. They report that up to 30 bases can be inserted into the small RNA genome of HDV without loss of replication. Future work may reveal that larger insertions are possible. While this size constraint disqualifies

HDV as a vector for expression of a protein-coding gene, 30 bases is sufficient for insertion of a hammerhead ribozyme construct[53b] (Hsieh and Taylor[54]).

Less work has been reported on the utility of several other, more exotic viruses for gene delivery. These include vaccinia, polio, and alphaviruses (of the Togaviridae, including Sindbis virus and Semliki Forest virus). Fuerst and Moss[55] used a recombinant vaccinia virus to express the bacteriophage T7 RNA polymerase in infected HeLa cells. T7 RNA polymerase can be used to drive cytoplasmic transcription of other genes, including ribozyme genes, within cells targeted by a T7 delivery vector (see L'Huiller et al.[56]). A potential advantage of the alphaviruses is that many of them mediate production of very high levels of (viral) protein within infected cells, accomplished in part by shutdown of host protein synthesis. This characteristic of alphavirus infection renders them particularly useful for cellular production of a desired protein or for protein-stimulated proliferation of specific subsets of T-cells. Another strong point of alphaviruses is their unusually broad host range. High levels of expression of heterologous genes, and broad host range, would be useful characteristics in a ribozyme delivery vector. However, two serious disadvantages which have yet to be fully addressed are cytopathicity and the risk of a potentially fatal infection (as has been reported: Willems et al.[57]). Semliki Forest virus vectors have recently been used (Paul et al.[58]) to express HIV-1 env protein and precursor in two human cell lines and a hamster cell line.

Some groups have investigated the delivery of DNA to target cells or tissues either directly (e.g., Wolff et al.59) or in a conjugate with some biological polymer, such as polylysine (Wu and Wu,[30] Gao et al.[31]). Nabel and colleagues[60] have reported whole-organism experiments in which direct delivery of DNA has accomplished the desired therapeutic goal. For example, introduction of a foreign MHC class I gene directly into tumors *in vivo* (in the form of a retroviral vector or complexed with liposome material) mediated expression of the gene and induced a cytotoxic T-cell response to the foreign antigen.[60] More recently, they report (Nabel et al.[61]) that direct gene transfer into porcine arteries mediated expression (by cells of the arterial intima) of fibroblast growth factor (FGF-1). Weiner and colleagues[66] have elicited immune responses in live mice against the HIV-1 envelope glycoprotein through direct intramuscular injection of the gene. Felgner and colleagues (Wolff et al.[59]) studied tissue-specific differences in efficacy of the direct delivery approach. Through direct injection of a reporter lacZ construct, they have demonstrated high expression in muscle tissue which lasted beyond a year, but which was promoter dependent (the Rous Sarcoma Virus (RSV) promoter, for example, mediated such long-lasting expression, whereas the cytomegalovirus (CMV-IE) promoter showed a drop in expression after 2 to 4 weeks postdelivery.) These studies suggest that stable ribozyme expression might well be feasible through direct delivery of a ribozyme-encoding gene to the tissue of interest, particularly if the tissue lends itself to such delivery methods.

IV. SELECTION OF REGULATORY SEQUENCES

Apart from the choice of vector, one must select regulatory sequences to mediate appropriate expression of the ribozyme. For transcriptional regulation the major decision is between a pol II or pol III transcriptional unit. The examples of *in vivo* ribozyme usage, detailed below, will illustrate advantages of both types of promoters. Some promoters enable expression of the ribozyme only under certain conditions (e.g., viral infection or the presence of an inducer); other strategies simply rely on constitutive promoters appropriate to the target cells. A further consideration is whether cell/tissue-specific expression will be obtained via the delivery method, or whether the expression cassette must provide this specificity (e.g., by a tissue-specific enhancer). The effectiveness of promoters selected for regulation of ribozyme expression can often be assessed empirically, at least in part, by transient transfection studies in cell culture. A common approach is to examine the levels of ribozyme expression obtained by several different promoter and enhancer combinations, and select for further development the combination yielding the highest levels of ribozyme RNA or the greatest inhibition of the targeted RNA.

The effectiveness of any particular promoter sequence, whether pol II or pol III transcribed, is highly dependent on the cell type(s) in which it will perform. Most of the pol II promoters presently used for ribozyme expression in mammalian cells are of viral origin; and a perusal of the literature reveals no single promoters which are clearly superior in all cell types. Numerous recent studies (e.g., Enssle et al.[62]) indicate that a given promoter sequence exerts a definite (and, currently, poorly understood) influence on the intracellular fate of RNA molecules transcribed under its control. As this influence is more fully elucidated investigators may harness the characteristics of different promoters to target trans-gene transcripts — including ribozymes — to desired processing pathways or subcellular domains. To some extent this is already possible: antisense and ribozyme molecules have been inserted into tRNA or snRNA genes. In addition to the advantages of high levels of transcription and intracellular stability, a pol III promoter might render a ribozyme more effectively localized within a target cell than a pol II promoter (e.g., tRNA localization would help when the target RNA is predominantly cytoplasmic). We will examine several instances of pol III promoters, and snRNA transcription units, for antisense and ribozyme expression.

Some expression parameters are unique to stable integration of the ribozyme gene, as is accomplished in retroviral delivery. One consideration is the poorly understood effect of the chromosomal integration site on the level of gene expression (for review, see Wilson et al.[63]). This "position effect" poses a problem for reliable expression of any transgene, including a ribozyme, from a retroviral vector. As mentioned earlier, the specificity of AAV vector integration is a boon in this regard, and would allow comparison of different constructs, in different cells, without the potential confounding effect of positional variation.

The post-transcriptional processing of a therapeutic ribozyme might play a major role in determining its effectiveness (or lack thereof) at reducing levels of the targeted RNA. The vast majority of *trans*-genes in general are expressed from RNA pol II promoters and therefore are processed as mRNA molecules — they are capped at the 5′ end with methylguanosine and polyadenylated at the 3′ end. As noted earlier, most of the ribozyme genes developed to date have also utilized pol II promoters. Often, this means that the ribozyme sequence is buried within a much larger transcript. In some cases, this seems beneficial to expression and stability. However, sometimes it seems to inhibit ribozyme function (e.g., Yu et al.[64]). This problem might be minimized through the use of "transcript-trimming" constructs, in which a *cis*-cleaving ribozyme, positioned upstream or downstream of the therapeutic ribozyme, removes the extraneous sequences to that side of the ribozyme (*in vitro* examples include Taira and Nishikawa,[65] Altschuler et al.,[66] Ventura et al.[67]). Application of such a strategy to *in vivo* ribozymes might increase their efficacy. In some cases, however, the shorter transcripts might be considerably less stable. Hence, a balance must be achieved between stability and effective cleavage capacity.

V. REVIEW OF INTRACELLULAR RIBOZYME STUDIES

Jennings and Molloy[68] used the lytic transforming virus SV40 to transfer to COS cells a pol III-driven antisense gene directed against the SV40 origin of replication. This antisense RNA inhibited replication of a plasmid dependent on the SV40 origin by 50% to almost 100%, and represents one of the earliest viral vector-delivered antisense constructs targeted to inhibition of a pathogenic process.

A later study by Cameron and Jennings[69] demonstrated the usefulness of SV40 early promoter and polyadenylation signals in directing ribozyme transcription. Hammmerhead ribozymes, directed against three target sites within chloramphenicol acetyltransferase (CAT) mRNA, were incorporated into the 3¢ untranslated region (UTR) of luciferase mRNA, since preliminary studies demonstrated that a gene encoding the ribozyme transcript alone was ineffectual in reducing CAT activity from even the weak HSV thymidine kinase (tk) promoter. (It is worth observing that this ineffectiveness of short ribozyme transcripts is not universal.) The three different ribozymes contained within the longer luciferase transcript decreased the level of CAT activity within transiently trans-fected COS cells by about 70% each. The authors note that an antisense mechanism cannot be excluded, since a noncleaving ribozyme control was not included (more recent studies generally include this important control). However, the CAT mRNA had been shown to be resistant to antisense inhibition (e.g., Kerr et al.),[3] suggesting that the cleavage capability of the ribozymes used may have made the difference.

HIV-1, the causative agent of AIDS, has become a major focus of therapeutic ribozyme development. Sarver et al.[70] reported the first intracellular ribozyme experiments target-ing the RNA of HIV. In these experiments, a hammerhead ribozyme gene (targeted to a GUC triplet in the gag region of HIV RNA) was placed downstream of the constitutive human ß-actin promoter, and an SV40 polyadenylation signal was located downstream to effect 3′ processing. They derived stable, ribozyme-expressing cell lines from HeLa CD4+ cells transfected with plasmid DNA carrying this ribozyme gene and a selectable marker. These cell lines were then challenged with HIV; levels of soluble p24 antigen were used to assess the efficacy of ribozyme protection at 7 d postviral challenge. Whereas untransformed cells secreted greater than 10 ng/ml, cloned and pooled transformants secreted 0.14 and 0.23 ng/ml, respectively, demonstrating a significant reduction in retroviral replication. Unfortunately, since a noncleaving ribozyme control was not included, antisense effects cannot be ruled out as the protective mechanism. However, the presence of cleaved *gag* RNA was indirectly demonstrated via PCR and the ability of ribozyme within total cellular RNA to cleave target RNA was demonstrated in a cell-free system (Chang et al.[71]).

Weerasinghe et al.[48] constructed retroviral vectors enabling expression of a hammer-head ribozyme cleaving a conserved sequence within the 5′ leader RNA of HIV-1. This ribozyme was expressed either as a portion of the 3′ UTR of neomycin phosphotransferase (neo) mRNA — under the control of the herpesvirus tk promoter — or as part of a shorter transcript generated from either the SV40 early promoter or the CMV-IE promoter. Another construct contained a tk promoter capable of both constitutive and tat-dependent transcription of the ribozyme, the latter due to the insertion of TAR sequences at the 5′ end of the gene. All retroviral constructs were used to stably transform the CD4+ lymphocyte cell line MT4; transformants were then analyzed for their resistance to HIV replication following virus challenge. The most effective was the tk-TAR promoter, suggesting that Tat-enhanced transcription would be a useful strategy for anti-HIV ribozymes transcribed from pol II promoters.

Lo et al.[72] also used the MoMLV system to deliver an anti-HIV ribozyme gene to cultured lymphocytes. A ribozyme targeted to the first exon of tat was inserted into the 3′ UTR of the neo mRNA; a corresponding antisense constuct, containing the 48 total pairing nucleotides, but lacking the catalytic domain, was also created. Transcription of the neo mRNA, containing the antisense or ribozyme sequence, was directed by the retroviral LTR promoter sequences. The anti-HIV activities of the ribozyme and antisense vectors were compared by HIV challenge of stably transformed Jurkat cells. Experiments done at different multiplicities of infection (m.o.i.) revealed that both constructs delayed shedding of HIV capsid protein; relative to the parental cell line p24 levels at 7 d post-challenge, ribozyme-expressing cells shed p24 at 15 d, and antisense-expressing cells at

19 d. Syncytium formation of infected Jurkat cells was similarly delayed. Levels of intracellular HIV proteins were analyzed, at 11 d postinfection, by metabolic labeling in all cell lines. No HIV proteins were detectable in antisense-expressing lines, whereas a reduced, but detectable level of viral proteins was observed in ribozyme-expressing clones. This study is especially useful in its direct comparison of antisense and ribozyme molecules targeted to the same RNA. However, the more effective inhibition obtained by the antisense molecule in this situation may not apply to other situations. The presence of large stretches of nontargeting RNA might have lessened the effectiveness of this particular ribozyme.

Ojwang et al.[73] designed a hairpin ribozyme targeted to a sequence within the 5′ leader of HIV RNA. Both a functional and a "disabled" (binding, but not cleaving) hairpin ribozyme were placed under the transcriptional control of the human β-actin promoter; plasmid DNA was used to effect transient transfection of CD4+ HeLa cells. Inhibitory activity of the ribozyme was assessed by its diminution of plasmid-based Tat production, as measured by a Tat-activated CAT reporter assay. The functional hairpin ribozyme reduced CAT activity by 70 to 80%, whereas the noncleaving ribozyme reduced activity by only 10%.

Yu et al.[64] reported further work with this hairpin ribozyme, cloned into retroviral vectors, within two different internal pol III transcription units (ribozyme is expressed from either the human tRNA^{Val} promoter or adenovirus VAI promoter). CD4+ HeLa cells were transfected with these vectors, and transient-transfection assays were used to assess the antiviral efficacy of the ribozyme-expressing constructs. The pol III ribozyme expression vectors inhibited the ability of several different HIV-1 isolates to produce p24 upon transfection of viral DNA. Interestingly, a pol II promoter ribozyme construct, tested simultaneously, produced significant amounts of ribozyme RNA (as did the pol III promoters), but did not inhibit HIV expression. The authors suggest that this relative ineffectiveness may be due to the position of the ribozyme sequence within a much longer transcript. It is possible, too, that the localization properties of the pol III-transcribed ribozymes increased their effectiveness.

Chen et al.[74] reported the development of multivalent ribozyme constructs, which incorporate up to nine different catalytic centers, each targeted to a different sequence within the env RNA of HIV-1. Genes encoding one, four, five, and nine ribozymes (under control of the β-actin promoter, SV40 late promoter, or HIV LTR) were tested for anti-HIV efficacy by cotransfection into CD4+ HeLa cells with DNA encoding the HIV clone pNL4-3. As measured by both syncytium formation and soluble p24 levels, the inhibitory effects of the "monoribozyme" were exceeded in every instance by those of the multiribozymes. The nine-ribozyme construct was most effective of all. This strategy might well be useful for enhancing the performance of therapeutic ribozymes in general.

Anti-HIV ribozymes have been utilized for investigative purposes, as well as therapeutic. Dropulic et al.[75] inserted a cis-cleaving hammerhead sequence into the nef region of an infectious HIV clone, in both a sense and an antisense (noncleaving) orientation. These clones were separately transfected into HeLa CD4+ cells, and virus production in cocultivated MT$_4$ lymphocytes (measured by reverse transcriptase activity) was used to assess the infectivity of both types of virus DNA. The sense-orientation ribozyme clones generated 100-fold less reverse transcriptase (RT) activity than the antisense clones over a period of 10 d. Since the folding characteristics (in the hybridizing arms) of the sense and antisense sequence would be equivalent, the reduction in RT activity argues that self-cleavage occurred. It is unknown how the ribozyme-containing virus escaped from this effect past the 1-week inhibition.

They next placed, into the nef gene of two different HIV isolates, a ribozyme targeted to the U5 region of the HIV LTR. (Both ribozyme and target sequences exist at different locations on the same RNA molecule, allowing either intra- or intermolecular RNA cleavage events.) Transfection into HeLa CD4+ cells, with MT4 cocultivation, was followed again by analysis of RT activity. The antisense ribozyme-containing virus, in both the pNL4-3 and LAI isolates, yielded RT patterns comparable to those of control virus lacking the insert. The sense ribozyme virus showed a delay in peak virus production by about 1 week as measured by RT activity and syncytium formation. The same U5 ribozyme sequence was then inserted into an amphotropic MuLV retroviral vector. Experiments with this vector demonstrated both a reduction in HIV RT production from a chronically infected cell line and a partial protective effect for transduced MT4 cells challenged with HIV. The authors suggest that the incomplete protection might be due to the percentage of nonvector-infected cells in the culture, perhaps as high as 50%.

Zhou et al.[75a] designed two hammerhead ribozymes targeting the coding sequences for the HIV regulatory proteins, tat and rev. Using amphotropic virions to deliver MoMLV-based vectors, they obtained stable CEM cell lines expressing functional and inactive versions of both ribozymes. The ribozymes were expressed as part of a single proviral transcript, in sequences of the neomycin resistance gene. On HIV challenge of the different cell lines, they found that the functional ribozymes inhibited HIV replication, preventing viral outgrowth until at least day 20 (control cells transduced with the parent vector showed viral outgrowth at day 12). In contrast, the noncleaving ribozymes displayed little or no delay in outgrowth. These results strongly support target RNA cleavage as the mechanism of inhibition; too, they argue that the antisense effect of ribozymes with short hybridizing sequences (about 20 base pairs) is not large. Interestingly, a dual-ribozyme-expressing vector showed no greater viral inhibition in this assay than did either of the single-ribozyme vectors. The authors suggest that this may not be the case in whole organisms. They also note that, in their hands, this simple, single-transcript vector produced higher levels of ribozyme than dual-promoter vectors utilizing either an internal CMV promoter or an internal tRNA[Met] promoter to effect ribozyme transcription.

Other viral pathogens have been targeted by ribozymes. Xing and Whitton[76] designed both *cis*- and *trans*-acting hammerhead ribozymes against several target trinucleotides within the RNA genome of lymphocytic choriomeningitis virus (LCMV.) LCMV is an arenavirus with a two-segment single-stranded RNA genome whose life cycle (like that of all arenaviruses) contains no DNA phase. They discussed the results of *in vitro* experiments designed to identify the most efficiently-cleaved target sites within the LCMV genome. Perhaps the most important lesson from these initial studies is that a high degree of secondary structure involving the target site (whether native to the sequence or induced by distal positioning, in *cis*, of complementary sequences) can strongly inhibit ribozyme function. Unfortunately, computer prediction of RNA secondary structure is still an inexact proposition. In the absence of phylogenetic or physical data to confirm such a predicted structure, it is probably best to analyze the function of several ribozyme-target RNA combinations *in vitro* and allow the comparison to guide selection of target sites for further work.

Xing and Whitton[77] have recently extended their work to stable cell lines expressing the various anti-LCMV ribozymes described above. Transcription of these ribozymes is driven by a combination RSV/mouse mammary tumor virus (MMTV) enhancer and promoter which yields increased levels of transcript in the presence of glucocorticoids. In viral challenge experiments they demonstrated that basal levels of ribozyme transcription prevent accumulation of viral RNA — even more so when transcription is enhanced with dexamethasone. Also, the yields of infectious virus are much diminished in ribozyme-expressing cells relative to parental cells. Although they do not report use of an inactive ribozyme control, they observed that the inhibitory effects of the ribozyme do not extend to a related arenavirus with eight "mismatches" in the target sequence.

L'Huillier et al.[56] designed five different ribozymes against five sites within the α-lactalbumin mRNA. Expression in mouse mammary cells and monkey kidney cells was accomplished using vaccinia virus to deliver T7 RNA polymerase to the cells, which were transfected with plasmid DNA bearing ribozyme and target RNA genes under the control of the T7 promoter. The intracellular experiments revealed (1) that *in vitro* ribozyme performance did not always reflect *in vivo* efficiency, (2) that a target site near the start codon was very poorly cleaved *in vivo*, and (3) that a site within the 3′ UTR of the target transcript was the most effective target *in vivo*. A catalytically-inactive form of one of the ribozymes was used to demonstrate that a majority of the reduction observed in target RNA and protein levels was due to cleavage and not to antisense function of the ribozyme RNA. The T7 RNA polymerase system is an alternative to the pol II and pol III transcription units utilized in many other studies; T7-generated transcripts are primarily cytoplasmic and can represent up to 30% of cytoplasmic RNA. Another useful attribute of T7 transcription is inducibility: without expression of the T7 RNA polymerase, no transcript will be generated from the T7 promoter.

VI. RIBOZYMES TARGETED AGAINST CANCER

Although most of the published work on therapeutic ribozymes has focused on antiviral applications, ribozymes have also been developed to decrease expression of genes involved in oncogenesis. Scanlon et al.[78] constructed a hammerhead ribozyme against the c-fos mRNA, whose gene product has been implicated in several aspects of cell growth in both normal and malignant cells. In particular, elevated levels of c-fos seem to confer heightened resistance to cisplatin, an important anticancer drug. To examine the dependence of cisplatin resistance on the c-fos gene product, the ribozyme was expressed from the dexamethasone-inducible MMTV promoter in an ovarian carcinoma cell line selected for cisplatin resistance. Negative controls were an antisense-orientation ribozyme gene, as well as a mutant ribozyme which lacked cleavage activity. Stably transformed cell lines were derived, and after exposure to dexamethasone, levels of c-fos expression and of cisplatin resistance were examined. The active ribozyme reduced both c-fos RNA and c-fos protein to levels found in noncisplatin-resistant cells, whereas the mutant ribozyme reduced fos protein levels by less than 50%. In addition, both the morphology and the cisplatin resistance phenotypes of the cells were similarly reversed by the active ribozyme. Since c-fos has been postulated to act as a transcriptional activator, they also examined the levels of several other gene products involved in cisplatin resistance. They found that levels of expression of two genes, dTMP synthetase and topoisomerase I, were markedly diminished by dexamethasone induction of the anti-c-fos ribozyme, suggesting a potential mechanism for the resistance conferred by c-fos. They also found that the ribozyme made the (cisplatin-resistant) cells more sensitive to three other antineoplastic agents, 5-fluorouracil, azidothymidine and camptothecin, as well as to cisplatin.

Koizumi et al.[79] availed themselves of the hammerhead target site created in the c-Ha-ras mRNA (the mutation is GGU to GUU, enabling cleavage of the oncogenic mutant, but not the wild-type sequence). They designed two functional ribozymes against this site, two differing only in the length of their hybridizing flanks, and a third with a G-C rich stem II. To assess the effect of these ribozymes on the target RNA and its transforming activity, NIH3T3 cells were transfected with two plasmids, one bearing a ribozyme gene and another bearing the c-Ha-ras gene (ribozyme expression was effected by placement downstream of the Rous sarcoma virus LTR promoter). Ribozymes one and two both reduced the number of transformed cell foci by approximately 50%; the third ribozyme did not show any effect, possibly due to disruption of its secondary structure by the changes made in one of the stems. In addition, RNA was isolated from these cells and analyzed by Northern hybridization; in cells cotransfected with ribozyme DNA, ras mRNA was not detectable. The argument that the inhibition was due to cleavage, and not antisense effects, was strengthened by use of a noncleaving ribozyme-substrate combination: in this case, a different c-Ha-ras mutant gene was used. The mRNA of this gene carries a point mutation in codon 16, but retains the GGU triplet and cannot be cleaved, though the ribozymes will bind to it. Its ability to induce foci in transfected NIH3T3 cells was undiminished by cotransfected ribozyme DNA.

The molecular event underlying creation of the Philadelphia chromosome (Ph) in chronic myelogenous leukemia (CML) is a transposition which fuses the coding regions of the c-abl and bcr genes in one of two ways. The protein(s) encoded by the bcr-abl gene are thought to be critical to the pathogenesis of Ph+ CML. Snyder et al.[80] devised a hammerhead ribozyme with flanking arms which span the fusion joint of the bcr exon 3-abl exon 2 oncogene mRNA. Hence, the ribozyme will fully bind only to the mutant mRNA and not to cellular RNA. A hybrid ribozyme molecule was synthesized with DNA in the flanking arms and RNA in the catalytic core; this hybrid molecule was encapsulated in a liposomal reagent and added to the culture medium of EM-2 cells, a Ph+ cell line. The functional ribozyme reduced bcr-abl mRNA levels to ~50% of levels observed in control cells, whereas neither a sense oligo nor an antisense oligo (bearing the same DNA nucleotides as the hybrid ribozyme) diminished the levels of bcr-abl mRNA observed on Northern hybridization. Levels of the bcr-abl fusion protein (p210) were markedly reduced by ribozyme transfection; the antisense oligo reduced protein levels slightly, and the sense oligo did not alter protein levels. Growth of EM-2 cells was inhibited 84% by the ribozyme and 71% by the antisense oligo.

VII. A RIBOZYME IN DROSOPHILA

Increasing awareness of the utility of ribozymes has prompted their application to purely investigative ends. A timely example is the work of Zhao and Pick,[81] who expressed a hammerhead ribozyme against the fushi tarazu (ftz) mRNA in developing Drosophila larvae transgenic for the ribozyme gene. (The ftz protein is involved in the control of segmentation and neurogenesis in Drosophila development.) Ribozyme expression from a heat-inducible promoter (hsp 70) allowed them to examine the consequences of ftz gene inactivation at various developmental stages. They showed that ribozyme activation (by heat pulse) at the blastoderm stage produced segmentation defects similar to those obtained in temperature-sensitive ftz mutants. Also, they demonstrated that expression of the ribozyme later in development disrupted nervous system development. These findings, since they mimic the phenotypes of known ftz mutations, confirm that the ribozyme inactivated the function of its target gene. Parallel experiments with both an inactive ribozyme and an antisense gene (lacking only the catalytic core sequences) strongly suggested that the majority of the inhibition was due to cleavage activity: only weak ftz mutant phenotypes were observed in larvae bearing either of the two control genes, whereas the active ribozyme induced strong phenotypes in a significant percentage of larvae (e.g., 5% of larvae transgenic for the active ribozyme exhibited a "pair-rule" phenotype indicative of disrupted segmentation). This study should be the harbinger of more widespread applications of ribozymes to study specific gene inactivation during development.

VIII. INTRACELLULAR TARGETING

Most early antisense experiments utilized a standard pol II expression strategy: the antisense molecule (including ribozymes) was treated as a mRNA. However, large excesses of antisense RNA over target RNA were almost invariably required for effective inhibition of target gene expression, and still the strategy often failed. Such cases prompted exploration of other expression systems. An example is the use of "chimeric" antisense/ribozyme constructs embedded within the genes for structural RNAs, such as tRNA genes (transcribed by pol III) or snRNA genes. Use of these promoter systems results in abundant and stable transcripts which often localize to the same intracellular compartment as the native RNA. (Abundance of transcripts is especially marked for pol III promoters; tRNA is up to 100 times more abundant in mammalian cells than polyadenylated RNA.) Another advantage, in many cases, is the facilitated interaction between the chimeric molecule and its intended target RNA.

As mentioned earlier, Jennings and Molloy[68] reported the use of an adenovirus pol III promoter linked to 163 nt of antisense targeted to the SV40 large T-antigen gene. Transfection of COS1 cells with this pol III antisense construct resulted in a substantial, though transient, inhibition of T-antigen-dependent DNA replication.

Cotten and Birnstiel[82] used the pol III transcription unit of a tRNA gene to express a hammerhead ribozyme-tRNA^{Met} chimeric molecule in microinjected Xenopus oocytes. (Their ribozyme-tRNA was targeted to a sequence near the 5′ end of U7 snRNA, a molecule essential to 3′ processing of histone mRNAs.) They found that the majority of the chimeric transcripts localized to the oocyte nucleus and remained in unprocessed, pre-tRNA form, which might account for the lesser stability of the chimeric molecule relative to the wild-type tRNA. Even with this handicap, the chimeric ribozyme in oocyte extract was as active as synthetic linear ribozyme at cleaving substrate molecules *in vitro*. If the (predominantly pre-tRNA) chimera is modified as are wild-type tRNA molecules, then the modifications did not hinder the cleavage ability of the ribozyme-tRNA.

Izant and colleagues (reviewed in Izant[83]) created antisense chimeric genes based on both tRNA and U2 snRNA genes for inhibition of a CAT reporter gene in Xenopus oocytes. They found that the U2 gene would accommodate inserts of up to 260 nucleotides (at an XhoI site near the 5′ end of the gene) without disruption of normal levels of transcription or normal subcellular localization and ribonucleoprotein formation. Microinjection of reporter and antisense-chimera plasmids into oocytes demonstrated up to 80% inhibition of CAT gene expression; however, this degree of inhibition required a large molar excess of antisense plasmid. When the antisense-chimera plasmid was injected 24 h prior to the reporter plasmid, inhibition of up to 98 to 100% was observed with only a fivefold excess of antisense plasmid.

The antisense-tRNA chimeras created by Izant and colleagues had antisense inserts placed either within the anticodon loop, or appended to the 3′ end, of the Xenopus tRNA^{Arg} molecule, obviating the antisense strategy. However, the 3′-extended constructs were always processed to normal-length molecules, obviating the antisense strategy. Work with the loop-embedded antisense genes showed that inserts of up to 80 bases were tolerated, and the resulting transcripts were present at very high levels, often representing the most abundant newly transcribed RNA in the oocytes. Up to 50% of the tRNA-antisense chimeras were localized in the germinal vesicle, 24 h after microinjection of plasmid, whereas native tRNAs are ~95% cytoplasmic. Extending this work in oocytes, Izant and colleagues studied the expression and activity of these constructs in transgenic mice. They found that the chimeric genes were expressed in multiple tissues, although at lower levels than the endogenous tRNA^{Arg} gene. (They note that viral LTR promoters, in *general, suffer from greater tissue variability in expression.*) When mice transgenic for the chimera gene were crossed with mice transgenic for a CAT reporter gene, they found significant inhibition of the reporter gene (up to 70%, as an average of littermates). No effect of the chimeric gene on weight, growth, or fertility of transgenic mice was noted, indicating the probable safety of such constructs. However, there was significant variation of the level of CAT gene inhibition between individuals, possibly a result of position effects on transcriptional levels.

The above-mentioned tRNA-antisense constructs were primarily studied in transient transfection assays (excepting the transgenic mouse work). Using a retroviral vector to obtain stable NIH3T3 cell lines, Sullenger et al.[85a] succeeded in appending MoMLV antisense constructs to the 3′ end of a tRNA$_i$^{Met} gene. This chimeric gene was efficiently expressed and effectively inhibited MoMLV replication only when placed within the 3′ LTR of a double-copy vector. (Constructs targeting the gag gene were much more effective than those designed against pol gene transcripts.) Inhibition seemed to be exerted through translational inhibition.

Colocalization of antisense/ribozyme RNA and target RNA may be enhanced by the incorporation of specific localization signals present in the target RNA. (Another advantage of such colocalization is the potential for reduced nonspecific activity of the ribozyme or antisense molecule.) For example, packaging of retroviral genomic RNAs requires a cis-acting sequence within the RNA itself, near the 5' terminus. Sullenger and Cech[85] designed a pair of hammerhead ribozymes targeted to a lacZ reporter gene; these constructs were expressed from MoMLV vectors in a packaging cell line. Both the ribozyme-expressing and the lacZ-expressing vectors generated genome-length transcripts bearing a retroviral packaging signal, which mediates incorporation of genomic RNAs into viral particles assembling at the cell membrane. Sullenger and Cech[85] reported that the titer of lacZ-containing viral particles was reduced by 90%, while the translation of lacZ mRNA (the pool of nonpackaged RNA, not colocalized with the ribozyme but still potentially cleavable) was unaffected.

Another example of the use of a cis-acting RNA sequence to colocalize ribozyme and target is the attachment of a ribozyme to the tRNA^{Lys} molecule required for synthesis of HIV DNA. (This strategy also affords the high expression levels of pol III genes, discussed earlier.) The human tRNA^{Lys} hybridizes to the primer binding site (PBS) of HIV RNA and primes synthesis of DNA by RT. Larson and Rossi (personal communication) have appended a hammerhead ribozyme to the 3' end of a modified tRNA^{Lys} gene. This ribozyme is targeted to a site just downstream of the PBS. This chimeric tRNA-ribozyme is efficiently recognized by HIV RT and can therefore become encapsidated in the virion. Early results indicate markedly reduced infectivity of HIV virions produced from tRNA-ribozyme-expressing cells.

IX. CONCLUSION

In conclusion, we might remark upon the relatively brief time in which the hammerhead and the hairpin have been ushered from their origins, in plant virioids, into the front lines of experimental therapy. Ribozyme applications have benefited greatly from earlier studies of antisense gene regulation, which laid the foundations of vector development and promoter usage. In this review, we have attempted to pull together diverse topics which converge in the effective cellular delivery of ribozymes. Currently, several different ribozyme expression vectors are nearing clinical application for such diseases as HIV infection and cancer. It is to be hoped that the delivery strategies outlined in this review will augment the potential of ribozymes as an effective disease intervention.

ACKNOWLEDGMENTS

This work was supported by the American Foundation for AIDS Research (AmFAR) Grant 01917-14-RG, the American Foundation for Pediatric AIDS Research (PAF/AmFAR) Grant 500331-14-PG, and the National Institute of Health Grants AI25959 and AI29329.

REFERENCES

1. Cech, T.R. and Bass, B.L. Biological catalysis by RNA. *Annu. Rev. Biochem.*, 55, 599, 1986.
2. Symons, R.H. Small catalytic RNAs. *Annu. Rev. Biochem.*, 61, 641, 1992.
3. Kerr, S.M., Stark, G.R., and Kerr, I.M. Excess antisense RNA from infectious recombinant SV40 fails to inhibit expression of a transfected, interferon-inducible gene. *Eur. J. Biochem.*, 175, 65, 1988.

4. Castgnotto, D., Rossi, J.J., and Deshler, J.O. Biological and functional aspects of catalytic RNAs. *Crit. Rev. Eukaryotic Gene Expression*, 2, 331, 1992.

5. Haseloff, J. and Gerlach, W.L. Simple RNA enzymes with new and highly specific endonuclease activity. *Nature*, 334, 585, 1988.

6. Burke, J.M. and Berzal-Herranz, A. *In vitro* selection and evolution of RNA: applications for catalytic RNA, molecular recognition, and drug discovery. *FASEB J.*, 7, 106, 1992.

7. Beaudry, A.A. and Joyce, G.F. Directed evolution of an RNA enzyme. *Science*, 257, 635, 1992.

8. Usman, N. and Cedergren, R. Exploiting the chemical synthesis of RNA. *Trends Biochem. Sci.*, 17, 334, 1992.

9. Perreault, J., Wu, T., Cousineau, B., Ogilvie, K.K., and Cedergren, R. Mixed deoxy- and ribo-oligonucleotides with catalytic activity. *Nature*, 344, 565, 1990.

10. Taylor, N.R., Kaplan, B.E., Swiderski, P., Li, H., and Rossi, J.J. Chimeric DNA-RNA hammerhead ribozymes have enhanced *in vitro* catalytic efficiency and increased stability *in vivo*. *Nucleic Acids Res.*, 20, 4559, 1992.

11. Hendry, P., McCall, M.J., Santiago, F.S., and Jennings, P.A. A ribozyme with DNA in the hybridising arms displays enhanced cleavage agility. *Nucleic Acids Res.*, 20, 5737, 1992.

12. Shimayama, T., Nishikawa, F., Nishikawa, S., and Taira, K. Nuclease-resistant chimeric ribozymes containing deoxyribonucleotides and phosphorothioate linkages. *Nucleic Acids Res.*, 21, 2605, 1993.

13. Pieken, W.A., Olsen, D.B., Benseler, F., Aurup, H., and Eckstein, F. Kinetic characterization of ribonuclease-resistant 2′-modified hammerhead ribozymes. *Science*, 253, 314, 1991.

14. Agrawal, S., Ikeuchi, T., Sun, D., Sarin, P.S., Konopka, A., Maizel, J., and Zamecnik, P.C. Inhibition of human immunodeficiency virus in early infected and chronically infected cells by antisense oligodeoxynucleotides and their phosphorothioate analogues. *Proc. Natl. Acad. Sci., U. S. A.*, 86, 7790, 1989.

15. Stein, C.A., Neckers, L.M., Nair, B.C., Mumbauer, S., Hoke, G., and Pal, R. Phosphorothioate oligodeoxycytidine interferes with binding of HIV-1 gp120 to CD4. *J. Acquired Immune Deficiency Syndromes*, 4, 686, 1991.

16. Slim, G. and Gait, M.J. Configurationally defined phosphorothioate-containing oligoribonucleotides in the study of the mechanism of cleavage of hammerhead ribozymes. *Nucleic Acids Res.*, 19, 1183, 1991.

17. Heidenreich, O. and Eckstein, F. Hammerhead ribozyme-mediated cleavage of the Long Terminal Repeat RNA of Human Immunodeficiency Virus type 1. *J. Biol. Chem.*, 267, 1904, 1992.

18. Paolella, G., Sproat, B.S., and Lamond, A.I. Nuclease resistant ribozymes with high catalytic activity. *EMBO J.*, 11, 1913, 1992.

19. Keller, T.H. and Haner, R. Synthesis and hybridization properties of oligonucleotides containing 2′-O-modified ribonucleotides. *Nucleic Acids Res.*, 21, 4499, 1993.

20. Fu, D.-J., Rajur, S.B., and McLaughlin, L.W. Importance of specific guanosine N[7]-nitrogens and purine amino groups for efficient cleavage by a hammerhead ribozyme. *Biochemistry*, 32, 10629, 1993.

21. Tuschl, T., Ng, M.M., Pieken, W., Benseler, F., and Eckstein, F. Importance of exocyclic base functional groups of central core guanosines for hammerhead ribozyme activity. *Biochemistry*, 32, 11658, 1993.

22. Sioud, M., Natvig, J.B., and Forre, O. Preformed ribozyme destroys Tumor Necrosis Factor mRNA in human cells. *J. Mol. Biol.*, 223, 831, 1992.

23. Loke, S.L., Stein, C.A., Zhang, X.H., Mori, K., Nakanishi, M., Subasinghe, C., Cohen, J.S., and Neckers, L.M. Characterization of oligonucleotide transport into living cells. *Proc. Natl. Acad. Sci. U.S.A.*, 86, 3474, 1989.

24. Yakubov, L.A., Deeva, E.A., Zarytova, V.F., Ivanova, E.M., Ryte, A.S., Yurchenko, L.V., and Vlassov, V.V. Mechanism of oligonucleotide uptake by cells: involvement of specific receptors? *Proc. Natl. Acad. Sci. U.S.A.*, 86, 6454, 1989.

25. Juliano, R.L. and Akhtar, S. Liposomes as a drug delivery system for antisense oligonucleotides. *Antisense Res. Dev.*, 2, 165, 1992.

26. Leonetti, J.-P., Machy, P., Degols, G., Lebleu, B., and Leserman, L. Antibody-targeted liposomes containing oligodeoxyribonucleotides complementary to viral RNA selectively inhibit viral replication. *Proc. Natl. Acad. Sci. U.S.A.*, 87, 2448, 1990.

27. Wagner, E., Cotten, M., Foisner, R., and Birnstiel, M.L. Transferrin-polycation-DNA complexes: the effect of polycations on the structure of the complex and DNA delivery to cells. *Proc. Natl. Acad. Sci. U.S.A.*, 88, 4255, 1991.

28. Cotten, M. The *in vivo* applications of ribozymes. *Trends Biotechnol.*, 8, 174, 1990.

29. Wu, G.Y. and Wu, C.H. Receptor-mediated *in vitro* gene transformation by a soluble DNA carrier system. *J. Biol. Chem.*, 262, 4429, 1987.

30. Wu, G.Y. and Wu, C.H. Receptor-mediated gene delivery and expression *in vivo*. *J. Biol. Chem.*, 263, 14621, 1988.

31. Gao, L., Wagner, E., Cotten, M., Agarwal, S., Harris, C., Romer, M., Miller, L., Hu, P.-C., and Curiel, D. Direct *in vivo* gene transfer to airway epithelium employing adenovirus-polylysine-DNA complexes. *Hum. Gene Ther.*, 4, 17, 1993.

32. Morgan, R.M. and Anderson, W.F. Human gene therapy. *Annu. Rev. Biochem.*, 62, 191, 1993.

33. Mulligan, R.C. The basic science of gene therapy. *Science*, 260, 926, 1993.

34. Tolstoshev, P. Gene therapy, concepts, current trials and future directions. *Annu. Rev. Pharmacol. Toxicol.*, 32, 57, 1993.

35. Friedman, T. A brief history of gene therapy. *Nature Genetics*, 2, 93, 1992.

36. Miller, A.D. Human gene therapy comes of age. *Nature*, 357, 455, 1992.

37. Majors, J. Retroviral vectors — strategies and applications. Semin. Virol., 3, 285, 1992.

38. To, R.Y.-L. and Neiman, P.E., The potential for effective antisense inhibition of retroviral replication mediated by retroviral vectors. *Gene Regulation: Biology of Antisense RNA and DNA*. Raven Press, New York, 1992. pp. 261–27.

39. Miller, D.G., Adam, M.A., and Miller, A.D. Gene transfer by retrovirus vectors occurs only in cells that are actively replicating at the time of infection. *Mol. Cell. Biol.*, 10, 4239, 1990.

40. Miller, A.D. and Buttimore, C. Redesign of retrovirus packaging cell lines to avoid recombination leading to helper virus production. *Mol. Cell. Biol.*, 6, 2895, 1986.

41. Markowitz, D., Goff, S., and Bank, A. Construction and use of a safe and efficient amphotropic packaging cell line. *Virology*, 167, 400, 1988.

42. Donahue, R.E., Kessler, S.W., Bodine, D., McDonagh, K., Dunbar, C., Goodman, S., Agricola, B., Byrne, E., Raffeld, M., Moen, R., Bacher, J., Zsebo, K.M., and Nienhuis, A.W. Helper virus-induced T cell lymphoma in nonhuman primates after retroviral-mediated gene transfer. *J. Exp. Med.*, 176, 1125, 1992.

43. Emi, N., Friedmann, T., and Yee, L.-K. Pseudotype formation of murine leukemia virus with the G protein of vesicular stomatitis virus. *J. Virol.*, 65, 1202, 1991.

44. Burns, J.C., Friedmann, T., Driever, W., Burrascano, M., and Yee, J.-K. Vesicular stomatitis virus G glycoprotein pseudotyped retroviral vectors: concentration to very high titer and efficient gene transfer into mammalian and nonmammalian cells. *Proc. Natl. Acad. Sci. U.S.A.*, 90, 8033, 1993.

45. Poznansky, M., Lever, M., Bergeron, L., Haseltine, W., and Sodroski, J. Gene transfer into human lymphocytes by a defective Human Immunodeficiency Virus type 1 vector. *J. Virol.*, 65, 532, 1991.

45a. Sodroski, J., 3rd Int. Symp. on Ribozymes and Targeted Gene Therapy, San Diego, Dec. 1992.

46. Buchschacher, G.L., Jr. and Panganiban, A.T. Human Immunodeficiency Virus vectors for inducible expression of foreign genes. *J. Virol.*, 66, 2731, 1992.

47. Lusso, P., Veronese, F.D., Ensoli, B., Franchini, G., Jemma, C., DeRocco, S.E., Kalyanaraman, V.S., and Gallo, R.C. Expanded HIV-1 cellular tropism by phenotypic mixing with murine endogenous retroviruses. *Science*, 247, 848, 1990.

48. Weerasinghe, M., Liem, S.E., Asad, S., Read, S.E., and Joshi, S. Resistance to Human Immunodeficiency Virus Type 1 (HIV-1) infection in human CD4+ lymphocyte-derived cell lines conferred by using retroviral vectors expressing an HIV-1 RNA-specific ribozyme. *J. Virol.*, 65, 5531, 1991.

49. Rosenfeld, M.A., Siegfried, W., Yoshimura, K., Yoneyama, K., Fukayama, M., Stier, L.E., Paakko, P.K., Gilardi, P., Stratford-Perricaudet, L.D., Perricaudet, M., Jallat, S., Pavirani, A., Lecocq, J.-P., and Crystal, R.G. Adenovirus-mediated transfer of a recombinant α-1 antitrypsin gene to the lung epithelium *in vivo*. *Science*, 252, 431, 1991.

50. Lemarchand, P., Jaffe, H.A., Danel, C., Cid, M.C., Kleinman, H.K., Stratford-Perricaudet, L.D., Perricaudet, M., Pavirani, A., Lecocq, J.-P., and Crystal, R.G. 1992. Adenovirus-mediated transfer of a recombinant human α-1-antitrypsin cDNA to human endothelial cells. *Proc. Natl. Acad. Sci. U.S.A.*, 89, 6482, 1992.

51. Chatterjee, S. and Wong, K.K., Jr. Adeno-associated viral vectors for the delivery of antisense RNA. *Methods: A Companion to Methods Enzymol.*, 5, 51, 1993.

52. Chatterjee, S., Johnson, P.R., and Wong, K.K., Jr. Dual-target inhibition of HIV-1 *in vitro* by means of an adeno-associated virus antisense vector. *Science*, 258, 1485, 1992.

53. Fink, D.J., Sternberg, L.R., Weber, P.C., Mata, M., Goins, W.F., and Glorioso, J.C. *In vivo* expression of β-galactosidase in hippocampal neurons by HSV-mediated gene transfer. *Hum. Gene Ther.*, 3, 11, 1992.

53a. Glorioso, J., 3rd Int. Symp., Dec. 1992.

53b. Taylor, J., 3rd Int. Symp., Dec. 1992.

54. Hsieh, S.-Y. and Taylor, J. Delta virus as a vector for the delivery of biologically-active RNAs: possibly a ribozyme specific for chronic hepatitis B virus infection. *Adv. Exp. Med. Biol.*, 312, 125, 1992.

55. Fuerst, T.R. and Moss, B. Structure and stability of mRNA synthesized by vaccinia virus-encoded bacteriophage T7 RNA polymerase in mammalian cells. *J. Mol. Biol.*, 206, 333, 1989.

56. L'Huillier, P.J., Davis, S.R., and Bellamy, A.R. Cytoplasmic delivery of ribozymes leads to efficient reduction in α-lactalbumin mRNA levels in C1271 mouse cells. *EMBO J.* 11, 4411, 1992.

57. Willems, W.R., Kaluza, G., Boscehk, C.B., and Bauer, H. Semliki Forest virus: cause of a fatal case of human encephalitis. *Science*, 203, 1127, 1979.

58. Paul, N.L., Marsh, M., McKeating, J.A., Schulz, T.F., Liljestrom, P., Garoff, H., and Weiss, R.A. Expression of HIV-1 envelope glycoproteins by Semliki Forest virus vectors. *AIDS Res. Hum. Retroviruses*, 9, 963, 1993.

59. Wolff, J.A., Malone, R.W., Williams, P., Chong, W., Acsadi, G., Jani, A., and Felgner, P.L. Direct gene transfer into mouse muscle *in vivo*. *Science*, 247, 1465, 1990.

60. Nabel, E.G., Plautz, G., and Nabel, G.J. Transduction of a foreign histocompatibility gene into the arterial wall induces vasculitis. *Proc. Natl. Acad. Sci. U.S.A.*, 89, 5157, 1992.

61. Nabel, E.G., Yang, Z., Plautz, G., Forough, Z.R., Zhan, X., Haudenschild, C.C., Maciag, T., and Nabel, G.J. Recombinant fibroblast growth factor-1 promotes intimal hyperplasia and angiogenesis in arteries *in vivo*. *Nature*, 362, 844, 1993.

61a. Weiner, D. et al., 3rd Int. Symp., Dec. 1992.

62. Enssle, J., Kugler, W., Hentze, M.W., and Kulozik, A.E. Determination of mRNA fate by different RNA polymerase II promoters. *Proc. Natl. Acad. Sci. U.S.A.*, 90, 10091, 1993.

63. Wilson, C., Bellen, H.J., and Gehring, W.J. Position effects on eukaryotic gene expression. *Annu. Rev. Cell Biol.*, 6, 679, 1990.

64. Yu, M., Ojwang, J., Yamada, O., Hamper, A., Rapapport, J., Looney, D., and Wong-Staal, F. A hairpin ribozyme inhibits expression of diverse strains of human immuno-deficiency virus type 1. *Proc. Natl. Acad. Sci. U.S.A.*, 90, 6340, 1993.

65. Taira, K. and Nishikawa, S. Construction of several kinds of ribozymes. *Gene Regulation: Biology of Antisense RNA and DNA*. Raven Press, New York, 1992, pp. 35–53.

66. Altschuler, M., Tritz, R., and Hampel, A. A method for generating transcripts with defined 5' and 3' termini by autocatalytic processing. *Gene*, 122, 85, 1992.

67. Ventura, M., Wang, O., Ragot, T., Perricaudet, M., and Saragosdti, S. Activation of HIV-specific ribozyme activity by self-cleavage. *Nucleic Acids Res.*, 21, 3249, 1993.

68. Jennings, P.A. and Molloy, P.L. Inhibition of SV40 replication function by engineered antisense RNA transcribed by RNA polymerase III. *EMBO J.*, 6, 3043, 1987.

69. Cameron, F.H. and Jennings, P.A. Specific gene suppression by engineered ribozymes in monkey cells. *Proc. Natl Acad. Sci. U.S.A.*, 86, 9139, 1989.

70. Sarver, N., Cantin, E.M., Chang, P.S., Zaia, J.A., Ladne, P.A., Stephens, D.A., and Rossi, J.J. Ribozymes as potential anti-HIV-1 therapeutic agents. *Science*, 247, 1222, 1990.

71. Chang, P.S., Cantin, E.M., Saia, J.A., Ladne, P.A., Stephens, D.A., Sarver, N., and Rossi, J.J. Ribozyme-mediated site-specific cleavage of the HIV-1 genome. *Clin. Biotechnol.*, 2, 23, 1990.

72. Lo, K.M.S., Biasolo, M.A., Dehni, G., Palu, G., and Haseltine, W. Inhibition of replication of HIV-1 by retroviral vectors expressing tat-antisense and anti-tat ribozyme RNA. *Virology*, 190, 176, 1992.

73. Ojwang, J.O., Hampel, A., Looney, D.L., Wong-Staal, F., and Rappaport, J. Inhibition of human immunodeficiency virus type 1 expression by a hairpin ribozyme. *Proc. Natl. Acad. Sci. U.S.A.*, 89, 10802, 1992.

74. Chen, C.-J., Banerjea, A.C., Harmison, G.C., Haglund, K., and Schubert, M. Multitarget-ribozymes directed to cleave at up to none highly conserved HIV-1 env RNA regions inhibits HIV-1 replication — potential effectiveness against most presently sequenced HIV-1 isolates. *Nucleic Acids Res.*, 20, 4581, 1992.

75. Dropulic, B., Lin, N.H., Martin, M.A., and Jeang, K.-T. Functional characterization of a U5 ribozyme: intracellular suppression of Human Immunodeficiency Virus type 1 expression. *J. Virol.*, 66, 1432, 1992.

75a. Zhou, C., Bahner, I.C., Larson, G.P., Zaia, J.A., Rossi, J.J., and Kohn, D.B. *Gene,* 149, 33, 1994.

76. Xing, Z. and Whitton, J.L. Ribozymes which cleave arenavirus RNAs: identification of susceptible target sites and inhibition by target site secondary structure. *J. Virol.*, 66, 1361, 1992.

77. Xing, Z. and Whitton, J.-L. An anti-lymphocytic choriomeningitis virus ribozyme expressed in tissue culture cells diminishes viral RNA levels and leads to a reduction in infectious virus yield. *J. Virol.*, 67, 1840, 1993.

78. Scanlon, K.J., Jiao, L., Funato, T., Wang, W., Tone, T., Rossi, J.J., and Kashani-Sabet, M. Ribozyme-mediated cleavage of c-fos mRNA reduces gene expression of DNA synthesis enzymes and metallothionein. *Proc. Natl. Acad. Sci. U.S.A.*, 88, 10591, 1991.

79. Koizumi, M., Kamiya, H., and Ohtsuka, E. Ribozymes designed to inhibit transformation of NIH3T3 cells by the activated c-Ha-ras gene. *Gene*, 117, 179, 1992.

80. Snyder, D.S., Wu, Y., Wang, J.L., Rossi, J.J., Swiderski, P., Kaplan, B.E., and Forman, S.J. Ribozyme-mediated inhibition of bcr-abl gene expression in a Philadelphia chromosome-positive cell line. *Blood*, 82, 600, 1993.

81. Zhao, J.J. and Pick, L. Generating loss-of-function phenotypes of the fushi tarazu gene with a targeted ribozyme in Drosophila. *Nature*, 365, 448, 1993.

82. Cotten, M. and Birnstiel, M.L. Ribozyme mediated destruction of RNA *in vivo*. *EMBO J.*, 8, 3861, 1989.

83. Izant, J.G. Chimeric antisense RNAs. *Gene Regulation: Biology of Antisense RNA and DNA*. Raven Press, New York, 1992, pp. 183–195.

84. Symons, R.H. Small catalytic RNAs. *Annu. Rev. Biochem.*, 61, 641, 1992.

85. Sullenger, B.A. and Cech, T.R. Tethering ribozymes to a retroviral packaging signal for destruction of viral RNA. *Science*, 262, 1566, 1993.

85a. Sullenger, B.A., Lee, T.C., Smith, C.A., Ungers, G.E., and Gilboa, E. Expression of chimeric tRNA-driven antisense transcripts renders NIH 3T3 cells highly resistant to Moloney Murine Leukemia Virus replication. *Mol. Cell. Biol.*, 10, 6512, 1990.

86. Joseph, S., Berzal-Herranz, A., Chowrira, B.M., Butcher, S.E., and Burke, J.M. Substrate selection rules for the hairpin ribozyme determined by *in vitro* selection, mutation, and analysis of mismatched substrates. *Genes Dev.*, 7, 120, 1993.

Chapter 3

Improved Structural Design and Cellular Uptake of Triplex-Forming Oligonucleotides

Nilabh Chaudhary, Jeffrey S. Bishop, Krishna Jayaraman, and Judith K. Guy-Caffey

CONTENTS

I. INTRODUCTION

Triple-stranded nucleic acids were first described soon after the classical structure of duplex DNA was solved.[1] Subsequently, short oligonucleotides were shown to bind to double-helical DNA or RNA to form triple-helical structures or triplexes (Figure 1). Experiments in a number of laboratories demonstrated that triple helix-forming oligonucleotides (TFOs) can recognize and bind to the major groove of duplex DNA in a sequence-specific manner.[2-4] More recently, TFOs have been designed to bind to the promoter region of certain genes at physiological pH and temperature with the aim of interfering with interactions between various transcription or enhancer factors and their cognate DNA elements.[5-21] The biological and pharmacological implications of these developments are enormous. By targeting a few well-defined DNA duplex sites in the nucleus instead of the much larger populations of RNA or proteins, the TFO antigene technology offers the promise of unprecedented control over gene expression. TFOs can potentially be used as therapeutic agents to specifically modulate the expression of genes involved in the development of a broad range of diseases of neoplastic, endocrine, immunological, viral, or microbial origin (Table 1).

In this chapter, we will briefly discuss the fundamental nature of triplex formation which provides the theoretical basis for the use of TFOs as antigene therapeutics and review the current developments of TFO-mediated inhibition of gene expression. After

0-8493-4778-5/95/$0.00+$.50
© 1995 by CRC Press Inc.

Figure 1 Schematic diagram of a triple helix showing the third (darker) strand binding to duplex DNA in the major groove.

Table 1 **Potential disease targets of TFO therapeutics**

Viral diseases[a]
  Adenovirus, CMV, EBV, hepatitis A and B, herpes zoster, HIV, HPV, HSV-1, HSV-2, HTLV-1, influenza A and B, parainfluenza
Oncologic diseases
  Carcinoma (of colon, prostate, kidney, bladder, breast), glioblastoma, leukemia, lymphoma, melanoma, osteosarcoma
Autoimmune diseases
  Arthritis, lupus erythromatosus, myasthenia gravis, psoriasis
Other
  Allergy, drug resistance, cardiovascular diseases, inflammatory diseases

[a] Abbreviations: CMV, cytomegalovirus; EBV, Epstein-Barr virus; HIV, human immunodeficiency virus; HPV, human papilloma virus; HSV, herpes simplex virus; HTLV, human T-cell lymphotropic virus.

outlining the existing technological limitations of the triple helix strategy with regard to biological and clinical applications, we will discuss the multifaceted approaches that are being used to address many of those limitations. In particular, we will focus on the recent progress made in our laboratory for enhancing the cellular and nuclear delivery of potentially therapeutic 2'-deoxyguanosine (G)- and thymidine (T)-rich TFOs.

II. OVERVIEW OF TRIPLEX FORMATION

TFOs bind in the major groove of duplex nucleic acids in a sequence-specific manner to form three-stranded complexes, or triplexes (Figure 1). The first reported triplex structures were composed of RNA polynucleotides containing one strand of polyadenosine (poly rA) bound to two strands of polyuridine (poly U).[1] Subsequent studies described the formation of homopolymeric DNA complexes, such as poly(T)·poly(A)·poly(T) and poly(C+T)·poly(GA)·poly(TC).[22,23] Later, it was discovered that short oligonucleotides of mixed base composition and sequence can recognize and bind to specific sites on duplex DNA.[3-5] TFO binds to the target DNA in either a parallel or antiparallel direction, which describes the 5' to 3' orientation of the third strand relative to the purine-rich strand of the duplex. The stability and orientation of triplex formation is influenced by several factors, including characteristics of the TFO such as base composition and size, as well as environmental factors such as salt concentration and pH. Oligonucleotides composed of 2'-deoxycytidine (C) and T have been shown to bind to purine-rich regions of duplex DNA in a parallel orientation. The T and C residues are hydrogen bonded to the purines of the duplex DNA, resulting in T·AT and C+·GC triplets.[3,4] Triplex formation in this motif requires protonation of the third-strand C residues and is thus limited to acidic conditions. In contrast, oligonucleotides containing G and T can form triplexes at physiological pH.[5] In this case, the third strand is oriented antiparallel to the purine-rich strand of the duplex and is stabilized by T·AT and G·GC triplets (Table 2).[24,25]

Table 2   **Evolution in the design of GT-Rich TFOs**

Year	Triple-Helix-Forming Oligonucleotides		Ref.
1988	d($\leftarrow$T$-$T$-$C$-$T$-$C$-$T$-$) d($-$A$-$A$-$G$-$A$-$G$-$A$\rightarrow$)	Duplex	5
	d($\leftarrow$T$-$T$-$G$-$T$-$G$-$T$-$)	Third strand	
1992[a]	d($\leftarrow$T$-$C$-$C$-$C$-$C$-$T$-$) d($-$A$-$G$-$G$-$G$-$G$-$A$\rightarrow$)	Duplex	34-36
	d($\leftarrow$T$-$G$-$<u>^{6}G</u>$-$G$-$G$-$T$-$)	Third strand	
1993[b]	d($\leftarrow$T$-$C$-$T$-$C$-$C$-$T$-$) d($-$A$-$G$-$A$-$G$-$G$-$A$\rightarrow$)	Duplex	37
	d($\leftarrow$T$-$G$-$<u>X</u>$-$G$-$G$-$T$-$)	Third strand	
1993[c]	d($\leftarrow$T$-$C$-$C$-$C$-$C$-$T$-$) d($-$A$-$G$-$G$-$G$-$G$-$A$\rightarrow$)	Duplex	38
	d($\leftarrow$T$-$G$-$<u>^{9}G</u>$-$G$-$G$-$T$-$)	Third strand	
1994[d]	d($\leftarrow$T$-$T$-$C$-$A$-$C$-$T$-$) d($-$A$-$A$-$G$-$T$-$G$-$A$\rightarrow$)	Duplex	39
	d($\leftarrow$T$-$T$-$G$-$<u>I</u>$-$G$-$T$-$)	Third strand	
1994[e]	d($\leftarrow$T$-$C$-$G$-$C$-$C$-$T$-$) d($-$A$-$G$-$C$-$G$-$G$-$A$\rightarrow$)	Duplex	40
	d($\leftarrow$T$-$G$-$<u>D$_f$</u>$-$G$-$G$-$T$-$)	Third strand	

[a] <u>^{6}G</u>, 6-Thio-dG, to prevent G-tetrad formation

[b] <u>X</u>, 7-Deaza-deoxyocanthosine at a TA inversion site

[c] <u>G</u>, 9-Deaza-dG, to prevent G-tetrad formation

[d] <u>^{9}I</u>, Imidazole at a TA inversion site

[e] <u>D$_f$</u>, 2′-Deoxyformycin A at a CG inversion site

TFOs can bind to their targets with apparent dissociation constants (K_d) in the 10^{-7} to 10^{-9} M range, comparing favorably with the binding affinity of the *Escherichia coli* Trp repressor protein for its target sequence.[24] Binding is highly sequence dependent, as single-base mismatches can cause a 20- to 30-fold reduction in the affinity of TFOs for duplex DNA.[26] It is this stringent sequence requirement that gives rise to the possibility of achieving selective binding of TFOs to specific target sites as small as 15 base pairs

Table 3 **TFO-Mediated transcription inhibition**

Target	Ref.
c-*myc* oncogene	5,6,11
Interleukin-2 receptor subunit α	7,13,16
G-free cassette	8
Bacteriophage T7 promoter	9,14,18
Transposon Tn3 *bla* gene	10
Sp1 binding site	12
HIV-1	15
Progesterone response element	17
Interferon response element	19
FMLV[a] *gag* gene	20

[a] Abbreviation: FMLV, Friend Murine Leukemia Virus

in the DNA of human chromosomes.[4] The high affinity and remarkable specificity of binding of TFOs to target DNA have led to the investigation of TFOs as artificial, sequence-specific regulators of gene function.

III. TRIPLEX-FORMING OLIGONUCLEOTIDES AS ANTIGENE THERAPEUTICS

TFO-based therapeutics can be of potential value in treating a wide variety of diseases since, in principle, any disorder in which a gene is expressed aberrantly is a potential target for antigene compounds (Table 1). In viral infections, TFOs directed against key viral-specific gene products may prevent viral replication and production. In many types of cancer, TFOs could be targeted to specific oncogenes. TFOs may prove useful in suppressing production of autoimmune antibodies or receptors in certain autoimmune disorders. Resistant strains of bacteria, malaria, tuberculosis, and other infectious agents may be overcome by specific antigene therapy in which the sequences of the therapeutic oligonucleotides are revised as mutation of the microbes occurs.

The triple helix approach may also be used to directly regulate gene expression in gene therapy treatments. Although there are several procedures for introducing foreign genes into animal cells,[27,28] very few allow for specific modulation of transcription. Unique triplex-forming sites could be engineered ahead of the coding sequence of the therapeutic gene. Upon transfection into target cells, the expression of the gene could be modulated by treating patients with the appropriate TFO.

A. MODULATION OF GENE EXPRESSION BY TRIPLEXES

The antigene role of TFOs directed against natural and artificial targets has been demonstrated *in vitro* in a number of systems. In 1988, Cooney et al.[5] were the first to demonstrate that TFOs can specifically repress the transcription of a gene. A discrete 27-base G-rich TFO targeted against the purine-rich promoter of the c-*myc* oncogene was shown to interfere with transcription. Other investigators subsequently discovered that TFOs can be potent inhibitors of transcription of viral, cancer, and other genes (Table 3).[6-20] Typically, these TFOs bind to their respective targets with high affinity (K_d in nanomolar range) and compete effectively with transcription factors for binding to the promoter region. The *in vitro* findings have been extended to the whole cell system, and evidence for sequence-specific recognition has been obtained by DNase protection assays.[6]

Recently, a quantitative cell culture-based *functional assay has been established to* better understand how the TFO composition, modification, and formulation might affect the biological activity of the oligonucleotides *in vitro* and *in vivo*.[17] The basic design of this assay utilizes a progesterone-responsive cell line transfected with a plasmid containing a chloramphenicol acetyl transferase (CAT) reporter gene (Figure 2). The transcription of the CAT gene is under the control of a progesterone response element (PRE) containing two progesterone receptor complex binding sites, separated by a G-rich region that has been engineered to enable triplex formation with specific TFOs. Triple helix formation can block the binding of the progesterone receptor complex, thus preventing the transcription of CAT mRNA and production of CAT protein. Changes in CAT expression can be easily quantified using enzymatic or immunological detection methods.[17] The system is simple enough that individual parameters can be dissected and analyzed separately *in vitro* or *in vivo*. Binding studies *in vitro* near physiological conditions have shown that a TFO designed to bind to PRE can bind to its target with a K_d of 100 nM. In cell-free extracts, the same concentration of TFO can compete effectively with the progesterone complex for the PRE site and specifically block the transcription of a PRE-driven template.[17] However, in intact cells, addition of 3'-amine-modified TFOs to the extracellular medium in high micromolar concentrations had virtually no effect. When the TFO was derivatized with a cholesterol moiety to increase nuclear association (see below), a concentration of 20 μM was necessary to achieve 50% inhibition in CAT production (Figure 3).[17] The abrupt, several hundredfold increase in TFO concentration required to observe inhibitory activity *in vivo* suggests that in living cells, delivery of a TFO to its target is highly inefficient. Furthermore, at high TFO levels, the risk of observing nonspecific oligonucleotide effects is substantially increased.

B. HURDLES TO THERAPEUTIC TRIPLEX TECHNOLOGY

The above *in vitro* and *in vivo* studies have raised several important issues which must be addressed in order for triplex technology to attain therapeutic value (Table 4). Considering the vast range of potential antigene targets, it is critical to design TFOs that interact with unique sites. The ideal antigene TFO transcriptional inhibitors should bind with high affinity and specificity to a specific stretch of target duplex DNA under physiological conditions. The TFO should compete effectively with factors that normally bind to DNA, and the triplex structure must remain intact long enough to interfere with the expression of the target gene. There should be a mechanism to efficiently deliver the TFO to its site of action within the cell and the nucleus. The TFO should be nuclease resistant both outside and inside cells, but it should not significantly interfere with normal metabolic processes. If internalized by cells that are not therapeutic targets, the oligo should be eventually degraded into harmless components. The TFO should not mimic regulatory or signaling factors and it should not bind to proteins, RNA, or other cellular components. For long-term therapy, TFOs should not elicit a significant immune response or drug resistance. Finally, the oligonucleotide should be active *in vivo* at low doses; otherwise the cost of production and treatment may be prohibitive. The list of desirable properties in a therapeutic TFO is long, but not impossible to attain. The biochemical basis of specific DNA recognition and binding by TFOs has been firmly established.[3-21,24] In addition, many reports provide evidence that specific TFO-duplex DNA interactions can also occur in living cells.[6,7,15,17] The current challenge is to further improve the biological efficacy of TFOs in cell-based systems, and then to extend the utility of this technology to whole animals and clinically relevant pharmacological targets. Because of the significant hurdles that remain in the development of therapeutic oligonucleotides (Table 4), it is not unreasonable to expect that active TFO formulations will incorporate a variety of covalent and noncovalent modifications to improve affinity, specificity, cellular uptake, stability, and biological efficacy.

C. ADVANCES WITH ANTIPARALLEL TRIPLEXES
1. Affinity and Specificity of Triplex Formation

A central theme in TFO design is the continuing quest for oligonucleotides that will bind to any designated target with high affinity and specificity under physiological pH and salt

44

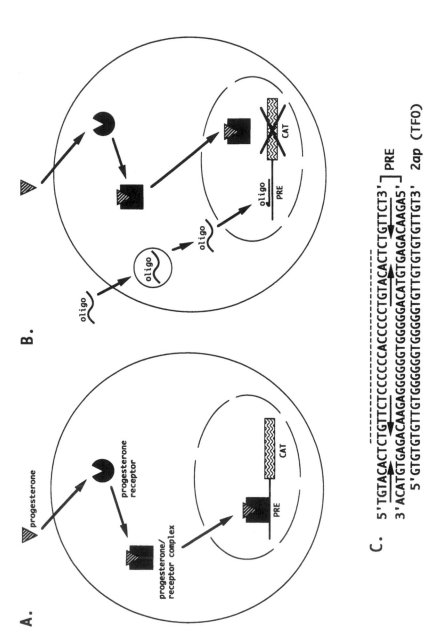

A.

progesterone

progesterone
receptor

progesterone/
receptor complex

PRE

CAT

B.

oligo

oligo

oligo

oligo

PRE

CAT

C.   5'TGTACACTCTGTTCTCCCCCACCCCTGTACACTCTGTTCT3'⎤ PRE
    3'ACATGTGAGACAAGAGGGGGTGGGGACATGTGAGACAAGA5'⎦
      5'GTGTGTTGTGTGGGGGTGGGGTGTTGTGTGTTGT3'   2αp (TFO)

    5'TTGTGGTGGTGGTGGGTTTGTGGTGGTGGTGT3'   CTR (control)

Figure 2 (facing page) Schematic diagram of the PRE assay for *in vivo* transcription inhibition. (A) In the absence of TFO, progesterone enters cells and binds to the cytoplasmic progesterone receptor, inducing a conformational change in the receptor. The progesterone/receptor complex enters the nucleus through a nuclear pore and binds to the progesterone response element (PRE) of the plasmid construct, thereby activating transcription of the CAT reporter gene. (B) In the presence of TFO, binding of the progesterone/receptor complex is blocked by the binding of the TFO at the PRE site, thereby inhibiting transcription of the CAT reporter gene. (C) The progesterone-responsive target sequence (PRE) contains two progesterone response elements from the tyrosine aminotransferase gene (half-sites indicated by arrows) separated by a linker with high purine/pyrimidine asymmetry. The 2ap TFO was designed to bind to the duplex target by Hoogsteen or reverse Hoogsteen base triplets to form an antiparallel DNA triplex. The extent of DNase I protection by the 2ap TFO is overlined on the target sequence. CTR is a control oligonucleotide.

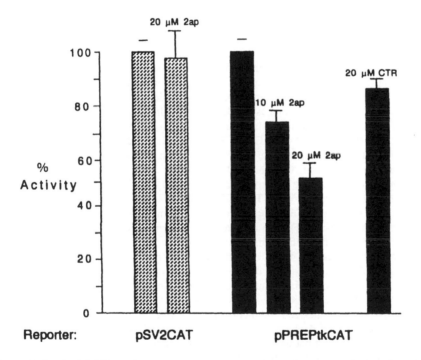

Figure 3 *In vivo* inhibition of progesterone receptor-dependent transcription. A 38-base cholesterol-modified GT-rich phosphodiester TFO (2ap) was added to CV-1 cells that were transiently transfected with either the pPREPtkCAT plasmid, with the PRE-containing TFO target (filled bars), or the control pSV2CAT plasmid (stippled bars). Full CAT activity from pPREPtkCAT was dependent upon cotransfection of a progesterone receptor expression vector and progesterone. The CAT activity of treated cells was calculated as a percentage of full CAT activity after subtraction of activity without progesterone. The average values from four experiments done in triplicate are shown with their standard errors. Cholesterol-derivatized oligonucleotides (PRE2ap and a control CTR) and their concentrations are indicated above the bars. (From Ing, N.H., Beckman, J.M., Kessler, D.J., Murphy, M., Jayaraman, K., Zendegui, J.G., Hogan, M.E., O'Malley, B.W., and Tesai, M.J., *Nucleic Acids Res.*, 21, 2789, 1993. With permission.)

Table 4 **Requirements for therapeutic triplex technology**

Uptake
- TFOs must be able to cross cellular membranes (plasma, endosomal membranes) to reach target in nucleus (intranuclear targeting?)

Stability
- TFOs must be resistant to degradation by nucleases both inside and outside cells

Specificity
- TFOs must be able to bind specifically to a unique duplex DNA target
- TFOs should not bind to proteins, RNA, or other cellular components

Affinity
- TFOs must be able to bind with high affinity to target duplex DNA under physiological conditions
- TFOs should compete effectively with factors that normally bind to target

Pharmacokinetics
- TFOs must have favorable absorption, distribution, metabolism, and excretion

concentrations. However, the actual "physiological" conditions in the microenvironment of the TFO target may never be known. To allow for this, it has been necessary to chemically modify bases in candidate TFOs to improve target recognition and binding over a wider range of salt and buffer conditions. In designing new phosphodiester TFOs, the general principle has been to retain the basic backbone of the molecules, to reduce the risk of introducing unknown and sometimes unfavorable characteristics into the modified molecule. Much progress on this aspect has been made in the subtle redesign of G-rich oligonucleotides which can form stable triplexes at physiological pH in the presence of divalent cations. One limitation of G-rich TFOs is the tendency to form intra- or intermolecular aggregates in the presence of high concentrations of sodium or potassium ions.[29,30] The guanine residues form "G-tetrads", structures that are formed by the coordination of the oxygen atoms of guanine with the alkali cation and stabilized by hydrogen bonds and strong stacking interactions.[30-33] G-Tetrad formation may reduce the bioavailability of the free, monomeric form of TFO, and the tetrad itself may have biological activities not normally associated with TFOs. In minimizing the formation of G-tetrads, the goal is to introduce modifications in the guanine residue which disrupt hydrogen bond formation and reduce self-association, without substantially altering the fundamental characteristics of the oligonucleotide. In addition, it is desirable to maintain the capacity for synthesizing the modified TFOs using existing automated oligonucleotide synthesizers. This approach to TFO design has recently been validated (Table 2).[34-40] Chemically modified guanine residues have been synthesized and incorporated into TFOs. The strategic replacement of a few G residues with 2'-deoxy-6-thioguanosine (6-thio-dG) disrupts multiple H-bonding, thereby reducing or eliminating the formation of G-tetrads, while preserving and enhancing the ability to form a triple helix in potassium-containing buffers.[34-36] Similarly, 2'-deoxy-7-deaza-xanthosine has been used successfully to reduce tetrad formation and enhance the binding affinity under physiological salt conditions (Table 2).[37] 9-Deaza-dG has been found to be equally successful for reducing G-tetrad formation but may decrease binding affinity.[38]

In vitro binding assays have shown that GT-rich TFOs can bind to duplex DNA in a sequence-specific manner, yet the requirement for relatively long stretches of purines (at least 85%) in one strand of the duplex significantly reduces the selection of biologically useful targets. To expand the range of potential targets, it is desirable to modify the TFOs to enable tight binding with high specificity to any sequence of duplex DNA. Triplex formation is severely disrupted as the number of pyrimidine interruptions in the target purine strand increases. The problem is referred to as base pair inversion or TA or CG inversion, depending on whether T or C is present in the purine-rich strand. The inversions cause steric repulsion, which prevents the stable association of the TFO with duplex DNA. The unfavorable interactions can be minimized by incorporating small base analogs in the third strand, at positions opposite the inversion site (Table 2).[39,40] For example, the use of simple azoles such as imidazole, pyrazole, and triazole can significantly reduce repulsion and maintain stacking interactions when TFOs are bound to duplex DNA containing CG inversions.[39] Imidazole also appears to enhance the specificity of binding at TA base pairs. An alternative approach involves the use of 2'-deoxyformycin A at CG inversion sites, which increases binding affinity tenfold as compared to its unmodified counterpart.[40]

When purine-rich targets are interrupted by a stretch of pyrimidine bases, binding affinity can be enhanced significantly by incorporating flexible polymeric linkers into the phosphodiester backbone.[41] Two different TFOs can be attached end-to-end, enabling each unit TFO to bind optimally to adjacent stretches of target duplex DNA that are separated by a duplex spacer region. Linkage of two 12-mers in this way has resulted in an increase in overall binding affinity of at least a factor of 10 ($K_d \sim 10^{-10} M$).[41] TFOs containing linkers have yet to be tested in biological systems, but offer the promise of more selective control over gene expression by binding two sites simultaneously.

2. Stability of Triplex-Forming Oligonucleotides

Phosphodiester oligonucleotides are highly susceptible to digestion by exonucleases and endonucleases inside and outside the cell. Blockage of the 3'-hydroxyl group, with a propanolamine moiety, for example, markedly increases the nuclease resistance of the TFOs, especially in serum.[42] Interestingly, blocking of the 5' end provides no significant protection. Base composition may also influence TFO stability, as oligonucleotides with stretches of G residues appear to be more stable than those with multiple pyrimidine residues.[42,57] Because of the constraints of triplex formation, some of the "nuclease-resistant" backbones may not be suitable for TFOs. As discussed below, TFOs may be further protected against nucleases by using a suitable formulation.

3. Nonspecific Actions of Oligonucleotides

There is a growing realization that antisense oligonucleotides, especially with modified backbones, may have nonantisense biological effects on living cells in the micromolar range.[43-48] Recent data from experiments using HIV-infected cells indicate that certain GT-rich phosphodiester oligonucleotides can have potent anti-HIV activity in the *submicromolar* range, with no evidence for antisense or triple-helix mechanism.[48] It is not known whether the antiviral activity is due to interference with targets inside or outside the cells.

It is likely that more examples of nonspecific effects of antisense or TFOs on living cells will be found. A simple 20-mer oligonucleotide composed of 4 bases may have $>10^{12}$ potential sequence and conformational variants, and it is possible that a candidate TFO may have a conformation that allows it to interact in an unpredictable manner, but with very high affinity, with proteins and cellular components other than duplex DNA. In fact, such "nonspecific" interactions are the basis of novel drug discovery programs initiated to identify oligonucleotides that bind to and interfere with the function of a wide variety of intracellular or extracellular targets at very low concentrations, but which do not inhibit other cellular functions. In these aptamer or SELEX-type approaches, massive libraries of oligonucleotides are screened for binding to a variety of targets by nonantisense, nontriplex mechanisms.[49-51]

To improve the specific activity of oligonucleotides and lower the effective dose, it is crucial to develop oligonucleotide delivery technology that leads to the specific transport of TFO to its intended target without any of the potential side effects. In addition, sequence similarity searches can reveal fortuitous sequence similarity to nontarget RNA or DNA sequences, which may lead to undesirable antisense-type associations.[52]

4. Cellular Uptake and Nuclear Delivery of Triplex-Forming Oligonucleotides

The recent advances in TFO base modifications are expected to greatly increase the selection of potential duplex DNA targets for TFOs, yet this approach addresses only half of the challenge in developing antigene therapeutics. The other half pertains to the availability of the TFO-binding site within the intact cell. The general consensus is that although oligonucleotides have the propensity to be taken up by cells (see Figure 4 for model), the internalized material tends to remain in cytoplasmic compartments of probable endocytic origin (Figure 9A,B).[53-56] Unaided, very little of the oligonucleotide seems to cross the endocytic membrane into the cytosol and the nucleus. The remaining material is believed to be degraded or perhaps returned to the extracellular medium by exocytosis. The barriers to cellular entry encountered by TFOs include those faced by antisense oligonucleotides, foremost among them being susceptibility to intra- and extracellular nucleases[57] and transport across the plasma and endosomal membranes (Table 4). There are additional transport problems specific to antigene TFOs, since these oligonucleotides must enter the nucleus and find their way to the transcription site. In addition, the intranuclear concentration of TFO must be high enough to allow it to compete effectively with the transcriptional complex for the same binding site on the gene.

Intuitively, it seems unlikely that foreign pieces of polyanionic phosphodiester nucleic acids can be delivered to cells, yet viruses and other microbes have been able to inject their unmodified genes into animal cells with high efficiency. Microbes that escape the immune response of the organism and enter intracellular compartments can remain intact and unscathed for days, weeks, and in the case of latent viruses, decades. Gene therapy protocols have attempted to take advantage of viral-mediated delivery systems, and in principle, the approach can lead to improved delivery of oligonucleotides into the nucleus.[27,28] However, the use of bulky viral components to achieve therapeutic doses of TFOs may not be a practical solution to the delivery problem. As discussed below, alternative delivery systems involving the use of covalent end modifications[17,58-61] as well as noncovalent modifications such as liposomes and cationic lipids/liposomes are being explored to optimize the targeting of oligonucleotides in a nonimmunogenic package.[62-68]

a. Covalent Modifications

One type of covalent modification which has been shown to enhance the intracellular delivery and efficacy of antisense and triplex-forming oligonucleotides is the attachment of lipophilic molecules such as cholesterol to oligonucleotides.[17,58-60] However, fluorescence studies using cholesterol-modified oligonucleotides show that much of the internalized material remains trapped within perinuclear compartments, most likely of endosomal origin.[58] The exact mechanism by which the lipid moiety enhances uptake is not known, but it is presumed that the presence of the lipid at one end enhances the membrane association and uptake of the oligonucleotide. An important constraint in the design of these conjugates is the linker that connects the lipid to the oligonucleotide. If the linker is too susceptible to in vivo degradation, then the lipid will detach from the oligonucleotide in the serum, and there will be little or no noticeable increase in cellular uptake. And if the bond between the lipid and the oligonucleotide is not biodegradable, the oligonucleotide may remain anchored to the lipid bilayer of cellular membranes through its lipid tail, unavailable for entry into the nucleus. Fluorescence microscopy analysis of a version of TFO-cholesterol (Figure 5A), used in the PRE assays described above, remained mostly sequestered in the endosomal compartment (data not shown), although there was some improvement in nuclear uptake.[17] To address this problem, a new generation of cholesterol-modified oligonucleotides has been synthesized in our laboratory, containing a potentially more biodegradable "labile" linker consisting of a triglycine moiety (Figure 5B).[69,70] According to the design, the lipid group should enhance uptake by cells,

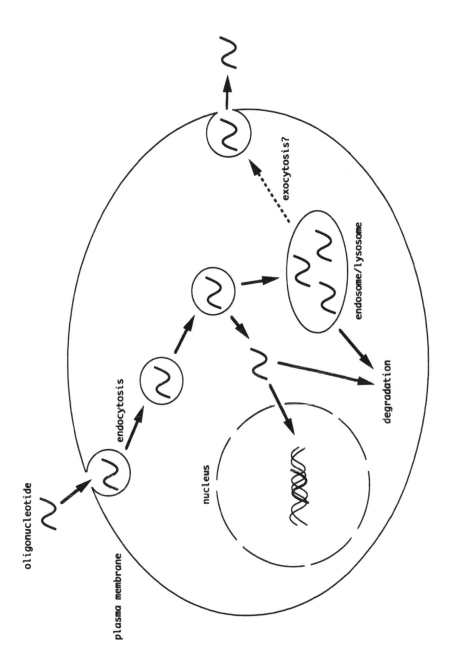

Figure 4 Model of cellular uptake of triple helix-forming oligonucleotides (TFOs). Oligonucleotides are thought to enter cells by endocytosis, during which they are enclosed within endosomes. Oligonucleotides may enter the cytoplasm by diffusion across the endosomal membrane, by leakage during fusion of endosomes, and/or by other unknown mechanisms. After release into the cytosol, TFOs can potentially enter the nucleus and bind to their duplex DNA target. Some proportion of intracellular oligonucleotides may be subject to degradation and/or possibly exocytosis.

A. Oligonucleotide

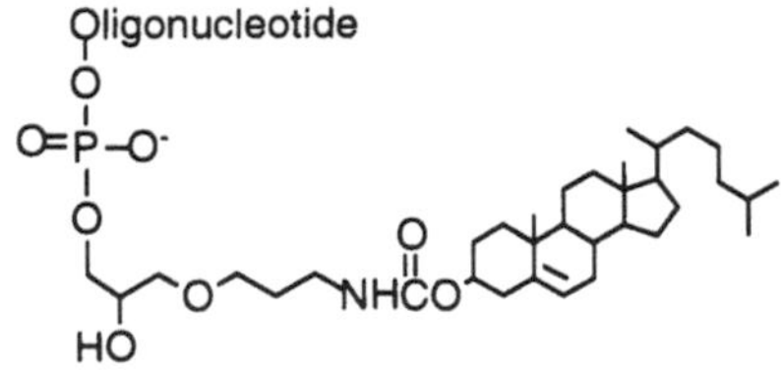

B. Oligonucleotide

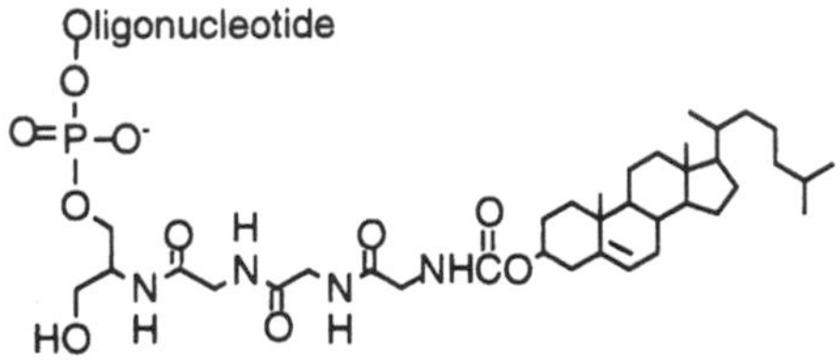

Figure 5 Structures of TFO-cholesterol linkages. (A) Cholesteryl-conjugated TFOs containing a 3-aminopropyl glycerol linker. (B) Cholesteryl-conjugated TFOs containing a 2-N-(glycylglycylglycyl)aminopropan-1,3-diol linker.

and upon exposure to cellular enzymes the tripeptide bond should be cleaved, releasing the free oligonucleotide. An experiment using a fluorescent-tagged 36-mer oligonucleotide indicates that this indeed appears to be the case. HeLa cells treated for up to 24 h with 0.1 μM TFO-3'-propanolamine internalize very small levels of the oligonucleotides (Figure 6A,B). However, if the cells are treated with 0.1 μM TFO-triglycyl-cholesterol, there is rapid uptake and delivery to the nucleus (Figure 6C,D). At time points up to 24 h, the fluorescence accumulated by endosomes rises relative to that in the nucleus, perhaps because the nuclear targets become saturated.[71] Experiments by Fisher et al.[57] show that nuclei retain fluorescein only if it is firmly attached to the oligonucleotide, suggesting that the nuclear fluorescence in Figure 6D is due to nuclear delivery of the TFO. In this import pathway, it is not clear how oligonucleotides manage to escape from endosomal to cytosolic compartments. Presumably, the oligonucleotides take advantage of transient openings in the endosomal membrane. One possibility is that the cholesterol-bound TFO may "flip" across the endosomal membrane through these fissures, but remain anchored to the lipid bilayer through cholesterol. Subsequently, a cytosolic peptidase could cleave the triglycyl linker, releasing the oligonucleotide into the cytosol. The utility of this labile linker approach needs to be validated in a variety of systems, but one benefit is already evident. The material is very easy to administer because only a single compound is present and no formulation is required. The peptide linker is short and not expected to be immunogenic. All components are natural and biodegradable, and no discernible toxicity has yet been seen in cell culture or animal systems. Moreover, the oligonucleotide can be synthesized using automated synthesizers.[69,70]

b. Noncovalent Uptake Enhancement

An alternative approach for improving the cellular and nuclear uptake of oligonucleotides is the use of uptake enhancers such as cationic lipids or liposomal agents.[61-68] These reagents are attractive because of their versatility. The same delivery agent may be coadministered with a variety of oligonucleotides, with little need to carry out complex chemistry each time a new oligonucleotide is designed.

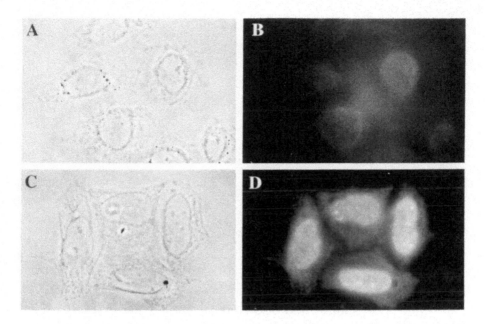

Figure 6 Enhanced uptake of cholesteryl-modified TFOs. HeLa cells were treated with fluorescent-tagged GT phosphodiester 36-mers blocked at the 3′ end with either propanolamine or cholesterol moieties. At various time points after addition of oligonucleotide, cells were fixed and processed for phase contrast (A,C) or fluorescence (B,D) microscopy. (A,B) TFO-amine, 24 h. (C,D) TFO-cholesterol, 40 min.

Since the introduction of N-[1-(2,3-dioleyloxy-propyl]-N,N,N-trimethylammonium chloride (DOTMA),[72] which is the active component of the cationic liposomal preparation Lipofectin® (containing 1:1 mass ratio of DOTMA and dioleoylphosphotidylethanolamine [DOPE]), there has been tremendous interest in the development of new reagents that will mimic the uptake enhancement effect of DOTMA, yet show reduced cytotoxicity. The design of these cationic lipids incorporates a positively charged head group that binds to the nucleic acid, and a membrane interactive tail that is proposed to interact with fusogenic lipids and/or destabilize cellular membranes. A number of other cationic lipid preparations are now available that can be used to enhance the uptake of oligonucleotides, as well as genes, into target cells.[65,66,73] The activity of these cationic lipids is influenced by several factors, including the composition and quantity of nucleic acid, cell type, and the concentration of serum in the medium. In addition, some preparations are cytotoxic. These constraints severely limit the utility of many of these cationic lipids as delivery agents for therapeutic TFOs in animal systems, and there continues to be a demand for novel uptake enhancers. Recently, a new polyaminolipid (PAL) family of uptake enhancers has been synthesized (Figure 7) by conjugating a polyamine (spermine or spermidine) to cholesterol through a carbamate linker. The PALs are designed to be biodegradable and nontoxic to cells through the use of components which are normally found in cells. Polyamines are involved in stabilizing nucleic acid structure and are found in the nucleus in millimolar range.[29,75] Cholesterol is a natural component of cellular membranes, involved in the modulation of membrane fluidity. In cytotoxicity assays, PALs, especially spermidine-cholesterol (SpdC), were found to be much less toxic than a DOTMA/DOPE preparation (Figure 8).

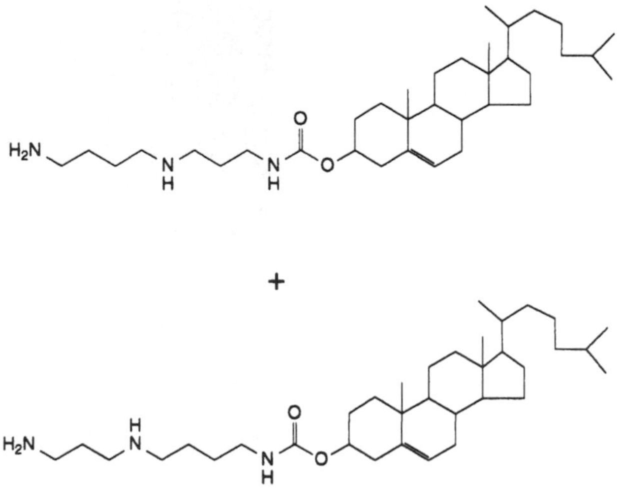

SPERMIDINE CONJUGATED TO CHOLESTEROL **(SpdC)**

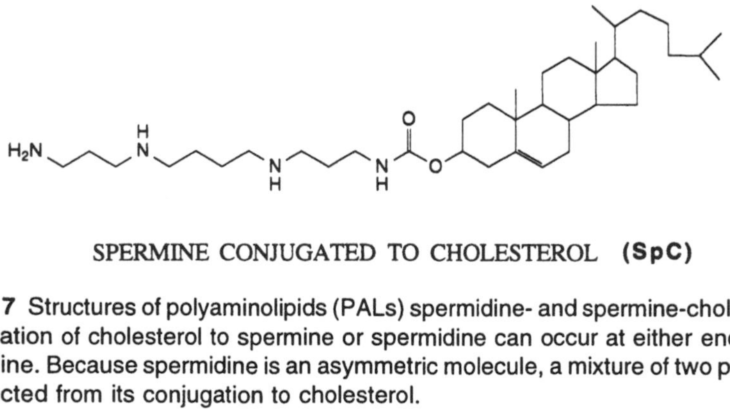

SPERMINE CONJUGATED TO CHOLESTEROL **(SpC)**

Figure 7 Structures of polyaminolipids (PALs) spermidine- and spermine-cholesterol. Conjugation of cholesterol to spermine or spermidine can occur at either end of the polyamine. Because spermidine is an asymmetric molecule, a mixture of two products is expected from its conjugation to cholesterol.

The PALs can improve the cellular and nuclear uptake of oligonucleotides in a variety of cell types in the presence of serum. Although the effect of these novel compounds on TFO activity has not been fully determined, there is sufficient evidence to indicate that PALs are excellent uptake enhancers. Cellular import of TFOs is increased by up to 100-fold.[71] Fluorescence microscopy studies indicate that cellular and nuclear uptake is discernible as early as 20 min after treatment (Figure 9C,D) and continues for more than 24 h. At later time points, nuclear uptake decreases somewhat and more of the oligonucle-otides accumulate in perinuclear cytoplasmic compartments.

Interestingly, the PALs can greatly increase the nuclease resistance of internalized oligonucleotides. For example, if phosphodiester TFOs (with free 3'-OH) are internally labeled with ^{33}P and added to Vero cells, no intact material can be extracted from the medium or the cells 20 min after treatment. Addition of a 3'-propanolamine group enhances the half-life to several hours, but even then, virtually no intact material can be extracted from cells 6 h after treatment (Figure 10A). In comparison, if the PAL SpdC is coadministered with the oligonucleotide, there is far greater uptake (~20-fold more) and

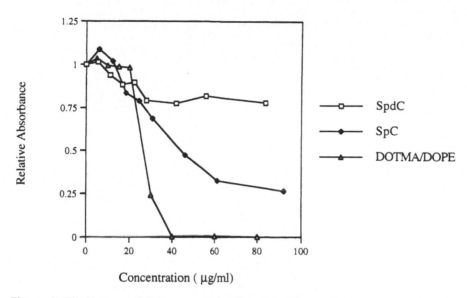

Concentration (μg/ml)

Figure 8 Effect of novel PALs on cell proliferation. Vero cells were exposed to the indicated concentrations of spermine-cholesterol (SpC), spermidine-cholesterol (SpdC), or DOTMA/DOPE (Lipofectin®; Gibco BRL, Gaithersburg, MD) for 4 d. Cell proliferation was measured by a nonradioactive cell-proliferation assay. The absorbance at 490 nm is directly proportional to the number of living cells. Data points are the average of four replicates for each test concentration.

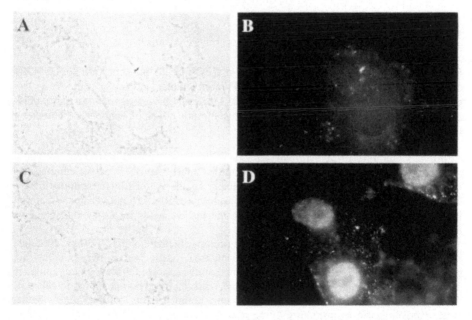

Figure 9 Enhanced uptake of TFOs in the presence of SpdC. Vero cells were treated with a fluorescent-tagged 3 G-base phosphodiester oligonucleotide blocked at the 3′ end with a propanolamine moiety in the presence or absence of SpdC. At various time points after addition of oligonucleotide, cells were fixed and processed for phase contrast (A,C), or fluorescence (B,D) microscopy. (A,B) TFO alone, 20 min. (C,D) TFO + SpdC, 20 min.

A. TFO only   B. TFO + SpdC

Time (hr)   0   3   6   24   M   0   3   6   24

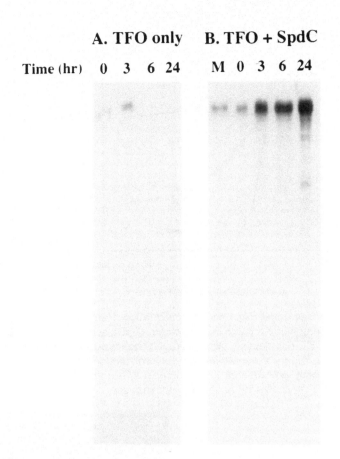

Figure 10 Increased intracellular stability of TFOs in the presence of SpdC. U251 cells were treated with phosphodiester 36-mers labeled internally with ^{33}P in the presence (B) or absence (A) of SpdC. At 0, 3, 6, or 24 h, cell-associated oligonucleotides were extracted and analyzed by polyacrylamide gel electrophoresis followed by autoradiography.

significant amounts of intact material can be isolated from cells treated for 24 h (Figure 10B). Quantitative estimates indicate that more than 80% of the internalized material is intact after 24 h. The TFO stability in the extracellular medium is also increased to the same extent. This remarkable ability of PALs to deliver intact material into cells indicates that they are binding to the nucleic acids in a way that prevents recognition and cleavage by nucleases.

The biochemical nature of the cytoplasmic compartments in which TFOs accumulate at later time points after coadministration with PALs is not yet known, but they are probably of endocytic origin. The data reported above suggests that oligonucleotides in these endocytic vesicles are not exposed to nucleases. It would be interesting to calculate the kinetics of TFO exit from these cytoplasmic compartments, to determine whether these deposits serve as miniature "time-release capsules" for TFOs, which may have utility in future development of TFOs as therapeutics. These nontoxic uptake enhancers may also have utility in other nucleic acid-based strategies, including gene therapy, antisense, and aptamer approaches.

c. Transport to the Nucleus

When fluorescent oligonucleotides are introduced into the cytosol of living cells by microinjection, the material is rapidly transported to the nuclei ($t_{1/2}$ ~5 min).[57,76,77] If injected directly into the nucleus fluorescence is visible until intranuclear nucleases digest the oligonucleotides.[57] The molecular reason for the swift delivery to the nucleus, and identity of the intranuclear retention sites are not known. The polyanionic oligonucleotides may be attracted to cationic proteins that are abundant in the nucleus. Nuclear entry does not appear to be rate-limiting in the uptake process, but it is not yet known whether the concentration of TFOs in the microenvironment of the transcription site can routinely reach levels high enough to displace the transcriptional complex.

The nuclear envelope of mammalian cells contains several hundred to thousands of nuclear pore complexes, through which the regulated exchange of macromolecules takes place.[78] Because only a small fraction of the nuclear DNA is actively transcribed, the genome is organized in three dimensions such that the transcriptionally active regions are accessible to one or more of the pore complexes, allowing for the transport of nascent RNA transcripts to the cytosol.[79,80] There is intriguing evidence for intranuclear shuttling systems along "tracks" that may be essential for nucleocytoplasmic exchange of RNA and proteins.[81] In agreement with this idea, *in situ* hybridization experiments suggest that nascent RNA transcripts are not randomly distributed in the nucleus. Instead, they seem to follow one or a few specific tracks from inside the nucleus to the nuclear surface.[80,82] These data support the logic of the TFO approach, because they suggest that transcriptional sites may indeed be accessible to incoming antigene TFOs. Although nuclear delivery does not appear to be a rate-limiting factor in the development of TFO technology, there is potential for further improvement in achieving superior targeting to the exact site of action. A number of sorting signals have been discovered that guide the movement of proteins and RNA inside and outside the nucleus.[78,83,84] These signals, such as the nucleolar protein localization signal or the cap structure of small nuclear RNAs, act as "zip codes" for sorting of the macromolecule within the numerous intranuclear and cellular compartments. If appropriate signals can be found, it is conceivable that short targeting elements (e.g., peptides or nucleic acids) may be tethered to the oligonucleotide to facilitate transport to the desired nuclear subcompartment.

IV. FUTURE DIRECTIONS

In the near future, there is likely to be continued emphasis on improving the biological efficacy of TFOs in cell culture systems. The recent developments in structural design that lead to improved TFO specificity, affinity, and stability, combined with advances in the delivery enhancement methods including tissue targeting, will greatly increase the range and utility of the TFO antigene technology. Continued success in a variety of cellular targets will provide incentive to optimize the approach against specific disease targets in whole-animal systems.

ACKNOWLEDGMENTS

We thank Drs. V. Bodepudi, P. Cossum, T. Hill, M. E. Hogan, J. O. Ojwang, R. Rando, T. S. Rao, G. R. Revankar, G. Reyes, H. Vu, T. Wallace, and J. G. Zendegui for helpful discussions.

REFERENCES

1. Felsenfeld, G., Davies, D. R., and Rich, A., Formation of a three-stranded polynucleotide molecule, *J. Am. Chem. Soc.*, 79, 2023, 1957.
2. Morgan, A. R. and Wells, R. D., Specificity of the three-stranded complex formation between double-stranded DNA and single-stranded RNA containing repeating nucleotide sequences, *J. Mol. Biol.*, 37, 63, 1968.

56

3. Le Doan, T., Perrouault, L., Praseuth, D., Habhoub, N., Decout, J.-L., Thuong, N. T., Lhomme, J., and Helene, C., Sequence-specific recognition, photo-crosslinking and cleavage of the DNA double helix by an oligo-[α]-thymidylate covalently linked to an azidoproflavine derivative, *Nucleic Acids Res.*, 15, 7749, 1987.
4. Moser, H. E. and Dervan, P. B., Sequence-specific cleavage of double helical DNA by triple helix formation, *Science*, 238, 645, 1987.
5. Cooney, M., Czernuszewicz, G., Postel, E. H., Flint, S. J., and Hogan, M. E., Site-specific oligonucleotide binding represses transcription of the human *c-myc* gene *in vitro*, *Science*, 241, 456, 1988.
6. Postel, E. H., Flint, S. J., Kessler, D. J., and Hogan, M. E., Evidence that a triplex-forming oligonucleotide binds to the *c-myc* promoter in HeLa cells, thereby reducing *c-myc* mRNA levels, *Proc. Natl. Acad. Sci. U.S.A.*, 88, 8227, 1991.
7. Orson, F. M., Thomas, D. W., McShan, W. M., Kessler, D. J., and Hogan, M. E., Oligonucleotide inhibition of IL2Rα mRNA transcription by promoter region collinear triplex formation in lymphocytes, *Nucleic Acids Res.*, 19, 3435, 1991.
8. Young, S. L., Krawczyk, S. H., Matteucci, M. D., and Toole, J. J., Triple helix formation inhibits transcription elongation *in vitro*, *Proc. Natl. Acad. Sci. U.S.A.*, 88, 10023, 1991.
9. Maher, L. J., Inhibition of T7 RNA polymerase initiation by triple-helical DNA complexes: a model for artificial gene repression, *Biochemistry*, 31, 7587, 1992.
10. Duval-Valentin, G., Thuong, N. T., and Helene, C., Specific inhibition of transcription by triple helix-forming oligonucleotides, *Proc. Natl. Acad. Sci. U.S.A.*, 89, 504, 1992.
11. Postel, E. H., Modulation of *c-myc* transcription by triple helix formation, *Ann. N.Y. Acad. Sci.*, 660, 57, 1992.
12. Maher, L. J., Dervan, P. B., and Wold, B., Analysis of promoter-specific repression by triple-helical DNA complexes in a eukaryotic cell-free transcription system, *Biochemistry*, 31, 70, 1992.
13. Grigoriev, M., Praseuth, D., Robin, P., Hemar, A., Saison-Behmoaras, T., Dautry-Varsat, A., Thuong, N. T., Helene, C., and Harel-Bellan, A., A triple helix-forming oligonucleotide-intercalator conjugate acts as a transcriptional repressor via inhibition of NF kappa B binding to interleukin-2 receptor alpha-regulatory sequence, *J. Biol. Chem.*, 267, 3389, 1992.
14. Ross, C., Samuel, M., and Broitman, S. L., Transcriptional inhibition of the bacteriophage T7 early promoter region by oligonucleotide triple helix formation, *Biochem. Biophys. Res. Commun.*, 189, 1674, 1992.
15. McShan, W. M., Rossen, R. D., Laughter, A. H., Trial, J., Kessler, D. J., Zendegui, J. G., Hogan, M. E., and Orson, F. M., Inhibition of transcription of HIV-1 in infected human cells by oligodeoxynucleotides designed to form DNA triple helices, *J. Biol. Chem.*, 267, 5712, 1992.
16. Grigoriev, M., Praseuth, D., Guieysse, A. L., Robin, P., Thuong, N. T., Helene, C., and Harel-Bellan, A., Inhibition of interleukin-2 receptor alpha-subunit gene expression by oligonucleotide-directed triple helix formation, *C.R. Acad. Sci. Iii*, 316, 492, 1993.
17. Ing, N. H., Beekman, J. M., Kessler, D. J., Murphy, M., Jayaraman, K., Zendegui, J. G., Hogan, M. E., O'Malley, B. W., and Tsai, M. J., *In vivo* transcription of a progesterone-responsive gene is specifically inhibited by a triplex-forming oligonucleotide, *Nucleic Acids Res.*, 21, 2789, 1993.
18. Skoog, J. U. and Maher, L. J., Repression of bacteriophage promoters by DNA and RNA oligonucleotides, *Nucleic Acids Res.*, 21, 2131, 1993.
19. Roy, C., Inhibition of gene transcription by purine rich triplex forming oligodeoxyribonucleotides, *Nucleic Acids Res.*, 21, 2845, 1993.
20. Rando, R., DePaolis, L., Durland, R. H., Jayaraman, K., Kessler, D. J., and Hogan, M. E., Inhibition of T7 and T3 RNA polymerase directed transcription elongation *in vitro*, *Nucleic Acids Res.*, 22, 678, 1994.

21. Blume, S. W., Gee, J. E., Shrestha, K., and Miller, D. M., Triple helix formation by purine-rich oligonucleotides targeted to the human dihydrofolate reductase promoter, *Nucleic Acids Res.*, 20, 1777, 1992.

22. Arnott, S. and Selsing, E., Structures for the polynucleotide complexes poly (dA)·poly (dT) and poly(dT)·poly (dA)·poly (dT)., *J. Mol. Biol.*, 88, 509, 1974.

23. Lee, J. S., Johnson, D. A., and Morgan, A. R., Complexes formed by (pyrimidine)n·(purine)n DNAs on lowering the pH are three stranded, *Nucleic Acids Res.*, 6, 3073, 1979.

24. Durland, R. H., Kessler, D. J., Gunnell, S., Duvic, M., Pettitt, B. M., and Hogan, M. E., Binding of triple helix forming oligonucleotides to sites in gene promoters, *Biochemistry*, 30, 9246, 1991.

25. Beal, P. A. and Dervan, P. B., Second structural motif for recognition of DNA by oligonucleotide-directed triple-helix formation, *Science*, 251, 1360, 1991.

26. Helene, C., The anti-gene strategy: control of gene expression by triplex-forming-oligonucleotides, *Anti-Cancer Drug Design*, 6, 569, 1991.

27. O'Malley, B. W. and Ledley, F. D., Somatic gene therapy: methods for the present and future, *Arch. Otolaryngol. Head Neck Surg.*, 119, 1100, 1993.

28. Mulligan, R. C., The basic science of gene therapy, *Science*, 260, 926, 1993.

29. Saenger, W., *Principles of Nucleic Acid Structure*, Springer-Verlag, New York, 1986, 238.

30. Jin, R., Breslauer, K. J., Jones, R. A., and Gaffney, B. L., Tetraplex formation of a guanine-containing nonameric DNA fragment, *Science*, 250, 543, 1990.

31. Sen, D. and Gilbert, W., Formation of parallel four-stranded complexes by guanine-rich motifs in DNA and its implications for meiosis, *Nature*, 334, 364, 1988.

32. Sundquist, W. I. and Klug, A., Telomeric DNA dimerizes by formation of guanine tetrads between hairpin loops, *Nature*, 342, 825, 1989.

33. Kim, J., Cheong, C., and Moore, P. B., Tetramerization of an RNA oligonucleotide containing a GGGG sequence, *Nature*, 351, 331, 1991.

34. Rao, T. S., Jayaraman, K., Durland, R. H., and Revankar, G. R., A novel synthesis of S⁶-cyanoethyl-2'-deoxy-6-thioguanosine and its incorporation into triple helix-forming oligonucleotides, *Tetrahedron Lett.*, 33, 7651, 1992.

35. Rao, T. S., Durland, R. H., Myrick, M. A., Seth, D. M., Bodepudi, V., and Revankar, G. R., Incorporation of 2'-deoxy-6-thioguanosine (S⁶-dG) into G-rich oligonucleotides inhibits G-tetrad formation and facilitates triple helix formation (Abstract No. CARB 55), in *207th ACS National Meeting Book of Abstracts*, American Chemical Society, San Diego, CA, 1994.

36. Rao, T. S., Durland, R. H., Seth, D. M., Myrick, M. A., Bodepudi, V., and Revankar, G. R., Incorporation of 2'-deoxy-6-thioguanosine (S⁶-dG) into G-rich oligodeoxyribonucleotides inhibits G-tetrad formation and facilitates triplex formation, *Biochemistry*, in press.

37. Milligan, J. F., Krawczk, S. H., Wadwani, S., and Matteucci, M. D., An anti-parallel triple helix motif with oligodeoxynucleotides containing 2'-deoxyguanosine and 7-deaza-2'-deoxyxanthosine, *Nucleic Acids Res.*, 21, 327, 1993.

38. Rao, T. S., Lewis, A. F., Durland, R. H., and Revankar, G. R., A total synthesis of 2'-deoxy-9-deazaguanosine (9-deaza-dG) and its incorporation into triple helix forming oligodeoxyribonucleotides with antiparallel motif, *Tetrahedron Lett.*, 34, 6709, 1993.

39. Rao, T. S., Durland, R. H., Jayaraman, K., and Revankar, G. R., Use of simple azoles in oligodeoxynucleotides containing CG or TA inversions improves triple helix formation (Abstract No. CARB 57), in *207th ACS National Meeting Book of Abstracts*, American Chemical Society, San Diego, CA, 1994.

40. Rao, T. S., Hogan, M. E., and Revankar, G. R., Synthesis of triple helix forming oligonucleotides containing 2'-deoxyformycin A, *Nucleosides Nucleotides*, 13, 95, 1994.

58

41. Kessler, D. J., Pettitt, B. M., Cheng, Y.-K., Smith, S. R., Jayaraman, K., Vu, H. M., and Hogan, M. E., Triple helix formation at distant sites: hybrid oligonucleotides containing a polymeric linker, *Nucleic Acids Res.*, 21, 4810, 1993.

42. Zendegui, J. G., Vasquez, K. M., Tinsley, J. H., Kessler, D. J., and Hogan, M. E., *In vivo* stability and kinetics of absorption and disposition of 3' phosphopropyl amine oligonucleotides, *Nucleic Acids Res.*, 20, 307, 1992.

43. Woolf, T. M., Melton, D. A., and Jennings, C. G. B., Specificity of antisense oligonucleotides *in vivo*, *Proc. Natl. Acad. Sci. U.S.A.*, 89, 7305, 1992.

44. Clarenc, J.-P., Lebleu, B., and Leonetti, J.-P., Characterization of the nuclear binding sites of oligoribonucleotides and their analogs, *J. Biol. Chem.*, 268, 5600, 1993.

45. Stein, C. A., Cleary, A. M., Yakubov, L., and Lederman, S., Phosphorothioate oligodeoxynucleotides bind to the third variable loop domain (v3) of human immunodeficiency virus type 1 gp120, *Antisense Res. Dev.*, 3, 19, 1993.

46. Praseuth, D., Guieysse, A.-L., Itkes, A. V., and Helene, C., Unexpected effect of an anti-human immunodeficiency virus intermolecular triplex-forming oligonucleotide in an *in vitro* transcription system due to RNase H-induced cleavage of the RNA transcript, *Antisense Res. Dev.*, 3, 33, 1993.

47. Wyatt, J. R., Vickers, T. A., Roberson, J. L., Buckheit, R. W., Jr., Klimkait, T., DeBaets, E., Davis, P. W., Rayner, B., Imbach, J. L., and Ecker, D. J., Combinatorially selected guanosine-quartet structure is a potent inhibitor of human immunodeficiency virus envelope-mediated cell fusion, *Proc. Natl. Acad. Sci. U.S.A.*, 91, 1356, 1994.

48. Ojwang, J., Elbaggari, A., Marshall, H. B., Jayaraman, K., McGrath, M. S., and Rando, R. F., Inhibition of human immunodeficiency virus type 1 activity *in vitro* by oligonucleotides composed entirely of guanosine and thymidine (GTOs), *J. A.I.D.S.*, 7, 560, 1994.

49. Tuerk, C. and Gold, L., Systematic evolution of ligands by exponential enrichment: RNA ligands to bacteriophage T4 DNA polymerase, *Science*, 249, 505, 1990.

50. Ellington, A. D. and Szostak, J. W., Selection *in vitro* of single stranded DNA molecules that fold into specific ligand-binding structures, *Nature*, 355, 850, 1992.

51. Jenison, R. D., Gill, S. C., Pardi, A., and Polisky, B., High-resolution molecular discrimination by RNA, *Science*, 263, 1425, 1994.

52. Mitsuhashi, M., Cooper, A., Ogura, M., Shinagawa, T., Yano, K., and Hosokawa, T., Oligonucleotide probe design — a new approach, *Nature*, 367, 759, 1994.

53. Akhtar, S. and Juliano, R. L., Cellular uptake and intracellular fate of antisense oligonucleotides, *Trends Cell Biol.*, 2, 139, 1992.

54. Budker, V. G., Knorre, D. G., and Vlassov, V. V., Cell membranes as barriers for antisense constructions, *Antisense Res. Dev.*, 2, 177, 1992.

55. Wickstrom, E., Strategies for administering targeted therapeutic oligodeoxynucleotides, *TIBTECH*, 10, 281, 1992.

56. Jaroszewski, J. W. and Cohen, J. S., Cellular uptake of antisense oligodeoxynucleotides, *Adv. Drug Del. Rev.*, 6, 235, 1991.

57. Fisher, T. L., Terhorst, T., Cao, X., and Wagner, R. W., Intracellular disposition and metabolism of fluorescently-labeled unmodified and modified oligonucleotides microinjected into mammalian cells, *Nucleic Acids Res.*, 21, 3857, 1993.

58. Letsinger, R. L., Zhang, G., Sun, D. K., Ikeuchi, T., and Sarin, P. S., Cholesteryl-conjugated oligonucleotides: synthesis, properties, and activity as inhibitors of replication of human immunodeficiency virus in cell culture, *Proc. Natl. Acad. Sci. U.S.A.*, 86, 6553, 1989.

59. Boutorin, A. S., Gus'kova, L. V., Ivanova, E. M., Kobetz, N. D., Zarytova, V. F., Yurchenko, L. V., and Vlassov, V. V., Synthesis of alkylating oligonucleotide derivatives containing cholesterol or phenazinium residues at their 3'-terminus and their interaction with DNA within mammalian cells, *FEBS Lett.*, 254, 129, 1989.

60. Shea, R. G., Marsters, J. C., and Bischofberger, N., Synthesis, hybridization properties, and antiviral activity of lipid-oligodeoxynucleotide conjugates, *Nucleic Acids Res.*, 18, 3777, 1990.
61. Clarenc, J. P., Degols, G., Leonetti, J. P., Milhaud, P., and Lebleu, B., Delivery of antisense oligonucleotides by poly(L-lysine) conjugation and liposome encapsulation, *Anti-Cancer Drug Design*, 8, 81, 1993.
62. Juliano, R. L. and Akhtar, S., Liposomes as a drug delivery system for antisense oligonucleotides, *Antisense Res. Dev.*, 2, 165, 1992.
63. Duzgunes, N., Goldstein, J. A., Friend, D. S., and Felgner, P. L., Fusion of liposomes containing a novel cationic lipid, N-[2,3-(dioleyloxy)propyl]-N,N,N-trimethyl-ammonium: induction by multivalent anions and asymmetric fusion with acidic phospholipid vesicles, *Biochemistry*, 28, 9179, 1989.
64. Bennett, C. F., Chiang, M. Y., Chan, H., Shoemaker, J. E., and Mirabelli, C. K., Cationic lipids enhance cellular uptake and activity of phosphorothioate antisense oligonucleotides, *Mol. Pharmacol.*, 41, 1023, 1992.
65. Behr, J.-P., Demeneix, B., Loeffler, J.-P., and Perez-Mutul, J., Efficient gene transfer into mammalian primary endocrine cells with lipopolyamine-coated DNA, *Proc. Natl. Acad. Sci. U.S.A.*, 86, 6982, 1989.
66. Gao, X. and Huang, L., A novel cationic liposome reagent for efficient transfection of mammalian cells, *Biochem. Biophys. Res. Commun.*, 179, 280, 1991.
67. Wagner, R. W., Matteucci, M. D., Lewis, J. G., Gutierrez, A. J., Moulds, C., and Froehler, B. C., Antisense gene inhibition by oligonucleotides containing C-5 propyne pyrimidines, *Science*, 260, 1510, 1993.
68. Gershon, H., Ghirlando, R., Guttman, S. B., and Minsky, A., Mode of formation and structural features of DNA-cationic liposomes complexes used for transfection, *Biochemistry*, 32, 7143, 1993.
69. Vu, H., Singh, P., Joyce, N., Hogan, M. E., and Jayaraman, K., Synthesis of cholesteryl supports and phosphoramidites containing a novel peptidyl linker for automated synthesis of triple-helix forming oligonucleotides (TFOs), *Nucleic Acids Symp. Ser.*, 1993, 19, 1993.
70. Vu, H. and Jayaraman, K., Synthesis and properties of cholesteryl modified triple helix forming oligonucleotides (TFOs) containing a triglycyl linker, *J. Bioconjugate Chem.*, in press.
71. Guy-Caffey, J. K., Bodepudi, V., Vu, H., Joyce, N. J., Jayaraman, K., Bishop, J. S., and Chaudhary, N., Improved cellular uptake and membrane permeability of oligonucleotides, *Mol. Biol. Cell*, 4, 324a, 1993.
72. Felgner, P. L., Gadek, T. R., Holm, M., Roman, R., Chan, H. W., Wenz, M., Northrop, J. P., Ringold, G. M., and Danielsen, M., Lipofection: a highly efficient, lipid-mediated DNA-transfection procedure, *Proc. Natl. Acad. Sci. U.S.A.*, 84, 7413, 1987.
73. Leventis, R. and Silvius, J. R., Interactions of mammalian cells with lipid dispersions containing novel metabolizable cationic amphiphiles, *Biochim. Biophys. Acta*, 1023, 124, 1990.
74. Felgner, J. H., Kumar, R., Sridhar, C. N., Wheeler, C. J., Tsai, Y. J., Border, R., Ramsey, P., Martin, M., and Felgner, P. L., Enhanced gene delivery and mechanism studies with a novel series of cationic lipid formulations, *J. Biol. Chem.*, 269, 2550, 1994.
75. Hampel, K. J., Crosson, P., and Lee, J. S., Polyamines favor DNA triplex formation at neutral pH, *Biochemistry*, 30, 4455, 1991.
76. Leonetti, J. P., Mechti, N., Degols, G., Gagnor, C., and Lebleu, B., Intracellular distribution of microinjected antisense oligonucleotides, *Proc. Natl. Acad. Sci. U.S.A.*, 88, 2702, 1991.
77. Chin, D. J., Green, G. A., Zon, G., Szoka, F. C., and Straubinger, R. M., Rapid nuclear accumulation of injected oligodeoxyribonucleotides, *New Biol.*, 2, 1091, 1990.

78. Miller, M., Park, M. K., and Hanover, J. A., Nuclear pore complex: structure, function, and regulation, *Physiol. Rev.*, 71, 909, 1991.
79. Blobel, G., Gene gating: a hypothesis, *Proc. Natl. Acad. Sci. U.S.A.*, 82, 8527, 1985.
80. Lawrence, J. B. and Singer, R. H., Spatial organization of nucleic acid sequences within cells, *Semin. Cell Biol.*, 2, 83, 1991.
81. Meier, U. T. and Blobel, G., Nopp140 shuttles on tracks between nucleus and cytoplasm, *Cell,* 70, 127, 1992.
82. Lawrence, J. B., Singer, R. H., and Marselle, L. M., Highly localized tracks of specific transcripts within interphase nuclei visualized by *in situ* hybridization, *Cell*, 57, 493, 1989.
83. Izaurralde, E. and Mattaj, I. W., Transport of RNA between nucleus and cytoplasm, *Semin. Cell Biol.*, 3, 279, 1992.
84. Zapp, M. L., RNA nucleocytoplasmic transport, *Semin. Cell Biol.*, 3, 289, 1992.

Chapter 4

Pharmacokinetics and Delivery of Oligonucleotides to the Brain

Leonard M. Neckers, Daniel A. Geselowitz, and Christine Chavany

CONTENTS

I. INTRODUCTION

The central nervous system (CNS) is conceptually a very exciting target for antisense therapy. The potential specificity of antisense oligonucleotides might open avenues for treatment of a number of diseases associated with overexpressed or defective genes. The relative lack of toxicity of oligonucleotides could minimize toxicity to nontargeted brain tissue and allow for a good therapeutic index. This latter point could be very significant in the treatment of brain cancers. Glioblastomas, in particular, are lethal and intractable, and current chemotherapy is of minimal value. To date, there have been numerous *in vitro* studies using antisense oligonucleotides in neuronal and glial cells,[1-11] but *in vivo* use of oligonucleotides in this setting is just beginning.

The rush to develop therapeutic uses of oligonucleotides has, of course, been tempered somewhat by the difficulties associated with systemic administration. The problems of degradation by blood nucleases, sequestering and clearance by liver and kidney,[12-14] and the large quantities of material necessary to achieve therapeutic concentrations may each require customized solutions before systemic delivery is generally practical.

If the target of the oligonucleotide is in the CNS, the problems of intravenous administration are confounded by the nature of the blood-brain and blood-cerebral spinal fluid (CSF) barriers. The blood-brain barrier is located at the cerebral capillary epithelium, and the blood-CSF barrier is located at the choroid plexus epithelium and arachnoid membrane.[15] The barrier is composed of a basement membrane covered with a sheet of cells connected by tight junctions, and has been described as essentially a modified, tight epithelium.[16] It was originally detected by Ehrlich and others, who found that anionic dyes injected into the bloodstream failed to penetrate and stain the brain.[17]

Among the properties of the blood-brain barrier are an extremely low permeability to ions, with passive permeability occurring mainly at intracellular junctions. Lipid-soluble molecules can passively cross the barrier, but passage of nonlipid-soluble molecules is extremely poor. Most of the nonlipid molecules which pass readily into the brain, such as glucose, cross the barrier by active transport.[18]

Thus, it would be expected that anionic oligonucleotides should penetrate the barrier quite poorly, and this is borne out in labeling studies of healthy animals. Hogan and co-workers[19] studied a 38-mer phosphodiester oligonucleotide with a phosphopropyl group at the 3' end injected i.v. into mice. With a peak plasma level of 1 mg/ml, the brain concentration paralleled the plasma level for the first 15 min after injection, reaching a maximum level of 10 mg/kg, and maintaining a concentration of 3 mg/kg for several hours. This concentration was some 50-fold less than that in the kidney, 30-fold less than liver, and even 10-fold less than skin. Similar brain levels were found with intraperitoneal administration. Agrawal and co-workers[12] studied distribution of a phosphorothioate 20 mer in mice. They found that the level of oligonucleotide in brain after 2 h was about 500-fold less than in the kidney, and about 100-fold less than in the liver, and indeed was far less than all other organs and tissues sampled. Interestingly, a similar distribution has been found with the neutral methylphophonate oligonucleotides, which might be expected to show some passive diffusion. Chen and co-workers[20] found that total brain concentration of a 12-mer methylphosphonate with a single phosphodiester linkage was roughly 2% of the plasma level if autopsied 2 to 5 min after injection, but shortly afterward fell to nearly undetectable levels as the oligonucleotide was cleared and taken up by other organs. It would seem that direct penetration of oligonucleotide from the bloodstream into the brain does not yield practical concentrations.

There are two conceptually different approaches to overcoming the blood-brain barrier and achieving a therapeutic level of an oligodeoxynucleotide (ODN) in the brain or CSF. One approach is to breach the barrier by any of several means concomitant with systemic administration of oligonucleotides; the other is to avoid the blood-brain barrier entirely by direct administration to the CNS. In this paper we will review these approaches. We will also discuss what we have learned in our laboratory in studies of oligonucleotide administration by continuous infusion into the rat brain.

II. BREACHING THE BLOOD-BRAIN BARRIER

A. DISRUPTION IN DISEASE STATES

As we have pointed out, the blood-brain barrier of a healthy animal is a formidable obstacle to intravenously administered oligonucleotides. One might hope that certain disease states, such as brain tumors, might weaken the barrier and allow direct access of oligonucleotides from the bloodstream. The increased vasculature of brain tumors, for example, allows their imaging with MRI contrast agents and PET radiopharmaceuticals. It was formerly thought that brain tumors lack an effective blood-brain barrier. However, this has been shown to be incorrect.[18] Although disease can perturb the nature of the blood-brain barrier, it rarely eliminates it. The presence of a tumor is unlikely to provide automatic access of a drug from the bloodstream, a fact which undoubtedly contributes to the failure of systemically administered conventional chemotherapeutics in patients with brain tumors.

In one oligonucleotide therapeutic study, however, there is some suggestion that intravenous oligonucleotides can have an effect in the CSF. Gewirtz and co-workers[21] have treated immunodeficient mice bearing a human leukemia xenograft with intravenous phosphorothioate oligonucleotides antisense to c-myb and have achieved a considerable increase in survival time. Interestingly, the authors found that the treatment leads to a great reduction of the disease in the CNS. The mechanism of action there remains unclear. Although it is established that in some disease states leukocytes actively infiltrate the blood-brain barrier,[16] it is unclear if this would permeabilize the barrier to oligonucleotide drugs in the bloodstream. The levels of oligonucleotide in the CSF have not been established in these animals, making it unclear if the effect observed in the CSF is due to drug action *in situ* or in the bloodstream.

B. INTENTIONAL DISRUPTION OF THE BLOOD-BRAIN BARRIER

The blood-brain barrier can be disrupted directly by hyperosmolar shock, allowing molecules to cross.[18] This method, usually involving a bolus of mannitol, has been used to deliver such chemotherapeutics as nimustine to patients with brain tumors,[22] and methotrexate and cylcophosphamide to patients with non-AIDS-related primary CNS lymphoma.[23] Although one study suggests that hyperosmolar disruption of the blood-brain barrier can permit passage of viral particles into the brain,[24] there are apparently no reports of this being tried with oligonucleotides.

Another approach to breaching the barrier is to conjugate the oligonucleotide to a molecule which is actively transported across the blood-brain barrier, thus "piggybacking" it into the brain. This general concept is being exploited commercially by workers at Alkermes, Inc.,[25] using antibodies to the transferrin receptor. These antibodies cross the blood-brain barrier in a receptor-mediated fashion and have been found to enhance delivery to the brain of molecules, such as AZT, conjugated to them.[26] Pardridge[27] has also suggested coupling oligonucleotides to blood-brain barrier-crossing proteins such as transferrin. However, there are no studies reported using such conjugated oligonucleotides.

In summary, it is unlikely that it will be possible to achieve effective brain concentrations of pure ODNs injected directly into the bloodstream. The strategies of piggybacking or hyperosmotic disruption of the blood-brain barrier may have considerable potential, and should be pursued. Any intravenous administration method will have to cope, however, with the inherent distribution and stability problems associated with the bloodstream.

III. DIRECT ADMINISTRATION OF NUCLEIC ACIDS TO THE CENTRAL NERVOUS SYSTEM

A. RATIONALE

As we have seen, if intravenous administration of oligonucleotides to the CNS is to work, fairly complex manipulations will be necessary. On the other hand, a simple way to assure a high concentration of oligonucleotide in the brain would appear to be to inject it there directly into cerebrospinal fluid (CSF). This administration method has several other theoretical reasons to recommend it. The first is that the biological makeup of the CSF differs enormously from that of blood. CSF contains few cells, and *in vitro* appears to have essentially no nuclease activity.[28] One might expect oligonucleotides to be very stable in the CSF. The second reason is that the volume of CSF is much less than that of the bloodstream. The possibility of maintaining a high CSF concentration of a desired drug without diluting it into the bloodstream would greatly reduce the amount of drug needed. Given the costliness of oligonucleotides, this could be an important consideration. Moreover, the CSF is cleared from the body by a different mechanism than the bloodstream: CSF is produced at the choroid plexus in the cerebral ventricles and cleared at the arachnoid granulations. The average turnover time is about 2 h in the rat. This also suggests that it might be easier to maintain high concentrations by direct infusion into CSF than by intravenous administration.

B. DIRECT INFUSION INTO THE RAT CENTRAL NERVOUS SYSTEM

It is clear that an understanding of the disposition of oligonucleotides in the CNS is of crucial importance to proper development of therapeutics. In this vein, we have undertaken a series of studies on pharmacokinetics of oligonucleotides in a rat CNS model. Some of these results have already been reported.[28,29]

Our studies initially focused on infusion of ODNs into the CSF of rats. We hoped, for the reasons stated above, that the oligonucleotide would be stable in CSF and that a moderate concentration could be maintained. We were surprised to find, however, that the CSF concentration of labeled phosphodiester oligonucleotides fell rapidly relative to a control substance (inulin) after a bolus injection into the rat cerebral ventricle. Gel electrophoretic analysis indicated that the oligonucleotides were being rapidly degraded. Phosphorothioate oligonucleotides, however, were not degraded and were seen at levels consistent with bulk clearance of the CSF. Corresponding results were seen when the oligonucleotides were continuously infused over several days: phosphodiester oligonucleotides were substantially degraded, while phosphorothioate oligonucleotides were seen at expected concentrations.

Thus, despite the absence of nucleases in CSF, intracerebroventricular oligonucleotides are rapidly degraded. The most likely explanation appears to be a substantial exchange of oligonucleotide with the brain parenchyma, where the nuclease level is quite high. This is corroborated by our observation of a substantial uptake of phosphorothioate oligonucleotide into the brain following intraventricular administration.[28] Presumably, phosphodiester oligonucleotide is also taken up in this way, but is rapidly degraded in the brain. Further evidence for the ability of the brain to degrade unprotected oligonucleotides was seen when CSF solutions of the ODNs were incubated with brain slices: again, phosphodiester oligonucleotides were degraded, but phosphorothioate ODNs remained intact.[28] Most of our subsequent studies have been based on the assumption that phosphodiester oligonucleotides will have a very short half-life in the brain parenchyma.

Our next studies focused on the direct injection of oligonucleotides into the caudate region of the brain. These studies again demonstrated the presence of significant nuclease activity in the brain parenchyma, and even some degradation of phosphorothioate oligonucleotides was observed. However, when phosphorothioates were used, we were struck by the high concentration achieved in the vicinity of the infusion, as evidenced by the fluorescence of the tagged oligonucleotide.[30] Moreover, intact oligonucleotide was recovered by extraction from distal parts of the brain, and fluorescence microscopy revealed small amounts of the fluorescent label in cells in the opposite hemisphere. Interestingly, the cellular localization away from the infusion site appeared to be primarily nuclear.

Quantitation of the fluorescence after continuous infusion of fluorescent oligonucleotides (Table 1) indicated that more than half of the total fluorescence was in the region within 2 mm of the injection slice with a phosphorothioate, and approximately 86% with a phosphodiester oligonucleotide. The concentration difference between the site of injection and the remainder of the brain was much higher with phosphodiester than phosphorothioate oligonucleotides. Moreover, when equal amounts of the fluorescent extracts were examined electrophoretically, little intact phosphodiester oligonucleotide was found, while phosphorothioate oligonucleotide was only partially degraded. McCarthy et al.[31] also reported that upon injection of a ^{32}P-labeled phosphorothioate oligonucleotide into the brain parenchyma, the oligonucleotide is largely intact after 5 h and the radioactivity can be seen sequestered in cells. Injection of rhodamine-labeled phosphodiester oligonucleotides revealed nonnuclear cellular uptake of fluorescence, although intactness of the oligonucleotide was not demonstrated.

We thus suspect that the higher clearance of phosphodiester oligonucleotides reflects their instability. Phosphorothioate oligonucleotides are also cleared from the brain and degraded somewhat, but detectable quantities manage to diffuse into cells all around the brain. Either with phosphorothioate or phosphodiester oligonucleotides, the continuous infusion method clearly has the potential to deliver focally high concentrations of oligonucleotide, with a strong concentration gradient to the rest of the brain.

We have recently begun to examine administration of oligonucleotides to growing glioblastoma tumors by continuous infusion. In two models, one involving a human glioblastoma (U87-MG) xenograft in athymic rats and the other a rat glioma (C6) tumor in Fisher rats, we have been able to achieve a similar sort of focal distribution in the

Table 1   **Distribution of fluorescent oligonucleotide in rat brain after continuous infusion**[a]

	Right Hemisphere		Left Hemisphere[b]
	Inf. Site[b]	Remainder[b]	
Amount/h-equiv[c]			
P(O)	1.9 (0.11)	0.22 (0.12)	0.05 (0.02)
P(S)	3.5 (2.2)	2.3 (1.1)	0.91 (0.18)
Relative amount[d]			
P(O)	100	15. (7.)	3.3 (1.)
P(S)	100	75. (18.)	37. (18.)
Relative concentration[e]			
P(O)	100	3.7	0.6
P(S)	100	21	6.3

[a] Oligonucleotide was infused into right caudate nucleus at 0.5 nmol/h for approximately 130 h. Values are for total fluorescence in aqueous layer after phenol/chloroform extraction, and represent upper limit of amount of intact oligonucleotide.

[b] Three brain regions were assayed: "Infusion site" was right hemisphere sliced coronally +/–2 mm around cannula site, "remainder" represents the remainder of the right hemisphere, and left hemisphere is the entire hemisphere, which was contralateral to the infusion.

[c] Amount of fluorescein divided by 0.5 nmol, to give number of hours necessary to infuse that amount. Standard errors in parentheses.

[d] Amounts of fluorescein relative to amount in infusion site in that rat; infusion site = 100. Standard errors in parentheses.

[e] Amounts of fluorescein divided by approximate size of brain sample. Infusion site = 100.

growing tumor as we saw in the healthy caudate.[30] We believe that it should be possible to achieve a very high ratio of concentration in tumor relative to the remainder of the brain, which should give a very positive contribution to the therapeutic index.

IV.   *IN VIVO* BIOLOGICAL EFFECTS

Several studies have now appeared which report biological effects upon administration of nucleic acids to the brain, both in the antisense and gene therapy literature. Antisense therapy is, of course, the other side of the coin of gene therapy. While gene therapy uses nucleic acids to introduce new genes, antisense therapy uses nucleic acids to control existing genes. While most of the gene therapy studies to date have focused on viral vectors, there are several reports of injections of naked RNA or DNA into the brain. For example, Jirikowski et al.[32] reported that injection of 5 μg of total cytoplasmic RNA of normal rat hypothalami, or 50 ng of synthetic mRNA for arginine vasopressin, into the hypothalami of Brattleboro rats (which lack expression of this gene) led to expression of the RNA and to reversal of low urine osmolality. Moreover, injection of labeled mRNA revealed radioactivity selectively in the magnocellular perikarya of the supraoptic and paraventricular nuclei, while injection of the antisense strand led to no accumulation in the hypothalamic perikarya. Thus there was not only expression, but apparent selective uptake or retention of the sense sequence.

Brain tissue has also been demonstrated to take up and express plasmids upon direct injection in association with lipids. Zhang et al.[33] injected a plasmid-Lipofectin®

formulation (10 µg) intracerebroventricularly (i.c.v.) into rats and reported expression by the lac reporter gene, mainly in the pia mater and ependymal cells, lasting for several days. The authors also reported a suppression of audiogenic epileptic seizures by a cholecystokinin octapeptide expressed by the plasmid.

There are also a few reports utilizing direct i.c.v. injection of antisense oligonucleotides into the brain. Wahlestedt et al.[34] injected an 18-mer phosphodiester ODN targeted against the rat neuropeptide Y1 receptor (50 µg i.c.v. twice a day for 2 d, approx. 32 nmol overall). When assayed 12 h later, these rats were found to have increased anxiety and, upon autopsy, cortical Y1 receptor level was reduced by 60%.

In a subsequent study, Wahlestedt and co-workers[2] used an 18-mer phosphodiester oligonucleotide antisense to rat N-methyl-D-aspartate (NMDA) receptor. Glutamate, the natural ligand of this receptor, is believed to be important in neuronal death following cerebral vascular occlusion. The antisense oligonucleotide reduced the amount of NMDA receptor *in vitro* and *in vivo*, although no effect on the level of NMDA-R1 mRNA was detected. The oligonucleotide was administered to rats i.c.v. (15 nmol in 5 µl) twice a day for 2 d, then the middle cerebral arteries were surgically occluded. A day later, the animals were sacrificed and the brains analyzed for infarctions. Antisense treatment reduced lesion volume by 44% over the controls, with the region of salvage being predominantly the posterior area, including portions of the occipital cortex and lateral caudate nucleus.

There are also reports of injections directly into specific regions of the rat brain. Chiasson and co-workers[35] injected 15-mer phosphorothioate oligonucleotides which were antisense or sense to c-fos into opposite striata of a rat (2 µl of a 1-mM solution). The animals were injected 10 or 22 h later with d-amphetamine to activate immediate-early genes and then sacrificed after 2 h. *Fos*-like immunoreactivity was virtually eliminated from 20 to 50% of the striatum at 10 h, but was back to control levels by 22 h.

McCarthy and co-workers[36] used a sesame oil vehicle to administer oligonucleotides (1 µg in 1 µl, bilaterally) to the ventral medial hypothalamus of 3-d-old female rat pups, which are at a crucial stage in sexual differentiation. The phosphodiester oligonucleotide was antisense to the estrogen receptor, and the dose was followed 6 h later by a lightly androgenizing subcutaneous dose of testosterone in some of the animals. The animals were allowed to mature, and were assayed for several effects of androgenization. There was no effect of oligonucleotide treatment on the animals not receiving testosterone, but in the testosterone-treated animals the antisense treatment significantly reduced androgenization. This was also apparent upon autopsy in the size of the sexually dimorphic nuclei of the brain. Based on these findings, McCarthy has made recommendations for the use of antisense oligonucleotides in the CNS.[37]

We have also been able to achieve biological and behavioral effects by unilateral administration to the caudate nucleus of rats of phosphorothioate oligonucleotides antisense to tyrosine hydroxylase.[38] Within 24 to 48 h of the initiation of continuous infusion (30 nmol/d for 7 d), the animals treated with the antisense oligonucleotide began to rotate toward the treated hemisphere, a behavior characteristic of unilateral reduction in striatal dopamine content. Animals treated with scrambled-sequence oligonucleotides did not show this behavior. Western blot analysis of striatal tissue taken from the treated and contralateral hemispheres revealed a 60 to 80% loss of tyrosine hydroxylase protein in the treated hemisphere. The control oligonucleotide had no effect on the tyrosine hydroxylase content.

These studies are intriguing, as they imply fairly significant antisense effects, sometimes reaching out some distance from the site of infusion and persisting for as long as 10 h. Oligonucleotides can apparently have a strong, but transient effect, suggesting a half-life of oligonucleotide long enough to achieve a significant antisense effect, yet short enough to have that effect gone within another 12 h. It is interesting that intracerebroventricular infusion of phosphodiester oligonucleotides can have a biological effect despite the degradation which we have observed with this administration method.[28] More work on the distribution of oligonucleotides and on nuclease activity in different parts of the brain is essential.

These early succeses hint at an enormous potential for use of antisense oligonucleotides in the brain. One can envision spatial and temporal control of oligonucleotide concentration not possible in other regions of the body. Focal targeting of oligonucleotides may be achievable simply by stereotaxic injection, given the unique distribution mechanisms in the brain. And on a temporal level, it appears that the clearance mechanisms of the brain allow for reversible biological effects with very small doses of oligonucleotide. The key to future success will be the coupling of the vast array of targets for antisense in the CNS with a delivery strategy designed to place the oligonucleotide where and when it is needed.

REFERENCES

1. Osen-Sand, A., Catsicas, M., Staple, J. K., Jones, K. A., Ayala, G., Knowles, J., Grenningloh, G., and Catsicas, S., Inhibition of axonal growth by SNAP-25 antisense oligonucleotides *in vitro* and *in vivo*, *Nature (London)*, 364, 445, 1993.
2. Wahlestedt, C., Golanov, E., Yamamoto, S., Yee, F., Ericson, H., Yoo, H., Inturrisi, C. E., and Reis, D. J., Antisense oligodeoxynucleotides to NMDA-R1 receptor channel protect cortical neurons from excitotoxicity and reduce focal ischaemic infarctions, *Nature (London)*, 363, 260, 1993.
3. Ulloa, L., Diaz-Nido, J., and Avila, J., Depletion of casein kinase II by antisense oligonucleotide prevents neuritogenesis in neuroblastoma cells, *EMBO J.*, 12, 1633, 1993.
4. Holopainen, I. and Wojcik, W. J., A specific antisense oligodeoxynucleotide to mRNAs encoding receptors with seven transmembrane spanning regions decreases muscarinic m_2 and gamma-aminobutyric acid$_B$ receptors in rat cerebellar granule cells, *J. Pharmacol. Exp. Ther.*, 264, 423, 1993.
5. Eng, L. F., Current antisense nucleic acid strategies for manipulating neuronal and glial cells, *Res. Publ. Assoc. Res. Nerv. Ment. Dis.*, 71, 293, 1993.
6. Lawlor, K. G. and Narayanan, R., Persistent expression of the tumor suppressor gene DCC is essential for neuronal differentiation, *Cell Growth Differ.*, 3, 609, 1992.
7. Caceres, A., Mautino, J., and Kosik, K. S., Suppression of MAP2 in cultured cerebellar macroneurons inhibits minor neurite formation, *Neuron*, 9, 607, 1992.
8. Wright, E. M., Vogel, K. S., and Davies, A. M., Neurotrophic factors promote the maturation of developing sensory neurons before they become dependent on these factors for survival, *Neuron*, 9, 139, 1992.
9. Ferreira, A., Niclas, J., Vale, R. D., Banker, G., and Kosik, K. S., Suppression of kinesin expression in cultured hippocampal neurons using antisense oligonucleotides, *J. Cell Biol.*, 117, 595, 1992.
10. Manev, H., Caredda, S., and Grayson, D. R., Nonselective inhibition by antisense oligonucleotides of cytosine arabinoside action, *NeuroReport*, 2, 589, 1991.
11. Caceres, A., Potrebic, S., and Kosik, K. S., The effect of tau antisense oligonucleotides on neurite formation of cultured cerebellar macroneurons, *J. Neurosci.*, 11, 1515, 1991.
12. Agrawal, S., Temsamani, J., and Tang, J. Y., Pharmacokinetics, biodistribution, and stability of oligodeoxynucleotide phosphorothioates in mice, *Proc. Natl. Acad. Sci. U.S.A.*, 88, 7595, 1991.
13. Goodarzi, G., Watabe, M., and Watabe, K., Organ distribution and stability of phosphorothioated oligodeoxyribonucleotides in mice, *Biopharm. Drug Dispos.*, 13, 221, 1992.

14. Inagaki, M., Togawa, K., Carr, B. I., Ghosh, K., and Cohen, J. S., Antisense oligonucleotides: inhibition of liver cell proliferation and in vivo disposition, *Transpl. Proc.*, 24, 2971, 1992.

15. Smith, Q. R., Quantitation of the blood-brain barrier permeability, in *Implications of the Blood-Brain Barrier and Its Manipulation*, Vol. 1, Neuwelt, E. A., Ed., Plenum Press, New York, 1989, chap. 4.

16. Crone, C., The blood-brain barrier; a modified tight epithelium, in *The Blood-Brain Barrier in Health and Disease*, Suckling, A. J., Rumsby, M. G., and Bradbury, M. W. B., Eds., VCH-Ellis Horwood, Chichester, England, 1986, 17.

17. Davson, H., History of the blood-brain barrier concept, in *Implications of the Blood-Brain Barrier and Its Manipulation*, Vol. 1, Neuwelt, E. A., Ed., Plenum Press, New York, 1989, chap. 2.

18. Neuwelt, E. A. and Frenkel, E. P., The challenge of the blood-brain barrier, in *Implications of the Blood-Brain Barrier and Its Manipulation*, Vol. 1, Neuwelt, E. A., Ed., Plenum Press, New York, 1989, chap. 1.

19. Zendegui, J. G., Vasquez, K. M., Tinsley, J. H., Kessler, D. J., and Hogan, M. E., In vivo stability and kinetics of absorption and disposition of 3' phosphopropyl amine oligonucleotides, *Nucleic Acids Res.*, 20, 307, 1992.

20. Chen, T.-L., Miller, P. S., Ts'o, P. O. P., and Colvin, O. M., Disposition and metabolism of oligodeoxynucleoside methylphosphonate following a single iv injection in mice, *Drug Metab. Dispos.*, 18, 815, 1990.

21. Ratajczak, M. Z., Kant, J. A., Luger, S. M., Hijiya, N., Zhang, J., Zon, G., and Gewirtz, A. M., In vivo treatment of human leukemia in a SCID mouse model with c-myb antisense oligodeoxynucleotides, *Proc. Natl. Acad. Sci. U.S.A.*, 89, 11823, 1992.

22. Miyagami, M., Tsubokawa, T., Tazoe, M., and Kagawa, Y., Intra-arterial ACNU chemotherapy employing 20% mannitol osmotic blood-brain barrier disruption for malignant brain tumors, *Neurol. Med. Chir. (Tokyo)*, 30, 582, 1990.

23. Neuwelt, E. A., Goldman, D. L., Dahlborg, S. A., Crossen, J., Ramsey, F., Roman-Goldstein, S., Braziel, R., and Dana, B., Primary CNS lymphoma treated with osmotic blood-brain barrier disruption: prolonged survival and preservation of cognitive function, *J. Clin. Oncol.*, 9, 1580, 1991.

24. Neuwelt, E. A., Pagel, M. A., and Dix, R. D., Delivery of ultraviolet-inactivated ^{35}S-herpesvirus across an osmotically modified blood-brain barrier, *J. Neurosurg.*, 74, 475, 1991.

25. Friden, P. M., Walus, L. R., Musso, G. F., Taylor, M. A., Malfroy, B., and Starzyk, R. M., Anti-transferrin receptor antibody and antibody-drug conjugates cross the blood-brain barrier, *Proc. Natl. Acad. Sci. U.S.A.*, 88, 4771, 1991.

26. Tadayoni, B. M., Friden, P. M., Walus, L. R., and Musso, G. F., Synthesis, *in vitro* kinetics, and *in vivo* studies on protein conjugates of AZT: evaluation as a transport system to increase brain delivery, *Bioconj. Chem.*, 4, 139, 1993.

27. Pardridge, W. M., Recent developments in peptide drug delivery to the brain, *Pharmacol. Toxicol.*, 71, 3, 1992.

28. Whitesell, L., Geselowitz, D., Chavany, C., Fahmy, B., Walbridge, S., Alger, J. R., and Neckers, L. M., Stability, clearance, and disposition of intraventricularly administered oligodeoxynucleotides: implications for therapeutic application within the central nervous system, *Proc. Natl. Acad. Sci. U.S.A.*, 90, 4665, 1993.

29. Neckers, L., Whitesell, L., Rosolen, A., and Geselowitz, D. A., Antisense inhibition of oncogene expression, *Crit. Rev. Oncog.*, 3, 175, 1992.

30. Geselowitz, D., Neckers, L., and Chavany, C., unpublished observations, 1993.

31. McCarthy, M. M., Brooks, P. J., Pfaus, J. G., Brown, H. E., Flanagan, L. M., Schwartz-Giblin, S., and Pfaff, D. W., Antisense oligodeoxynucleotides in behavioral neuroscience, *Neuroprotocols*, 2, 64, 1993.

32. Jirikowski, G. F., Sanna, P. P., Maciejewski-Lenoir, D., and Bloom, F. E., Reversal of diabetes insipidus in Brattleboro rats: intrahypothalamic injection of vasopressin mRNA, *Science (Washington, D.C.)*, 255, 996, 1992.
33. Zhang, L. X., Wu, M., and Han, J.-S., Suppression of audiogenic epileptic seizures by intracerebral injection of a CCK gene vector, *NeuroReport*, 3, 700, 1992.
34. Wahlestedt, C., Pich, E. M., Koob, G. F., Yee, F., and Heilig, M., Modulation of anxiety and neuropeptide Y-Y1 receptors by antisense oligodeoxynucleotides, *Science (Washington, D.C.)*, 259, 528, 1993.
35. Chiasson, B. J., Hooper, M. L., Murphy, P. R., and Robertson, H. A., Antisense oligonucleotide eliminates *in vivo* expression of c-*fos* in mammalian brain, *Eur. J. Pharmacol.*, 227, 451, 1992.
36. McCarthy, M. M., Schlenker, E. H., and Pfaff, D. W., Enduring consequences of neonatal treatment with antisense oligodeoxynucleotides to estrogen receptor messenger ribonucleic acid on sexual differentiation of rat brain, *Endocrinology (Baltimore)*, 133, 433, 1993.
37. McCarthy, M. M., Use of antisense oligodexoynucleotides to block gene expression in the central nervous system, in *Neurobiology of Steroids: Methods in Neurosciences*, de Kloet, E. R. and Sutanto, W., Eds., Academic Press, New York, in press, 1994.
38. Chavany, C., Geselowitz, D., and Neckers, L., unpublished observations, 1993.

Chapter 5

In Vivo Pharmacokinetics of Oligonucleotides Following Administration by Different Routes

Valentin V. Vlassov, Leonid A. Yakubov, Valery Karamyshev,
Ludmila Pautova, Elena Rykova, and Marina Nechaeva

CONTENTS

I. INTRODUCTION

Exciting results of studies on suppression of virus multiplication and attenuation of expression of deleterious genes with oligonucleotide derivatives[1-3] suggest that the compounds may become efficient therapeutics. For future therapeutic purposes, pharmacokinetic investigation of oligonucleotide behavior *in vivo* is needed for elucidation of optimal administration routes and estimation of therapeutic doses and frequency of administration of the compounds.

Studying on the pharmacokinetics of oligonucleotides just have begun, and most of the experiments have been done with phosphodiester oligonucleotides and phosphorothioate oligonucleotide analogs which are already being used in animal and human trials. These compounds behave similarly in animal organisms. A different behavior can be expected for the heavily modified and nonionic analogs of oligonucleotides; however, detailed pharmacological studies with these compounds have yet to be reported.

This chapter summarizes data on the *in vivo* pharmacokinetics of deoxyribo-oligonucleotides introduced by traditional intravenous and intraperitoneal injection and results of our recent studies on delivery of oligonucleotides in animals by various nondamaging (noninvasive) routes.

II. FACTORS AFFECTING DISTRIBUTION OF OLIGONUCLEOTIDES IN ORGANISMS

Distribution and fate of oligonucleotides in whole organisms are affected by a few factors. Oligonucleotides are relatively large and complex molecules and may be expected to interact with some macromolecules present in the bloodstream and at the cell surface. These interactions will affect distribution of the compounds in the organism, decrease bioavailability, and can cause toxic and other effects. It should be noted that oligonucleotides are not unknown foreign molecules to the organism, because the appearance of nucleic acids in the circulation can be caused by many events. They are detected in healthy patients and can appear in large amounts as a result of pathological disease (e.g., systemic lupus erythematosus), apoptosis, after trauma, and during hemodialysis. Studies in experimental animals have shown that DNA is removed rapidly and efficiently from circulation by liver.[4]

A. CELLULAR UPTAKE

The data on the *in vitro* pharmacokinetics of oligonucleotides and various oligonucleotide analogs were reviewed recently,[5,6] and more information concerning the interaction of oligonucleotides with cells can be found in other chapters of this volume. In this section we consider briefly the main features of the oligonucleotide-cell interaction, which can affect distribution of the compounds in organism and which should be taken into account when developing approaches for administration of the compounds.

In tissue culture experiments, it was found that eucaryotic cells take up deoxyribooligonucleotides by a mechanism compatible with receptor-mediated endocytosis.[5-7] The uptake process is saturable, specific, and dependent on temperature and concentration of the compounds. Plateau binding level is achieved by 1 to 2 h of incubation of cells with oligonucleotides at 37°C. Known inhibitors of endocytosis reduce uptake.

It was found that the cells can release the oligonucleotides taken up.[8] Efflux of the compounds occurs with a rate similar to that of the binding process. This phenomenon suggests that the occurrence of transcytosis and transportation of oligonucleotides through cell layers in tissues is a possibility.

In affinity labeling experiments and by affinity chromatography, 75 to 80 kDa cellular surface proteins capable of specific binding oligonucleotides were detected at the surface of various cells.[8-11] These proteins were suggested to be receptors specific for nucleic acids and are believed to play an important role in oligonucleotide internalization. The number of these putative receptors was calculated to be approximately 10^5 per one L929 cell. Also, a 35-kDa protein on the lymphocyte membrane was found to bind oligonucleotides, although interaction with this protein occurs only at acidic pH.[12] Recently we have found that phosphodiester oligonucleotides and phosphorothioate oligonucleotide analogs can bind to cellular CD4 receptors.[13] Interaction with cell-surface proteins can play an important role in the distribution of the compounds in organisms. Thus, the number of the 75- to 80-kDa proteins at the cells of different organs of mouse correlate with the quantity of the labeled oligonucleotide internalized by the organs.[14]

Phosphorothioate oligonucleotides enter cells in a manner similar to the phosphodiester oligonucleotides; however, they bind more tightly to the cellular receptors, and the uptake process occurs somewhat more slowly.[5,9]

Methylphosphonate analogs of oligonucleotides also were suggested to be taken up by cells by absorptive and fluid phase endocytosis.[15]

Numerous modifications were introduced in oligonucleotides for increasing their affinity to cell membranes and targeting to specific cells, taking advantage of receptor-mediated endocytosis.[3] Thus, conjugation of cholesterol residues to oligonucleotides increased efficiency of uptake of the compounds by cells in culture by more than an order of magnitude.[16] Conjugation to polycations[17] and absorption at the positively charged carriers[18] considerably increased the cellular uptake and allowed delivery of the compounds in the cell cytoplasm rather than in the endosomes. Apparently, the mentioned modifications of oligonucleotides and use of delivery vehicles will affect the fate of oligonucleotides in the organism.

B. INTERACTION OF OLIGONUCLEOTIDES WITH BLOOD PROTEINS

To investigate molecular interactions of oligonucleotides in the bloodstream, we have performed affinity labeling experiments with alkylating derivatives of oligonucleotides bearing an aromatic 2-chloroethylamino group capable of crosslinking to nucleic acids and proteins. The group was conjugated to the terminal phosphate, which is the least important position for interactions of oligonucleotides with macromolecules. We incubated the whole human blood serum with the $[^{32}P]$-labeled alkylating derivative of $(pT)_{16}$, and analyzed proteins reacted with the oligonucleotides. We have found that immunoglobulins M and G (IgM, IgG) and serum albumin are the major oligonucleotide-binding proteins in the blood. The reactivity decreased in the order: IgM>IgG>albumin; corresponding dissociation constants were estimated to be 4, 6, and 20 μM for the mentioned proteins, respectively.[19] Analysis of the labeled Ig revealed that the oligonucleotide derivatives react with both heavy and light chains of the proteins at, or in close proximity to, the antigen binding site. The interaction of oligonucleotides with the proteins should be taken into account when considering the fate and interactions of oligonucleotides in an organism, because the complexes can form in the organism at therapeutic concentrations of oligonucleotides *in vivo*. It also raises the question of other potential biological effects which may be caused by this interaction.

Specific immune response also can be a factor affecting the fate of oligonucleotides in an organism, although normal nucleic acids are known to be poor immunogens.[28] However, some oligonucleotide derivatives can form conjugates with blood proteins which can cause an immune response, as in the case of repeating administration of the reactive derivatives and affect biodistribution of the compounds. Thus, alkylating derivatives of oligonucleotides react with serum albumin and yield immunogenic conjugates.[20]

As expected, conjugation of oligonucleotides to groups facilitating cellular uptake of the compounds considerably affects the fate of the oligonucleotides in whole organism. Experiments with the cholesterol-conjugated oligonucleotides have shown that the compounds bind to low-density lipoproteins in the bloodstream, and this results in longer circulation and survival time of the compounds.[21]

III. PHARMACOKINETICS OF OLIGONUCLEOTIDES

A. ADMINISTRATION BY INTRAVENOUS, INTRAPERITONEAL, AND SUBCUTANEOUS ROUTES

1. Accumulation and Degradation of Oligonucleotides in Organs

Pharmacokinetics, biodistribution, and excretion of oligodeoxynucleotides and oligodeoxynucleotide phosphorothioates were investigated in experiments with different animals.[7,22-25] We investigated the fate in mice of the 5'-$[^{32}P]$-labeled oligonucleotides, bearing a benzylamine residue at the 5'-phosphate, for protection from enzymatic dephosphorylation. It was found that after intravenous administration, phosphodiester oligonucleotides rapidly distribute in mice and reach most of the animal organs.[7,24] In the case of intraperitoneal (i.p.) injection, the compounds appeared in the bloodstream after a

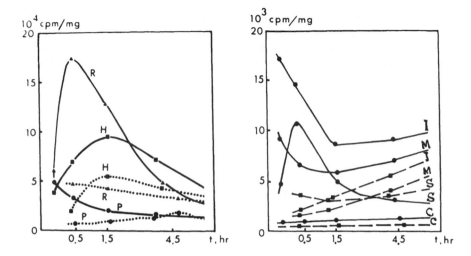

Figure 1 Kinetics of oligonucleotide accumulation in mice organs after the intraperitoneal and subcutaneous (dashed and dotted lines) administration. 2 nmol of the [^{32}P]-5'benzylphosphoramide derivative of the oligonucleotide pTGACCCTCTTCCCATC (2.5 Ci/mmol) was administered. After certain time points the mice were decapitated, and the radioactivity of blood and sample organs was determined in a liquid scintillation counter. S, blood; R, kidney; I, spleen; H, liver; P, pancreas; M, muscles; C, brain.

delay of about 30 min and tissue distributions were similar to the i.v. route (Figure 1a,b). Subcutaneously injected oligonucleotides distributed in an organism considerably slower. Accumulation of the label was more efficient in excreting organs and reticuloendothelial tissue in liver, spleen, and kidneys.[24] The brain was the least accessible for the compounds. When administered subcutaneously the compound concentrated more in the spleen and muscles.

A significant amount of the deoxyribooligonucleotide and degradation products was excreted with urine: 30 and 50% at 4 and 24 h postinjection, respectively. Similar results were obtained with phosphorothioate oligonucleotide derivatives.[22] Within the first 24 h, about 30% of the phosphorothioate oligonucleotides was excreted with urine and an additional 15% was excreted by 48 h. The excreted material was represented by intact and partially degraded oligonucleotides. Electrophoretic analysis of deoxyribooligonucleotides extracted from mice tissues after i.p. injection showed that half-life times of oligonucleotides vary in organs, from 15 min or less in liver to 30 min in blood and pancreas.

Phosphorothioate oligonucleotide analogs were much more stable in mice. By 30 min post-i.p. injection, the derivatives were practically undegraded in blood and pancreas; in most other organs the degradation rate was also diminished by a factor of 1.5 to 2, as compared to the phosphodiester oligonucleotides. By 48 h, degradation of phosphorothioate oligonucleotides in tissues varied from 50% in liver and kidney to 15% in plasma, stomach, intestine, and heart. The degradation pattern was characteristic of the 3'-exonuclease digestion. Surprisingly, oligonucleotides were extended in some organs such as liver, kidney, and intestine.[22] Accumulation of phosphorothioate oligonucleotides in the blood plasma of different animals (rat, rabbit, monkey) after single intravenous and intraperitoneal injections of the compound has been investigated.[25] It was found that peak plasma concentration was achieved a few minutes after the intravenous injection. Intravenously administered phosphorothioate oligonucleotide at doses of 5 mg/kg and 50 mg/kg resulted in the rat plasma concentrations of the compounds of 3 μM and 20 to 60

μM, respectively. The half life time of the compounds was approximately 70 h for the 50-mg/kg dose.[23,25] When the compounds were administered intraperitoneally, the plasma concentration-time curve was very similar to that for the intravenous administration, although the peak concentration was achieved 30 min later and the peak was lower. In accordance with the experiments with mice, it was found that in rabbit and monkey organisms oligonucleotides accumulate most intensively in the kidney and liver, and the lowest concentration was detected in the brain. In monkey, low accumulation of phosphorothioate compounds was observed in brain, cartilage, thyroid, and prostate. Moderate accumulation was detected in muscle, tongue, bladder, esophagus, duodenum, fat, gallbladder, and trachea. A high accumulation was characteristic of thymus, kidney, lymph nodes, liver, adrenal gland, lung, aorta, pancreas, bone marrow, heart, and salivary gland. The efficiency of accumulation of oligonucleotides in various tissues was highly correlated with the binding of oligonucleotides to membranes of the corresponding cells.[24,25]

In rats, up to 70% of the material present in plasma 20 h after intraperitoneal injection was represented by intact oligonucleotide, while in rabbits oligonucleotides survived for a longer time than in rats; degradation in monkeys was more rapid. It was concluded that daily injections are frequent enough to maintain therapeutic concentrations of phosphorothioate oligonucleotides in plasma.

2. Effect of Modifications and Delivering Devices on Oligonucleotide Pharmacokinetics

To elucidate the role of some oligonucleotide modifications on their fate in an organism, we injected the animals with [^{32}P]-labeled derivatives of phosphodiester oligonucleotides carrying an aromatic 2-chloroethylamino group, a phenazinium group, and a cholesterol group. We also investigated the fate of oligonucleotides with alternating phosphodiester and methylphosphonate linkages. The distribution of the compounds among animal organs was similar; however, a threefold greater amount of the last two compounds was bound to blood cells, as compared to the first one.[7] Electrophoretic analysis, performed 1 h postinjection, has revealed that the modifications provided a significant protection of the oligonucleotide moiety from degradation. To investigate the effect of incapsulation of the oligonucleotide derivatives in membrane carriers on their distribution in the organism, oligonucleotides were incorporated into multilamellar liposomes and into Sendai virus envelopes.[7] When the preparations were injected in mice, an enhanced delivery of the labeled oligonucleotides into lymphatic nodi and spleen was observed.

The pharmacokinetics of methylphosphonate oligonucleotide analog injected into the tail vein of a mouse has been investigated.[26] Within a few minutes the compound distributed among all animal tissues, with the lowest level found in the brain. Elimination of the compound from the circulation was rapid (half-life was 17 min). Within 2 h postinjection, up to 70% of the total amount of the oligomer injected was excreted with urine. It should be noted that the studies were performed with a chimeric oligonucleotide with a single phosphodiester linkage at the 5' end. One can expect that the behavior of the charged oligonucleotide and electroneutral methylphosphonate analogs may be different.

B. PHARMACOKINETICS OF OLIGONUCLEOTIDES INTRODUCED BY NONDAMAGING ROUTES

1. Comparative Efficiency of Oligonucleotide Delivery by Different Administration Routes

To optimize therapy, it is essential to search for nondamaging routes of administration which would allow delivery of the compounds in a controlled release fashion. Oligonucleotides could be expected to be poorly absorbed by the oral route due to degradation in the gastrointestinal tract. Because of their polyanionic nature, they are hardly expected

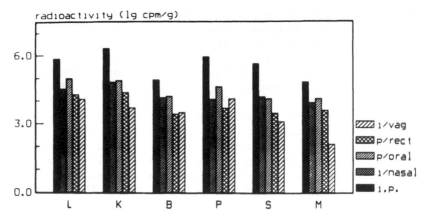

Figure 2 Distribution of oligonucleotides in mice organism after administration by different routes. 2 nmol of the [^{32}P]-5'benzylphosphoramide derivative of the oligonucleotide pTGACCCTCTTCCCATC (2.5 Ci/mmol) was administered. The mice were decapitated 20 min after the administration and tissue samples were weighed and counted. L, liver sample; K, kidney sample; B, blood sample; P, pancreas sample; S, spleen sample; M, muscle sample. I.p., intraperitoneal injection; i/nasal, intranasal injection; p/oral, per os injection; i/vag, oligonucleotide was introduced into the vagina of mice; p/rect, oligonucleotide was introduced into the rectum of mice. (From Vlassov, V.V., Karamyshev, V.N., and Yakubov, L.A., FEBS Lett., 327, 271, 1993. With permission.)

to penetrate through mucosa or nondamaged skin. However, keeping in mind the possibility of transcytosis and in hope for the existence of natural transport mechanisms, we investigated systematically the various routes of administration of phosphodiester oligonucleotides.

The experiments were performed with the 5'-^{32}P-labeled oligodeoxynucleotides The 5'-phosphate of the oligonucleotide was protected from dephosphorylation by conjugation to benzylamine. Equal amounts (2 nmol, 2×10^7 cpm) of the ^{32}P-labeled oligonucleotide derivatives were introduced in Balb mice. The compounds were dissolved in physiological salt solution and introduced intraperitoneally, intranasally (dropwise into the nose holes, total volume 5 µl), per os (dropwise into the mouth of mice, 50 µl), intravaginally (by Gilson® pipette into the vagina of mice, 10 µl), and rectally (in the distal part of the rectum, 10 µl). Application onto skin was performed as follows: 10 µl of the oligonucleotide solution was applied onto the skin of ear helices of mice. Special attention was paid to preventing the mice from licking the applied solution. In the case of ocular delivery, oligonuclotides were introduced in 2 ul of aqueous solution in mouse eye with a pipette.

Typically, mice were decapitated 20 min after the administration of oligonucleotides. Radioactivity of different organs was determined, and radioactive material extracted from the organs was subjected to analysis by electrophoresis. Figure 2 shows distribution of the label among mice organs after administration of the oligonucleotide derivative by different methods. Intranasal and oral routes provide 5 to 20% of the delivery efficiency achieved by the i.p. injection, i.e., in the experimental conditions mean concentration of the oligonucleotides in the blood was 3 µM in the former vs. 15 µM in the latter case. These concentrations fit the concentration range required for the oligonucleotide derivatives to affect target viral and cellular nucleic acids. In the cases of intravaginal and rectal administration of oligonucleotides, the concentrations of the labeled compound in organs was 5 to 10 times less than that achieved by intranasal and oral routes. Application of the aqueous oligonucleotide solution onto the skin resulted in much lower efficacy of

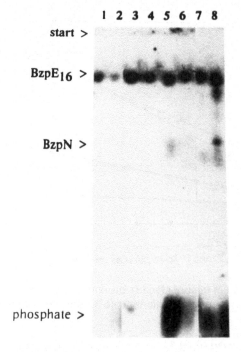

Figure 3 Stability of oligonucleotide derivatives introduced in mice by different routes. The experimental conditions were as described in the legend to Figure 2. The radioactivity of the blood samples (odd-numbered lanes on the gel) and pancreas samples (even-numbered lanes) was analyzed by electrophoresis under denaturing conditions (20% PAAG, 7 M urea). 1, 2, Intranasal administration; 3, 4, intavaginal administration; 5, 6, administration per os; 7, 8, i.p. injection. BzpE$_{16}$, position of the initial oligonucleotide derivative; BzpN, position of the mononucleotide derivative. (From Vlassov, V.V., Karamyshev, V.N., and Yakubov, L.A., FEBS Lett., 327, 271, 1993. With permission.)

delivery in the organism.[27] All the administration routes tested provided similar oligonucleotide distribution among organs. It suggests that there is one mechanism of transport of the oligonucleotides into organs: transportation by the bloodstream followed by uptake by cells.

Electrophoretic analysis has revealed that oligonucleotides reach various organs and tissues of the animals less degraded as compared to the i.p.-injected compounds.

Stability of the oligonucleotide derivatives in animal organisms depends considerably on the route of administration used: the oligonucleotide derivative tested remains practically undegraded in the organism for 30 min when being introduced intranasally, intravaginally, and per rectum, while in the case of intraperitoneal degradation of the compound reaches 50% at the same time.

Figure 3 shows results of electrophoretic analysis of the radioactive material isolated from blood and pancreas of mice 20 min after the administration of the labeled oligonucleotide derivatives. It is seen that in all the samples undegraded oligonucleotides are present in reasonable amounts. Accumulation of the labeled inorganic phosphate in the blood was observed when the oligonucleotides were administered intraperitoneally and per os (up to 70% of the total radioactivity 20 min postinjection). Approximately 15% of the recovered radioactivity in the case of intraperitoneal injection and 3% in the case of administration per os was associated with partially degraded oligonucleotides, whereas in the case of the intravaginal and intranasal injection, no products of the oligonucleotide

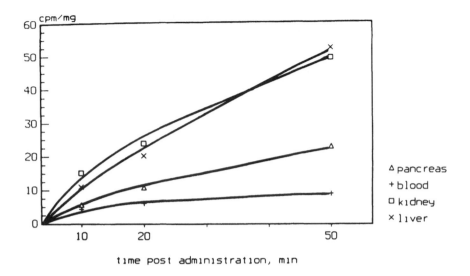

Figure 4 Time course of oligonucleotide accumulation in mice organs after the nasal administration. Experimental conditions were as described in the legend to Figure 2 except for the times.

degradation were detected (only 3% of the inorganic phosphate). This can be explained by the increased activity of macrophages and the presence of nucleases in the peritoneal cavity and intestines, which degrade oligonucleotides before they reach the bloodstream. In the pancreas, degradation of the oligonucleotides seems somewhat slower than in the blood: 30, 45, 70, and 95% of the oligonucleotides remained undegraded in the case of the i.p., peroral, intranasal, and intravaginal injections, respectively. This suggests that the pancreatic cells select preferentially undegraded oligonucleotides from the blood, perhaps by using some specific receptors. It was found that the CD4 receptor binds long oligo-nucleotides more tightly than the short ones, and that the oligonucleotides shorter than 5 mer bind to the receptor very poorly.[13] In the case of the intravaginal application, a less pronounced degradation of the oligonucleotides was observed. The obtained results have shown that the nasal and ocular routes can provide reasonable efficiency of delivery of oligonucleotides. We investigated administration of oligonucleotides by these and ocular routes in more detail. We also attempted to increase efficiency of the transdermal route by using detergent lotions and iontophoresis.

2. Intranasal Administration

The use of the nasal cavity as an alternative route of delivery has received much attention due to the ease of administration. We have found that oligonucleotides can be delivered by this route quite efficiently. Kinetics of the process (Figure 4) and the nonlinear dependence of the absorbed amount of the compound on the applied dose suggested that it is not a passive diffusion, but rather an active saturable transport through a biological barrier. At high oligonucleotide doses applied (5 nmol per animal) radioactivity continued to accumulate in animals for up to 2 h after the application.

3. Ocular Administration

We have found that administration of oligonucleotides in the form of eye drops was followed by absorption of the compounds and distribution in the organism. Figure 5 shows the kinetics of accumulation of the radiolabeled derivatives of an oligonucleotide in mouse organism after ocular administration. Concentration of the compound in the bloodstream

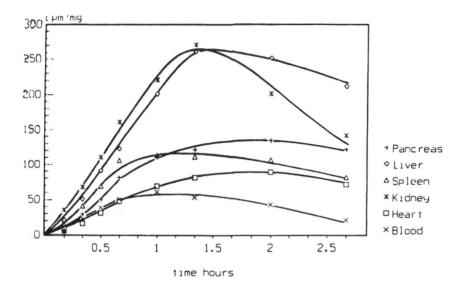

Figure 5 Kinetics of oligonucleotide accumulation in mice organs after the ocular administration. 2 µl 150 µM ^{32}P BzpE$_{16}$ were administered through mouse eye. After certain time points the mice were decapitated, and the radioactivity of blood and sample organs was determined in a liquid scintillation counter. X, blood; □, heart; *, kidney; △, spleen; ◊, liver; +, pancreas.

reaches its maximum within 1 h. The pattern of the oligonucleotide distribution after its absorption seems to be similar to that produced by other methods tested. The applied dose-absorbed amount curve showed a dependence similar to that observed for the intranasal route, also suggesting the existence of a saturable absorption mechanism.

To learn more about the mechanism of transportation of oligonucleotide through the eye we investigated the dependence of the process efficiency on the size of the oligo-nucleotide. We administered by ocular route derivatives of oligonucleotides of various length: 4, 10, 16, and 22 mers. Figure 6 shows the results of the experiments. Ocular route provides 10% of the penetration efficiency achieved by the intraperitoneal injection for all the oligonucleotides tested. The delay in the appearance of the compounds in the organism is an apparent result of the complicated pathway of the compounds through the mucous membrane in conjunctiva and in the nasolacrimal system: 30 min after adminis-tration only a small part of the material applied accumulates in the mice organism (Figure 5). Results of the experiment show that longer oligonucleotides penetrate by the ocular route more easily. This dependence is opposite to the one known for peptides introduced by the ocular route.[28] The fact may suggest the existence of the transcytosis mechanism of the oligonucleotide absorption through the eye, in contrast to the diffusion for peptides. Specific nucleic acid-binding receptors could provide such a mechanism, which can also explain the length dependence of the delivery efficiency because long oligonucleotides bind to the receptors more tightly than short ones.

It was found that simultaneous application of the oligonucleotides and competitors such as sonicated DNA (5 mg/ml), unlabeled 16-mer oligonucleotide (1 mM) and heparane sulfate (1 mM) inhibited the oligonucleotide uptake. These results are evidence in favor of the active oligonucleotide transportation through the eye.

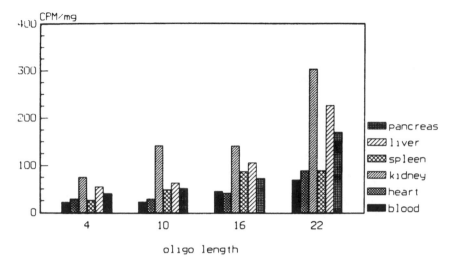

Figure 6 Dependence of delivery efficiency on the oligonucleotide length by the ocular route. 0.3 n*M* (4.4 mln cpm) of each oligonucleotide were administered in mice through the eye. After 30 min the radioactivity of the organ samples was determined.

4. Transdermal Administration

Permeation of oligonucleotides across the mammalian skin could be expected to be very poor. Indeed, when the compounds were applied in water, this route was found to be the least efficient. However, when a mild detergent was included in the solution and the compound was applied in the form of a lotion composed of equal volumes of water, glycerol, and Tween® 80, the penetration efficiency was increased by an order of magnitude and reached the level of the administration routes through mucosa.[27] Further modifications of the solution composition may result in further enhancement of the efficiency of the process. Penetration of ionic oligonucleotides through the unperturbed skin can take place only via hydrophilic pathways. Logically, it could then be influenced by an electric field because of the anionic nature of oligonucleotides, and we attempted to use iontophoresis for facilitating the percutaneous flux of oligonucleotides.

5. Iontophoresis

There are several advantages of iontophoresis over classical transdermal patches: iontophoresis yields much higher efficacy of drug delivery and it provides a means of manipulating the rate of the process. We performed iontophoresis across mouse skin in salt solution. An aluminum cathode was placed under the 1.5- × 1.5-cm Whatman® 3MM paper soaked with 100 μl of the solution containing 0.15 *M* NaCl and 2 nmol (2 × 10⁷ cpm) of radiolabeled oligonucleotide and was applied at the stomach of the immobilized mouse. An aluminum anode with the paper was put on the wet back of the mouse. Electrophoresis was run for 20 min at 2 mA. After the procedure, we observed a considerable amount of oligonucleotides in all organs of the animal. The distribution pattern was similar to that observed by using other routes tested. The efficiency of administration far exceeded (by a factor of 20 under the conditions used) that achieved by simple application of the oligonucleotide solution at the animal skin. It was inferred from this study that controlled transdermal administration is possible through this approach.[27]

We tried to deliver oligonucleotides by means of iontophoresis into a solid tumor of a living mouse. The experiment has been performed with C3H mice with spontaneous subdermal mammary gland tumors. The electrophoretic procedure was performed as described above, except that the cathode was applied on the region of the tumor and electrophoresis was run for 50 min. After the procedure, distribution of the compounds in the animal body and in different parts of the tumor was determined. It was found that the oligonucleotide entered the tumor and further distributed among the animal organs as well as in the healthy animals. The surface layers of the tumor which were close to the electrode accumulated more oligonucleotides as compared to its opposite part. The mean concentration of the oligonucleotide in the tumor was as high as in the liver, which retains the compounds actively. This result evidences that increased local concentration of oligonucleotides can be created in solid tumor tissue by means of iontophoresis directed in the tumor.

IV. CONCLUSIONS

The described experiments evidence that oligonucleotides can be delivired into mice by the intranasal, ocular, rectal, and vaginal routes and even through the skin. Taking into account the existence of specific nucleic acid-binding cellular receptors, we suggest that the mechanism of oligonucleotide penetration is transcytosis through the mucosa cells. Our studies have shown that it is possible by using alternative strategies to obtain clinically relevant levels of administration via the nasal and ocular routes without the necessity to change the structure of the oligonucleotides or destroying the epithelial membrane. We have found that the transdermal delivery of oligonucleotides can be improved considerably by using iontophoresis. Oligonucleotides delivered by the above mentioned administration routes reach animal organs in a biologically intact form. This makes one optimistic about the prospects of the *in vivo* use of natural oligonucleotides.

ACKNOWLEDGMENTS

Work in this chapter was supported in part by grants from the Foundation for Fundamental Researches of Russia and from the Russian Ministry of Sciences, as a project aimed at development of gene-targeted drugs.

REFERENCES

1. Cohen, J. S., Ed., *Oligodeoxynucleotides. Antisense Inhibitors of Gene Expression,* CRC Press, Boca Raton, FL, 1989.
2. Wickstrom, E., Ed., *Prospects for Antisense Nucleic Acids Therapy of Cancer and AIDS,* Wiley-Liss, New York, 1991.
3. Crooke, S. T. and Lebleu, B., Eds., *Antisense Research and Applications,* CRC Press, Boca Raton, FL, 1993.
4. Emlen, W., Rifai, A., Magilavy, D., and Mannik, M., Hepatic binding of DNA is mediated by a receptor on nonparenchymal cells, *Am. J. Pathol.,* 133, 54, 1988.
5. Crooke, R. M., Cellular uptake, distribution and metabolism of phosphorothioate, phosphodiester and methylphosphonate oligonucleotides, in *Antisense Research and Applications,* Crooke, S. T. and Lebleu B., Eds., CRC Press, Boca Raton, FL, 1993.
6. Neckers, L. M., Cellular internalization of oligodeoxynucleotides, in *Antisense Research and Applications,* Crooke, S. T. and Lebleu, B., Eds., CRC Press, Boca Raton, FL, 1993.
7. Vlassov, V. V. and Yakubov, L. A., Oligonucleotides in cells and in organisms: pharmacological considerations, in *Prospects for Antisense Nucleic Acid Therapy of Cancer and AIDS,* Wickstrom, E., Ed., Wiley-Liss, New York, 1991, 243–266.

8. Yakubov, L. A., Deeva, E. A., Zarytova, V. F., Ivanova, E. M., Ryte, A. S., Yurchenko, L. V., and Vlassov, V. V., Mechanism of oligonucleotide uptake by cells: involvement of specific receptors?, *Proc. Natl. Acad. Sci. U.S.A.*, 86, 6454–6458, 1989.

9. Loke, S. L., Stein, C. A., Zhang, X. H., Mori, K., Nakanishi, M., Subasinghe, C., Cohen, J. S., and Neckers, L. M., Characterization of oligonucleotide transport into living cells, *Proc. Natl. Acad. Sci. U.S.A.*, 86, 3474, 1989.

10. Geselowitz, D. A. and Neckers, L. M., Analysis of oligonucleotide binding, internalization, and intracellular trafficking utilizing a novel radiolabeled crosslinker, *Antisense Res. Dev.*, 2, 17, 1992.

11. Vlassov, V. V., Deeva, E. A., Nechaeva, M. N., Rykova, E. N., and Yakubov, L. A., Interaction of oligonucleotide derivatives with animal cells, *Nucleosides Nucleotides*, 10, 581, 1991.

12. Goodarzi, G., Watabe, M., and Watabe, K., Binding of oligonucleotides to cell membranes at acidic pH, *Biochem. Biophys. Res. Commun.*, 181, 1343, 1991.

13. Yakubov, L. A., Khaled, Z., Zhang, L.-M., Truneh, A., Vlassov, V., and Stein, C. A., Oligodeoxynucleotides interact with recombinant CD4 at multiple sites, *J. Biol. Chem.*, 268, 18818, 1993.

14. Bazanova, O. M., Vlassov, V. V., Zarytova, V. F., Ivanova, E. M., Kuligina, E. A., Yakubov, L. A., Abdukajumov, M. N., Karamyshev, V. N., and Zon, G., Oligonucleotide derivatives in organism: distribution among organs, rates of release and degradation, *Nucleotides Nucleosides*, 10, 523, 1991.

15. Shoji, Y., Akhtar, S., Periasamy, A., Herman, B., and Juliano, R., Mechanism of cellular uptake of modified oligodeoxynucleotides containing methylphosphonate linkages, *Nucleic Acids Res.*, 19, 5543, 1991.

16. Butorin, A. S., Gus'kova, L. V., Ivanova, E. M., Kobetz, N. D., Zarytova, V. F., Ryte, A. S., Yurchenko, L. V., and Vlassov, V. V., Synthesis of alkylating oligonucleotide derivatives containing cholesterol or phenazinium residues at their 3'-terminus and their interaction with DNA within mammalian cells, *FEBS Lett.*, 254, 129, 1989.

17. Clarenc, J. P., Degols, G., Leonetti, J. P., Milhaud, P., and Lebleu, B., Delivery of antisense oligonucleotides by poly(L-lysine)conjugation and liposome encapsulation, *Anti-Cancer Drug Design*, 8, 81, 1993.

18. Wagner, R. W., Matteucci, M. D., Lewis, J. G., Gutierrez, A. J., Moulds, C., and Froehler, B. C., Antisense gene inhibition by oligonucleotides containing C-5 propyne pyrimidines, *Science*, 260, 1510, 1993.

19. Rykova, E. Yu., Pautova, L. V., Yakubov, L. A., Karamyshev, V. N., and Vlassov, V. V., Interaction of oligonucleotides with blood serum proteins, *FEBS Lett.*, 344, 96, 1994.

20. Brossalina, E. B., Vlassov, V. V., and Ivanova, E. M., New methods for preparation and detection of antibodies to nucleotide sequences: evidence with oligonucleotide pApCpApC, *Biokhimija (Russia)*, 53, 18, 1988.

21. Smidt, P. C., Doan, T. L., Falco, S., and van Berkel, T. J. C., Association of antisense oligonucleotides with lipoproteins prolongs the plasma half-life and modifies the tissue distribution, *Nucleic Acids Res.*, 19, 4695, 1991.

22. Agrawal, S., Temsamani, J., and Tang, J. Y., Pharmacokinetics, biodistribution and stability of oligodeoxynucleotide phosphorothioates in mice, *Proc. Natl. Acad. Sci. U.S.A.*, 88, 7595, 1991.

23. Goodchild, J., Kim, B., and Zamecnik, P. C., The clearance and degradation of oligodeoxynucleotides following intravenous injection into rabbits, *Antisense Res. Dev.*, 1, 153, 1991.

24. Karamyshev, V. N., Vlassov, V. V., Zon, G., Ivanova, E. M., and Yakubov, L. A., Distribution of oligonucleotide derivatives and their stability in mouse tissues, *Biochimija (Russia)*, 86, 590, 1993.

25. Iversen, P., In vivo studies with phosphothioate oligonucleotides: rationale for systemic therapy, in *Antisense Research and Applications*, Crooke, S. T. and Lebleu, B., Eds., CRC Press, Boca Raton, FL, 1993.
26. Miller, P. S., Ts'O, P. O. P., Hogrefe, R. I., Reynolds, M. A., and Arnold, L. J., Jr., Anticode oligonucleoside methylphosphonates and their psoralen derivatives, in *Antisense Research and Applications*, Crooke, S. T. and Lebleu, B., Eds., CRC Press, Boca Raton, FL, 1993.
27. Vlassov, V. V., Karamyshev, V. N., and Yakubov, L. A., Penetration of oligonucleotides into mouse organism through mucosa and skin, *FEBS Lett.*, 327, 271, 1993.
28. Stollar, B.D., The experimental induction of antibodies to nucleic acids. *Meth. Enzymol.*, 70, 70, 1980.

Chapter 6

Nuclease-Resistant DNA Therapeutics

Eric Wickstrom

CONTENTS

I. ABSTRACT

Antisense and antigene DNA therapy of aberrant genes represents an exciting possibility. A variety of synthetic DNA derivatives have been applied to control many different pathogenic genes in cell culture, and a few in whole organisms, such as mice. The array of offshoots from the original phosphodiester backbone include methylphosphonate, phosphorothioate, phosphoramidate, α-phosphodiester, phosphorodithioate, boranophosphate, formacetal, and polyamide, to name only a few. In one example of an animal trial, transgenic mice bearing a murine immunoglobulin enhancer/c-*myc* fusion transgene (Eμ-*myc*) have been treated with antisense DNA methylphosphonates targeted against c-*myc* mRNA. A single intravenous dose of 300 nmol inhibited production of c-MYC protein in peripheral and splenic lymphocytes. In addition, DNA methylphosphonates did not induce acute toxicity following i.v. administration of a 300-nmol dose. An identically administered scrambled sequence oligomer did not decrease c-MYC protein or induce toxicity. Finally, recovery of DNA methylphosphonates from the blood plasma of treated mice indicated that the oligomers remained intact up to 3 h, while their concentrations decreased rapidly for the first hour, then slowly decreased over the

0-8493-4778-5/95/$0.00+$.50
© 1995 by CRC Press Inc.

next 2 h. Beyond this simple demonstration of antisense DNA therapy in an animal model, realistic design of DNA-based human therapeutic strategies requires many aspects of a candidate disease to be considered: disease prevalence, the number and nature of genes and mutations involved, and the tissues which must be targeted. For each DNA derivative intended for therapy, methods of targeting, mode of administration, pharmacokinetics, tissue distribution cellular uptake, toxicity, degradation, and excretion must be considered.

II. INTRODUCTION

Development of agents that will block replication, transcription, or translation in transformed cells, and at the same time defeat the ability of cells to become resistant has been the goal of many approaches to chemotherapy of cancer or AIDS. Many chemotherapeutic agents lack specificity, and will affect the growth and metabolism of nondiseased cells, also. This can result in strong adverse side effects, making the patients very ill and often compromising the immune system. In addition, the diseased cells may become resistant to the existing drugs, so new therapeutic approaches are needed. For diseases which result from the inappropriate expression of genes — either genes of the cell itself, or of an invading pathogen — specific prevention or reduction of the expression of such genes represents an ideal therapy.

One technique for turning off a single activated gene is the use of antisense DNAs and their analogs for inhibition of gene expression.[1] In principle, production of a particular gene product may be inhibited, reduced, or shut off by a single-stranded DNA complementary to an accessible sequence in the mRNA for that gene product,[2] or to some sequence in a transcript essential for pre-mRNA processing,[3] or to some sequence in the gene itself.[4] This paradigm for gene control is called antisense or antigene inhibition, and displays significant potential for functional probing, diagnosis, and therapy.[5-7] If indeed only the pathogenic gene is inhibited, and all other normal genes in both diseased and healthy cells are uninhibited, no toxicity should occur. This represents a most stringent and optimistic expectation, yet experiments to date have borne out the prediction of very low toxicity in cells and animals.[5-7]

The fundamental target is the gene, and within the gene one must choose a sequence target which is unique to the disease. This locus may be as simple as a single mutated base, or a junction between rearranged sequences, or a wild-type sequence in a gene which is normally silent. The nature of the unique target will usually dictate whether one must aim for the DNA of the pathogenic gene or its RNA transcript, in the nucleus, or for the processed messenger RNA, in the cytoplasm. The concept of target must also include the type of cell which is diseased, the tissue in which the cell is located, and the condition of the individual who requires treatment.

Because normal, unmodified DNA oligomers lack stability in the presence of serum nucleases[8] they are a poor choice for use in animal models. Hence, each application of DNA therapeutics will require consideration of which combination of backbone and terminal modifications of DNA (Figure 1) will be ideal for administration, distribution, cellular uptake, compartmentalization, and metabolism.

III. GENETIC DISEASE TARGETS

At their roots, all diseases are examples of inappropriate gene expression. This clearly applies to infectious agents and parasites, which bring exogenous genes into the body. It is also true for physiological diseases which arise due to imbalanced somatic gene expression, as well as inherited gene disorders. Investigations of the last decade have made clear that cancer itself stems from mutations of one kind or another in otherwise normal human genes which control proliferation.

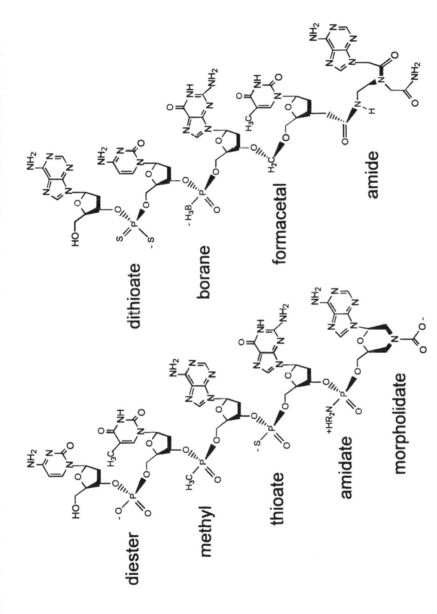

Figure 1 The most common oligodeoxynucleotide backbone modifications, arbitrarily displayed on a pair of random oligomers. (Modified from Wickstrom, E., et al., *Trends in Biotech*, 10, 281-287, 1992.)

A. INFECTIONS

Many infections may be expected to prove amenable to gene-specific therapy. In the first bacterial example, DNAs complementary to the 3′ terminus of prokaryotic 16S rRNA were synthesized, and found to inhibit mRNA translation by direct competition for ribosomes.[9] Hence, such DNAs may be exploited as a new class of antibiotics. These may be very valuable against strains of tuberculosis resistant to all known antibiotics[10] or mycoplasmas thought by some to be synergistic in human immunodeficiency virus (HIV) infection.[11] During viral infection of a cell, DNAs directed against viral-specific gene products may prevent viral replication and production, thereby stopping the disease in an unvaccinated individual. HIV,[12] herpes simplex virus,[13] and influenza virus[14] are good candidates, as they are not adequately dealt with by the immune system. Indeed, the first human trial of an antisense DNA therapeutic for HIV is underway.[15] Similarly, trypanosomes, the cause of sleeping sickness, evade the immune system and develop drug resistance. Their utilization of a common 35-nucleotide leader on their messengers provides an excellent opportunity for antisense DNA intervention.[16]

B. PHYSIOLOGICAL DISEASES

Restenosis frequently follows balloon angioplasty due to smooth muscle cell proliferation in response to injury. Inhibition of this phenomenon has been observed in cell culture following antisense DNA inhibition of c-*myc* expression[17] or in denuded carotid arteries of rats following antisense DNA inhibition of c-*myb*.[18] Furthermore, efficacy of normal antisense DNA administered into the cerebral ventricle of rats has been demonstrated against the Y-Y1 neuropeptide receptors, leading to an increase in anxiety.[19] As a corollary, one may expect that abnormal monoamine oxidase levels could be lowered in the management of schizophrenia, or of renin in hypertension.

C. CANCER

A wide variety of modes of uncontrolled cell growth fall under the umbrella name of cancer. Solid tumors and leukemias exhibit mutational activation or overexpression of a family of proliferative genes called oncogenes, as well as inactivation or underexpression of suppressor genes.[20] Antisense DNAs have proven effective in cell culture against a wide variety of oncogenes.[5-7] In those instances where the malignant state is maintained by identifiable oncogene or proto-oncogene products, cell transformation may be reversed, and the cell returned to normal, by antisense DNAs directed against the active oncogenes.[21-23] In particular, the aberrant expression of the viral v-*myc* oncogene, or the mouse or human c-*myc* proto-oncogene has been implicated in a number of leukemias and solid tumors in a variety of avian and mammalian species,[20] and antisense oligomer inhibition of human c-*myc* expression of the c-MYC p65 antigen has been effective in normal and transformed cells.[5-7] Transgenic mice which overexpress c-*myc* in their B-cells develop aggressive multifocal lymphoma/leukemia involving lymphoid organs.[24] The principal line examined of these Eμ-*myc* transgenic mice has been assigned the designation Tg(IgH,Myc)Bri157. Mice in these lines develop aggressive multifocal lymphoma/leukemia involving lymphoid organs, mice providing a well-characterized animal model for c-*myc* activation in B-cell lymphoma. Mice contain about 2 ml of blood, carrying up to 10^7 lymphocytes/ml.[25] Thus, for the purpose of studying antisense DNA inhibition of c-*myc* expression, the circulating immature lymphocytes in the bloodstream represent a close parallel in cell number and volume to cells in a culture dish.

IV. ANTISENSE DNA THERAPEUTICS

Gene-specific normal DNAs and their analogs are able to inhibit expression of specific proteins, inhibit growth, and inhibit viral replication in cultured cells.[5-7] Antisense DNA synthesized with normal phosphodiester linkages is useful as a research tool for probing biological function. Although such normal DNAs have been used successfully by many research groups to inhibit the expression of a wide variety of target genes, in viral, bacterial, plant, and animal systems, both in cell-free extracts and in whole cells, they are limited in their potential as therapeutic agents. Attempts to improve the therapeutic efficiency have involved many different approaches to derivatization (Figure 1) to improve nuclease resistance, specificity of binding to target nucleic acids, and cellular uptake.[26] Substitutions on the backbone include those which reduce nuclease sensitivity while maintaining a negative charge, and those which replace ionic moieties in order to allow the DNA to diffuse through cellular, organellar, or nuclear membranes.[5-7] Substituents at the 5' and 3' ends include reactive groups which allow covalent crosslinking of the antisense or antigene DNA to the target sequence in the RNA or DNA of interest, and bulky groups which improve cellular uptake.[5-7]

A. CELLULAR UPTAKE

A major issue in the development of DNAs as therapeutic agents is delivery of the antisense or antigene DNA into cells under *in vivo* conditions. Initially, it was considered unlikely that oligodeoxynucleotides could enter cells under any conditions, because as polyanionic molecules they would be unable to cross lipid membranes. However, after the appearance of convincing data on the antisense DNA inhibition of translation of specific cellular mRNAs and viral multiplication, studies of uptake in mammalian cells were initiated. Several laboratories found that DNAs can be taken up by the DNA binding to a surface protein and then internalized by active endocytosis, utilizing ATP hydrolysis. The putative import protein displays the electrophoretic mobility of an 80-kDa polypeptide,[27] though another group has characterized a 32-kDa cell-surface protein which operates in a similar manner;[28] both polypeptides may be involved, and may be fragments of a larger protein. This route of receptor-mediated endocytosis works for both normal and phosphorothioate DNAs,[27] though uptake of the phosphorothioate is much slower. The neutral methylphosphonates do not compete with this receptor, but recent results imply that their uptake and inhibitory effects occur more rapidly than would be possible solely by passive diffusion.[29]

A high-capacity natural mechanism of nonadsorptive endocytosis of DNAs also exists which is not limited by the number cell-surface receptors. The efficiency of uptake by this mechanism can be enhanced by coupling the DNAs to lipophilic groups. Cellular uptake and efficacy of DNA oligomers is enhanced by modification with polylysine[30] or cholesterol[31] at the 3' end, or with dimethoxytrityl at the 5' end.[32] Most cells carry the receptor for charged DNAs, and neutral DNAs enter all cell types which have been tested. Hence, targeting of DNA drugs based on cell type is unlikely to be successful, unless the DNA has been conjugated to a ligand or immunoglobulin which will bind specifically.

B. MODIFICATIONS TO STRENGTHEN BINDING

Independent of the question of nuclease resistance, potency of antisense or antigene DNAs may be enhanced by end modification with a variety of substituents. One such approach is to add an alkylating group, which results in permanent covalent binding of a DNA therapeutic to its target.[33] Similarly, the addition of an intercalating substituent which noncovalently strengthens a hybrid complex[34] may be used to enhance potency. Alternatively, one might add a moiety which induces chain scission in the target RNA.[33,34] However, from kinetic analysis, it is clear that any antisense method, including ribozymes, which depends on rapid cleavage or crosslinking following initial association, is limited by the simultaneous opportunity for nonspecific reaction with a vast excess of other available targets.[35] Hence, with any alkylating, intercalating, or cleaving derivative, it is essential that the dissociation rate of mismatched oligomers from mRNA targets be significantly faster than alkylation or chain scission.

C. NUCLEASE-RESISTANT DERIVATIVES

The relatively high concentrations of DNA necessary for activity in cell culture[5-7] and the rapid degradation of normal DNA in serum[8] make its use in therapy unrealistic. Hence, many backbone derivatives have been prepared (Figure 1). Uncharged methylphosphonate DNAs are unrecognized by nucleases, enter animal cells rapidly without utilizing the DNA receptor, and specifically inhibit expression of targeted genes, distinguishing even a single mismatch.[36] DNA phosphorothioates display a negatively charged backbone; the sulfur-modified phosphate linkage is much less sensitive to nucleases than normal DNA.[37] However, the phosphorothioates display significant nonsequence-specific background inhibition[5-7] relative to the other derivatives. To overcome the chiral problems of phosphorothioates, phosphorodithioates have been synthesized which are prochiral just like phosphodiesters.[38] However, one comparison of their antisense properties with phosphoromonothioates and normal DNA revealed lower melting temperatures and lower antisense efficacy or specificity.[39] The nuclease-resistant borane phosphonates are inter-mediate in their electronic properties between methylphosphonates and phosphorothioates;[40] their properties as antisense inhibitors remain to be explored.

A variety of phosphoramidates, morpholidates, and other cationic derivatives have been prepared, but not yet studied in great detail.[41,42] To bypass the stereochemical problems brought on by modifying the phosphate, some non-phosphate linkages have been synthesized, such as formacetal[43] and amide.[44] The amide derivatives hybridize to RNA more strongly than does normal DNA, but unfortunately they are not taken up by cells;[45] this may be a general problem of nonribose-phosphate backbones. Instead of modifying the backbone, the synthesis of oligomers of α-deoxynucleotides, rather than the normal β-deoxynucleotides, provides another way of achieving nuclease resistance without loss of base pairing effectiveness.[46] Similarly, replacement of deoxyribose with a 2'-modified ribose provides nuclease resistance and strong hybridization.[47] Surprisingly enough, the simple addition of a 3' terminal amine is sufficient to greatly extend the lifetime of otherwise unmodified DNA in blood.[48] This result is made plausible by the observation that 3'-exonucleases are responsible for most oligodeoxynucleotide degrada-tion.[49] In agreement with this model, both a 3' hairpin[50] and circularization of oligodeoxynucleotides[51] provide remarkable nuclease resistance.

These observations invited further studies of these compounds as potential therapeutic agents in animal systems. Mice can be used as a useful model to approach the conditions of the human environment. Before antisense DNA technology can be used therapeuti-cally, it must be determined whether or not the specific DNA oligomers are effective in the animal environment, that is less favorable than a culture dish, and that there are a minimum of toxic or harmful effects. Encouraging results have been observed in mice with antisense phosphorothioate therapy directed against c-*myb*,[18,52] NF-κB,[53] and the neuropeptide Y-Y1 receptor;[19] on the other hand, a lack of specificity was observed against tick-borne encephalitis virus in mice.[27] While no acute toxicity was noted in these studies, longer term administration revealed splenic hyperplasia and hypergamma-globulinemia.[54]

For a comparable test of methylphosphonate DNA efficacy in mice, the first steps were to determine whether or not DNA methylphosphonates were toxic in mice,[12] whether they could survive long enough in the mouse bloodstream to be taken up by tissue,[55] and whether they are capable of downregulating c-*myc* gene expression *in vivo*. These questions were addressed using mice bearing a murine immunoglobulin enhancer/c-*myc* fusion transgene.[24]

D. ANTI-C-*MYC* DNA METHYLPHOSPHONATES

For the test of antisense DNA therapy in c-*myc* transgenic mice,[56] two methylphosphonate DNA pentadecamers were synthesized[57] on a Millipore® 8750 DNA synthesizer and

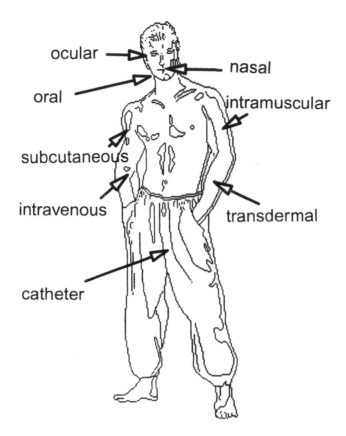

Figure 2 Schematic of routes of administration in a human.

purified by reversed-phase liquid chromatography[58] with a gradient from water to acetonitrile. Buffer salts were eliminated due to the lack of charge on DNA methylphosphonates that obviates the need for counterions. The first methylphosphonate was complementary to the initiation codon and the next 4 codons of the human c-*myc* mRNA (5'-dAACGTTGAGGGGCAT-3', antisense methylphosphonate), that were predicted to occur in a large loop in human c-*myc* mRNA.[59] As a normal oligodeoxynucleotide, this sequence has been observed to inhibit c-MYC protein expression and cellular proliferation in normal peripheral blood lymphocytes[60,61] and c-*myc*-transformed HL-60 cells,[58,62] and to induce terminal granulocytic differentiation of HL-60 cells.[22,63] The mouse and human sequences differ at the 3' end of the target sequence,[64] resulting in an A:G mismatch. The second was a scrambled version of the antisense oligomer (5'-dCTGAAGTGGCATGAG-3', scrambled methylphosphonate), that did not inhibit c-MYC expression in previous studies.[62] The scrambled oligomer was used as a control for specificity and toxicity. Several cycles of vortexing and warming to 56°C allowed the preparation of 1.2-mM samples of DNA methylphosphonates in Dulbecco's phosphate buffered saline (DPBS) solution.

V. ADMINISTRATION AND DISTRIBUTION

Drugs are given to patients (Figure 2) by a variety of modes: orally, subcutaneously, intramuscularly, intravenously, intraperitoneally, dermally, nasally, ocularly, and by catheter. Oral and topical administration are least invasive. If a therapeutic does not enter the body easily by these routes, administration by syringe or catheter is unavoidable. Though one's first impulse is to assume that DNAs must be given intravenously, all of the other routes are plausible for particular derivatives aimed at particular tissues.

A. PARENTERAL

Nevertheless, most of the published data to date in animal models report the results of intravenous administration. Normal and phosphorothioate DNAs were rapidly excreted by the kidney, but showed very low toxicity relative to concentrations which are effective in cell culture.[27,65,66] DNA methylphosphonates were also found to be nontoxic and removed quickly from circulation, but distributed well to tissues, including the brain.[55] Circulating peripheral lymphocytes presented the most favorable target cells for an intravenous oligomer in c-*myc*-transgenic mice.[56] Briefly, it was observed that DNA methylphosphonates inhibited MYC expression in B-cells with sequence specificity, as described below.

Abdominal injection of normal and phosphorothioate DNAs has been found to mimic intravenous administration in most respects, except that an extra delay of 10 to 20 min was observed before the DNAs reached peak concentrations in serum.[27,66] It is reasonable to predict that intraperitoneal administration will be less traumatic in general for animal models than intravenous injection, though in humans the intravenous route is better tolerated.

Administration of normal and phosphorothioate DNAs under the skin displayed similar distribution and excretion kinetics to those found with intraperitoneal injection.[27] With normal DNA, it has been found possible to inhibit the tumorigenesis of c-Ha-*ras*-transformed NIH3T3 cells implanted subcutaneously in athymic nude mice, but only by pretreatment of the cells with the antisense DNA prior to implantation.[23] In order to achieve normal antisense DNA inhibition of N-*myc* gene expression in human neuroepithelioma CHP-100 cells implanted subcutaneously in athymic nude mice, it was found necessary to administer the normal oligomers by continuous perfusion from a microosmotic pump implanted next to the tumor cells.[67]

B. DERMAL, NASAL, OCULAR, AND ORAL

In mice, DNA methylphosphonates have shown efficacy against herpes simplex virus 1 when applied to skin lesions in the ear.[13] This implies dermal penetration, which is indeed logical in view of the measured water:octanol partition coefficients on the order of 0.03 for dodecamers and pentadecamers.[29] For comparison, L-tyrosine and its *N*-acetyl derivative displayed similar partition coefficients and were absorbed nasally by rats.[68] Even 5'-benzylamino oligodeoxynucleotides have been found to distribute intact throughout the tissues of mice following administration to mucosal membranes and ear helices.[69] Thus, amphipathic compounds are often capable of absorption through the skin and nasal membranes. The molecular explanation for dermal availability of neutral DNA derivatives also allows the possibility of uptake by absorptive cells in the small intestine. The counterargument holds that DNA oligomers, even amphipathic ones, are too large to be absorbed by the intestinal lining. No examination of this question has been published. Oral administration under anesthesia of a measured dose, followed by measurements of oligomer levels in serum and bodily fluids and tissues, would quickly answer the question. The same rationale implies that DNA methylphosphonates or other neutral derivatives may also be available by eye drops. This route would be most applicable to ocular infections and also for neurological diseases.

C. INTRAVENOUS ANTI-C-*MYC* DNA METHYLPHOSPHONATES

In order to assess the practicality of antisense DNA therapy in c-*myc* transgenic mice,[56] the DNA methylphosphonates were administered to 4- to 8-week-old Eμ-*myc* transgenic mice, 17 to 25 g each[24] by injection into the tail vein. In each of the five trials, a mouse received 300 nmol of either antisense methylphosphonate or scrambled oligomer in 250 µl DPBS, or DPBS alone, from blind coded vials. In the fourth trial, a transgenic mouse received no treatment instead of DPBS. In each double-blind trial the dose administered corresponded to 50 to 75 mg/kg. Neither the antisense nor scrambled oligomers displayed any observable toxicity in any of the treated mice when compared with untreated or saline-injected control mice, in agreement with other reports.[12]

D. PHARMACOKINETICS AND METABOLISM

All backbone derivatives studied so far are rapidly excreted by the kidney; the α phase serum half-lives for normal, phosphorothioate, and methylphosphonate oligomers are all on the order of 10 to 20 min.[27,55,65] In addition, they do not exhibit toxicity at doses which are effective for downregulation of gene expression. The derivatives studied distribute themselves throughout body tissues; neutral derivatives, such as methylphosphonates, may cross the blood-brain barrier.[55] Furthermore, cholesterol derivatization markedly increases serum half-life,[48] presumably by increasing the hydrophobicity of the DNAs so that they bind to cell membranes and low-density lipoproteins. Paradoxically, optimal distribution and maximal serum half-lives may call for greater hydrophobicity in DNA derivatives, rather than less.

Normal DNA is readily degraded in physiological fluids,[8] primarily by 3'-exonucleases,[49] yielding deoxynucleotides which are scavenged by salvage pathways. DNA phosphorothioates are more slowly hydrolyzed, generating 5' deoxynucleoside phosphorothioates which will be incorporated into newly replicated DNA. The neutral DNA derivatives are virtually undegraded, unless some linkages are charged.[55] The predominant mode of loss of DNA oligomers, however, is excretion rather than degradation. Further modifications which slow clearance will increase the significance of metabolic detoxification pathways, particularly in the liver. No data exist at present which allow the significance of this effect to be evaluated.

E. SERUM LEVELS OF ANTI-C-MYC METHYLPHOSPHONATES

From Eµ-myc transgenic mice which had been given intravenous antisense DNA 4 h previously, blood was collected from the retroorbital sinus and the plasma isolated by sedimentation. Plasma supernatants were frozen and shipped over dry ice ($CO_2(s)$) from Philadelphia, PA to Tampa, FL. For analysis of DNA oligomers in serum, blind coded plasma samples were thawed, and aliquots of 200 µl were mixed with equal volumes of acetonitrile. Insoluble material, mainly precipitated protein, was removed by sedimentation.[70] The supernatant fractions were then lyophilized, redissolved in H_2O, filtered through 0.2-µm filters, and analyzed on a 4.6 × 250-mm Econosphere™ C_{18} reversed-phase analytical column (Alltech) using a 0 to 80% gradient from 100% H_2O to 80% acetonitrile/20% H_2O over 60 min, at a flow rate of 1.0 ml/min, delivered by an ISCO liquid chromatograph. Oligomers were detected by absorbance at 260 nm; baseline runs were subtracted from sample runs.

Blood was drawn from each mouse 3 to 4 h after administration of the DNA methylphosphonates, and analyzed for the presence of the oligomers in plasma. Plasma samples were deproteinized[70] and analyzed by reversed-phase liquid chromatography.[58] Plasma from mice treated with DNA methylphosphonates yielded peaks co-eluting with control oligomer (Figure 3). Shorter fragments were not observed, suggesting little or no breakdown of DNA methylphosphonate in the 3 h after injection. Oligomer levels in the bloodstream dropped rapidly for the first hour after injection, but leveled off to a much slower rate of decrease during the second and third hours, implying rapid removal of DNA methylphosphonates from the blood by kidneys and other tissues (Figure 4).

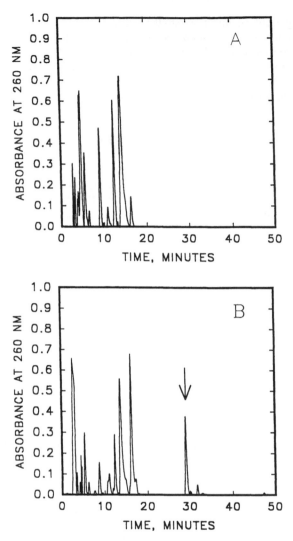

Figure 3 Reversed-phase liquid chromatograms of DNA methylphosphonates recovered from mouse plasma. (A) Chromatogram of extracted plasma from 250 μl of blood from a mouse treated with DPBS; (B) chromatogram of extracted plasma from a mouse treated with antisense oligomer for 1 h. Arrow shows normal elution time of antisense *oligomer*. (From Wickstrom, E., Bacon, T.A., and Wickstrom, E.L., *Cancer Res.*, 52, 6741-6745, 1992. With permission.)

VI. INTRAVENOUS THERAPY OF MICE WITH B-CELL LYMPHOMA

A. EFFECT ON ANTIGEN LEVELS

Peripheral lymphocytes were recovered from the blood samples of the mice treated above, or normal nontransgenic mice, by sedimentation on a Percoll® density gradient (Pharmacia). Cells were prepared for shipment by washing twice in DPBS supplemented with 0.1% (w/v) bovine serum albumin, pelleting, decanting, and resuspending in 900 μl of RPMI 1640 supplemented with 10% fetal calf serum. The cell suspensions were allowed to equilibrate for 5 to 10 min, and 100 μl of glycerol were added to each suspension in preparation for freezing. Samples were frozen and shipped as above. Upon

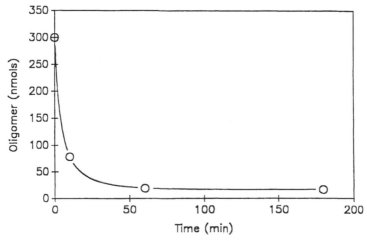

Figure 4 Recovery of methylphosphonates from plasma samples of c-*myc* transgenic mice. Samples of 250 µl removed at 3 min, 1 h, and 3 h were analyzed by liquid chromatography as in Figure 3, quantitated by integrated peak areas, then multiplied by 8 to estimate the number of nmol in the entire 2-ml bloodstream. (From Wickstrom, E., Bacon, T.A., and Wickstrom, E.L., *Cancer Res.*, 52, 6741-6745, 1992. With permission.)

receipt, blind coded cell samples were brought rapidly to room temperature in a water bath at 37°C, washed in 0.5 ml DPBS, and fixed in 1% paraformaldehyde in DPBS for 30 min. The cells were then pelleted, washed with DPBS, and preadsorbed with 1% goat serum in DPBS for 5 min followed by rabbit polyclonal anti-p65 IgG[71] for 1 h. The cells were then washed once in DPBS, preadsorbed with 1% goat serum in DPBS for 5 min, and a fluorescein-conjugated goat polyclonal anti-rabbit IgG was added for 30 min. The cells were then washed twice in DPBS. At least 50 cells were then visualized by light and fluorescence microscopy at ×40, and an average intensity estimated on an arbitrary geometric scale of 0 to 5. The fluorescence intensities from two separate fields were estimated for each sample. For flow cytometry, cell samples were passed through a Becton/Dickinson FACSCAN™ flow cytometer, and fluorescein fluorescence was quantitated from 10,000 cells. Mean channel fluorescence was linearized by dividing by 64 channels/decade, yielding a quotient that is the logarithm of the fluorescence. The antilog of this number represents the linear fluorescence. Background due to nonspecific binding by preimmune serum was subtracted. Investigators revealed sequence and cell codes to each other at the conclusion of each experiment.

Five double-blind studies were conducted to determine the effect of the DNA methylphosphonates on the level of c-MYC antigen in peripheral lymphocytes. Mice were administered 300-nmol doses of each oligomer, and blood samples were withdrawn 3 to 4 h later. In the first two trials, blood samples were also drawn prior to administration of DNA methylphosphonate in order to check for variability in unperturbed controls. White blood cells were isolated from each sample, and c-MYC was visualized by indirect immunofluorescence of cells in suspension on slides.[58] In the subsequent trials, blood was not withdrawn prior to treatment in order to maximize the number of cells that could be recovered after administration. In the third and fourth trials, white blood cells collected at 3 to 4 h were sedimented onto slides in order to allow viewing of a large number of cells in a single field. These cells were fixed with cold 95% ethanol prior to blocking.

In each of these trials, cells from untreated nontransgenic mice showed faint nuclear fluorescence, in contrast to cells from Eµ-*myc* transgenic mice 3 h after treatment with DPBS, that showed intense nuclear fluorescence. Cells isolated from transgenic mice 3 h after treatment with the scrambled DNA methylphosphonate retained intense nuclear

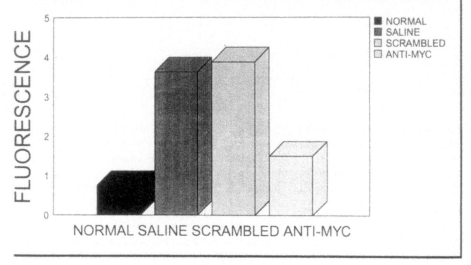

Figure 5 Inhibition of mouse c-MYC antigen levels in circulating lymphocytes collected from Eµ-*myc* transgenic mice treated with antisense DNA methylphosphonate, measured by indirect immunofluorescence, estimated visually on a geometrically increasing scale from 0 to 5. Bars, average of four experiments. (A) Untreated nontransgenic mouse cells; (B) transgenic mouse cells 3 h after administration of DPBS; (C) transgenic mouse cells 3 h after administration of scrambled DNA methylphosphonate; (D) transgenic mouse cells 3 h after administration of antisense DNA methylphosphonate. (From Wickstrom, E., Bacon, T.A., and Wickstrom, E.L., *Cancer Res.*, 52, 6741-6745, 1992. With permission.)

fluorescence. However, cells isolated from mice treated with the antisense oligomer showed nuclear fluorescence equal to or less than that of cells from nontransgenic mice, implying that the antisense DNA methylphosphonate inhibited c-MYC production.

Averaging c-MYC immunofluorescence levels among the four 300-nmol trials reported above according to the arbitrary geometrically increasing 0 to 5 scale (Figure 5), the estimated intensities of the four normal mice were 0.75 ± 0.14, transgenic mice that received saline only showed 3.63 ± 0.24, while transgenic mice that received the scrambled DNA methylphosphonate displayed 3.88 ± 0.59, and transgenic mice that received the antisense DNA methylphosphonate scored 1.50 ± 0.46. An overall analysis of the data set by the Kruskal-Wallis one-way analysis of variance for ordinal measurements[72] indicated that it was highly unlikely ($p < 0.01$) that all groups came from the same or equivalent populations. Therefore, it is probable that at least one group is statistically different from one other group. Comparisons among individual groups using the Mann-Whitney U test[72] indicated that the DPBS-treated samples were not significantly different from the scrambled-treated samples ($p = 0.343$). However, the DPBS-treated samples were found not likely to be equivalent to the antisense-treated samples ($p = 0.014$), nor were the scrambled-treated samples likely to be equivalent to the antisense-treated samples ($p = 0.014$). Levels of MYC antigen were quantitated by flow cytometry in one additional trial. Cells

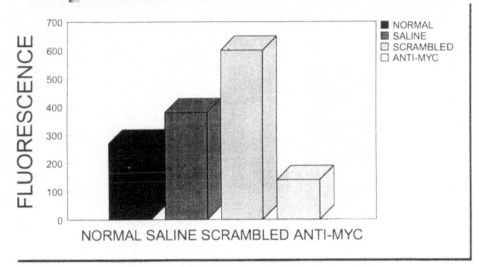

Figure 6 Fluorescence measurements of c-MYC antigen quantitated by flow cytometry of splenic lymphocytes from transgenic mice treated as in Figure 5. (A) Untreated nontransgenic mouse cells; (B) transgenic mouse cells 3 h after administration of DPBS; (C) transgenic mouse cells 3 h after administration of scrambled DNA methylphosphonate; (D) transgenic mouse cells 3 h after administration of antisense DNA methylphosphonate. Fluorescence y-axis is linear; 10,000 cells were measured; background due to nonspecific binding by preimmune serum was subtracted. (From Wickstrom, E., Bacon, T.A., and Wickstrom, E.L., *Cancer Res.*, 52, 6741-6745, 1992. With permission.)

were labeled with a fluorescein isothiocyanate-conjugated second antibody only, or with preimmune IgG first, or with polyclonal anti-c-MYC first, or with polyclonal antiactin first. No DNA oligomer had any impact on actin levels. When these profiles were compared by the Kolmogorov-Smirnov two-sample test,[73] each member of each pair of profiles was significantly different from the other member ($p < 0.001$). The linearized MYC fluorescence intensities (Figure 6) illustrate further the correlation of reduced MYC antigen with antisense DNA methylphosphonate treatment. While significant effects could be seen with doses of 300 nmol, a single trial with 50 nmol of each oligomer yielded no statistically significant diminution of c-MYC expression at 12 h after administration. Hence, there is at least some indication of dose or time dependence.

B. EFFECT ON MESSENGER LEVELS

Quantitation of c-*myc* transcripts required a larger number of cells than could easily be obtained from the peripheral lymphocyte samples utilized for indirect immunofluorescence. Spleens, however, contain many lymphocytes and are significantly enlarged in the transgenic mice,[24] and thus were used as a source of cells. Spleens were removed from mice treated with DNA methylphosphonates and homogenized in 1% NaDodSO$_4$, 5 m*M* EDTA, 10 m*M* Tris-HCl, pH 7.4, containing 100 μg/ml proteinase K. Spleen cell total nucleic acid was isolated and transgene mRNA was measured by solution hybridization[74]

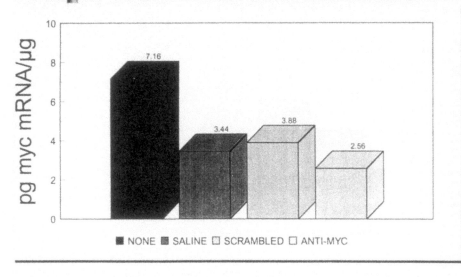

■ NONE ■ SALINE □ SCRAMBLED □ ANTI-MYC

Figure 7 Levels of c-*myc* mRNA in splenic lymphocytes from transgenic mice treated as in Figure 5. Messenger levels were quantitated by solution hybridization of cell total nucleic acid with an oligodeoxynucleotide specific for the ΦX174 bacteriophage sequences present in the transgene. (A) Untreated nontransgenic mouse cells; (B) transgenic mouse cells 3 h after administration of DPBS; (C) transgenic mouse cells 3 h after administration of scrambled DNA methylphosphonate; (D) transgenic mouse cells 3 h after administration of antisense DNA methylphosphonate. Values are reported as pg transgene mRNA/μg total nucleic acid.

using an oligodeoxynucleotide specific for the ΦX174 bacteriophage sequences present in the transgene.[24] Values are reported as pg transgene mRNA/μg total nucleic acid.

Levels of c-*myc* mRNA (Figure 7) were interpreted by averaging over the first four trials reported above. Normal mice were negative for the transgene-specific probe. Seven transgenic mice that received no injection displayed 7.16 ± 3.03 pg c-*myc* mRNA/μg total nucleic acid in their spleen cells. Three mice that received saline only showed 3.44 ± 0.46 pg/μg, while four mice that received the scrambled DNA methylphosphonate had 3.88 ± 1.88 pg/μg, and four mice that received the antisense DNA methylphosphonate yielded 2.56 ± 1.20 pg/μg. While the latter value was lower than the others, the difference was not statistically significant.

VII. DISCUSSION AND FUTURE GOALS

Given the need for nuclease resistance, antisense DNA oligomers with all methylphosphonate linkages were prepared for the first attempt at systemic antisense DNA treatment of c-*myc*-induced B-cell lymphoma in Eμ-c-*myc* transgenic mice.[56] The neutral methylphosphonate backbone was chosen to avoid nuclease hydrolysis in the blood and to maximize cellular uptake. The DNA methylphosphonates were soluble to >1 mM in DPBS, in agreement with partition coefficient measurements showing methylphosphonate preference for water over octanol.[29] The lack of acute toxicity in mice of DNA methylphosphonates at doses that are effective, 50 to 75 mg/kg, implies that further studies *in vivo* are worth pursuing. This observation agrees with an earlier study that reported no signs of toxicity below 140 mg/kg.[12] Similarly, it is encouraging that no breakdown of the DNA methylphosphonates was detected. In a previous study in mice,[55] DNA methylphosphonate undecamers with a normal phosphodiester linkage at the 5′ end were observed to be truncated in the blood to the pure methylphosphonate decamer that was not further degraded. This report also agrees with the latter work in finding rapid clearance of DNA methylphosphonates from the blood.

The observation of reduction in the level of c-MYC antigen in lymphocytes of transgenic mice treated with the anti-c-*myc* DNA methylphosphonate, but not with the scrambled sequence, is consistent with previous studies in cells.[58-63] This first positive result in an animal system implies that antisense DNA methylphosphonates, or further derivatives, may have the potential for therapeutic inhibition of c-*myc* expression in human leukemias and lymphomas. The lack of significant reduction in the level of c-*myc* mRNA in the same cell type agrees with the model that antisense DNA methylphosphonates act by inhibiting ribosomal translation of the targeted mRNA, rather than by inducing the breakdown of mRNA. This observation is consistent with the lack of RNase H degradation of mRNA/DNA methylphosphonate hybrids.[75]

It would be desirable to determine whether daily doses of antisense oligomer would actually reverse development of lymphomas. Ideally, one should target a unique sequence in the transgene that would allow downregulation of the transgene alone rather than the normal gene. To extend the serum half-life, end modification with lipophilic groups[31-34,48] or liposome encapsulation[29] might be attempted. The rather high dose required for effective inhibition may in principle be reduced by synthesis of stereospecific methylphosphonate oligomers with stronger hybridization.[76-78]

From some perspectives, the prospects for DNA therapeutics seem too good to be true. However, from a molecular point of view, their application appears logical for any disease which is caused by inappropriate gene expression. One must point out that diseases which are due to a lack of an active gene product will not be amenable to antisense or antigene DNA therapy, but will require gene *replacement* therapy.[79]

Furthermore, the ultimate challenge to DNA oligomer therapy will not be biological, but financial. How can such a regimen be carried out affordably? The answer lies in chemical modifications and synthetic routes which elevate potency to the point where milligram doses will be effective in a human patient.

ACKNOWLEDGMENTS

It is a pleasure to thank my colleagues who have worked with me on this problem: Thomas Bacon, Dr. Ralph L. Brinster, Audrey Gonzalez, Christine O'Connell, Dr. Richard D. Palmiter, Dr. Eric P. Sandgren, and Erica Wickstrom. I also thank Dr. Grace Ju, Dr. Robert Eisenman, and Dr. Gerard Evan for samples of polyclonal antisera against c-MYC protein. The following organizations contributed support for various aspects of the work: American Cancer Society, Leukemia Society of America, Biosearch, Inc., and the National Cancer Institute.

REFERENCES

1. Belikova, A. M., Zarytova, V. F., and Grineva, N. I. Synthesis of ribonucleosides and diribonucleoside phosphates containing 2-chloroethylamine and nitrogen mustard residues. *Tetrahedron Lett.*, 37, 3557, 1967.

2. Zamecnik, P. C. and Stephenson, M. L. Inhibition of Rous sarcoma virus replication and cell transformation by a specific oligodeoxynucleotide. *Proc. Natl. Acad. Sci. U.S.A.*, 75, 280, 1978.

3. Zamecnik, P. C., Goodchild, J., Taguchi, Y., and Sarin, P. S. Inhibition of replication and expression of human T-cell lymphotropic virus type III in cultured cells by exogenous synthetic oligonucleotides complementary to viral RNA. *Proc. Natl. Acad. Sci. U.S.A.*, 83, 4143, 1986.

4. Cooney, M., Czernuszewicz, G., Postel, E. H., Flint, S. J., and Hogan, M. E. Site-specific oligonucleotide binding represses transcription of the human c-*myc* gene *in vitro*. *Science*, 241, 456, 1988.

5. Wickstrom, E., Ed. *Prospects for Antisense Nucleic Acid Therapeutics for Cancer and AIDS*, Wiley-Liss, New York, 1991.

6. Erickson, R. P. and Izant, J. G., Eds. *Gene Regulation: Biology of Antisense RNA and DNA*, Raven Press, New York, 1992.

7. Murray, J. A. H., Ed. *Antisense RNA and DNA*, Wiley-Liss, New York, 1992.

8. Wickstrom, E. Oligodeoxynucleotide stability in subcellular extracts and culture media. *J. Biochem. Biophys. Methods*, 13, 97, 1986.

9. Jayaraman, K., McParland, K., Miller, P., and Ts'o, P. O. P. Selective inhibition of *Escherichia coli* protein synthesis and growth by nonionic oligonucleotides complementary to the 3′ end of 16S rRNA. *Proc. Natl. Acad. Sci. U.S.A.*, 78, 1537, 1981.

10. Marwick, C. Resurgence of tuberculosis prompts US search for effective drugs, expanded research effort. *J. Am. Med. Assoc.*, 269, 191, 1993.

11. Lo, S.-C., Tsai, S., Benish, J. R., Shih, J. W.-K., Wear, D. J., and Wong, D. M. Enhancement of HIV-1 cytocidal effects in CD4+ lymphocytes by the AIDS-associated mycoplasma. *Science*, 251, 1074, 1991.

12. Sarin, P. S., Agrawal, S., Civeira, M. P., Goodchild, J., Ikeuchi, T., and Zamecnik, P. C. Inhibition of acquired immunodeficiency syndrome virus by oligodeoxynucleoside methylphosphonates. *Proc. Natl. Acad. Sci. U.S.A.*, 85, 7448, 1988.

13. Kulka, M., Aurelian, L., Wachsman, M., Fishelevich, R., Miller, P., and Ts'o, P.O.P. Antiviral effect of an oligo(nucleoside methylphosphonate) to the acceptor splice site of HSV-1 IE mRNA 4. *Abstracts, 15th International Herpesvirus Workshop*, Georgetown University, Washington, D.C., 397, 1990.

14. Leiter, J. M. E., Agrawal, S., Palese, P., and Zamecnik, P. C. Inhibition of influenza virus replication by phosphorothioate oligodeoxynucleotides. *Proc. Natl. Acad. Sci. U.S.A.*, 87, 3430, 1990.

15. Alper, J. Oligonucleotides surge into clinical trials. *BioTechnology*, 11, 1225, 1993.

16. Cornelissen, A. W. C. A., Verspieren, M. P., Toulmé, J.-J., Swinkels, B. W., and Borst, P. The common 5′ terminal sequence on trypanosome mRNAs: a target for anti-messenger oligodeoxynucleotides. *Nucleic Acids Res.*, 14, 5605, 1986.

17. Biro, S., Fu, Y.-M., Yu, Z.-X., and Epstein, S. E. Inhibitory effects of antisense oligodeoxynucleotides targeting c-*myc* mRNA on smooth muscle cell proliferation and migration. *Proc. Natl. Acad. Sci. U.S.A.*, 90, 654, 1993.

18. Simons, M., Edelman, E. R., DeKeyser, J.-L., Langer, R., and Rosenberg, R. D. Antisense c-*myb* oligonucleotides inhibit intimal-arterial smooth muscle cell accumulation *in vivo*. *Nature*, 359, 67, 1992.

19. Wahlestedt, C., Pich, E. M., Koob, G. F., Yee, F., and Heilig, M. Modulation of anxiety and neuropeptide Y-Y1 receptors by antisense oligodeoxynucleotides. *Science*, 259, 528, 1993.

20. Bishop, J. M. Molecular themes in oncogenesis. *Cell*, 64, 235, 1991.

21. Daaka, Y. and Wickstrom, E. Target dependence of antisense oligodeoxynucleotide inhibition of c-Ha-*ras* p21 expression and focus formation in T24-transformed NIH3T3 cells. *Oncogene Res.*, 5, 267, 1990.

22. Bacon, T. A. and Wickstrom, E. Daily addition of an anti-c-*myc* DNA oligomer induces granulocytic differentiation of human promyelocytic leukemia HL-60 cells in both serum-containing and serum-free media. *Oncogene Res.*, 6, 21, 1991.

23. Gray, G., Hebel, D., Hernandez, O., Pow-Sang, J. M., and Wickstrom, E. Antisense DNA inhibition of tumor growth by T24 c-Ha-*ras*-transformed NIH3T3 cells in athymic nude mice, *Cancer Res.*, 53, 577, 1993.

24. Adams, J. M., Harris, A. W., Pinkert, C. A., Corcoran, L. M., Alexander, W. S., Cory, S., Palmiter, R. D., and Brinster, R. L. The c-*myc* oncogene driven by immunoglobulin enhancers induces lymphoid malignancy in transgenic mice. *Nature*, 318, 533, 1985.

25. Jacoby, R. O. and Fox, J. G. Biology and diseases of mice, in *Laboratory Animal Medicine*, Fox, J. G., Cohen, B. J., and Loew, F. M., Eds., Academic Press, New York, 1984, 31.

26. Agrawal, S. Antisense oligonucleotides as antiviral agents. *Trends Biotechnol.*, 10, 152, 1991.

27. Vlassov, V. V. and Yakubov, L. A. Oligodeoxynucleotides in cells and in organisms: pharmacological considerations, in *Prospects for Antisense Nucleic Acid Therapeutics for Cancer and AIDS*, Wickstrom, E., Ed., Wiley-Liss, New York, 1991, 243.

28. Bennett, R. M. As nature intended? The uptake of DNA and oligonucleotides by eukaryotic cells. *Antisense Res. Dev.*, 3, 235, 1993.

29. Akhtar, S., Basu, S., Wickstrom, E., and Juliano, R. L. Interactions of antisense DNA oligonucleotide analogues with phospholipid membranes (liposomes). *Nucleic Acids Res.*, 19, 5551, 1991.

30. Leserman, L., Degol, G., Machy, P., Leonetti, J.-P., Mechti, N., and Lebleu, B. Targeting and intracellular delivery of antisense oligonucleotides interfering with oncogene expression, in *Prospects for Antisense Nucleic Acid Therapeutics for Cancer and AIDS*, Wickstrom, E., Ed., Wiley-Liss, New York, 1991, 25.

31. Letsinger, R. L., Zhang, G., Sun, D. K., Ikeuchi, T., and Sarin, P. S. Cholesteryl-conjugated oligonucleotides: synthesis, properties, and activity as inhibitors of replication of human immunodeficiency virus in cell culture. *Proc. Natl. Acad. Sci. U.S.A.*, 86, 6653, 1989.

32. Farese, R. V., Standaert, M. L., Ishizuka, T., Yu, B., Hernandez, H., Waldron, C., Watson, J., Farese, J. P., Cooper, D. R., and Wickstrom, E. Antisense DNA downregulates protein kinase C isozymes (β and α) and insulin-stimulated 2-deoxyglucose uptake in rat adipocytes. *Antisense Res. Dev.*, 1, 35, 1991.

33. Knorre, D. G. and Zarytova, V. F. Novel antisense derivatives: antisense DNA intercalators, cleavers, and alkylatiors, in *Prospects for Antisense Nucleic Acid Therapeutics for Cancer and AIDS*, Wickstrom, E., Ed., Wiley-Liss, New York, 1991, 195.

34. Hélène, C. and Toulmé, J. J. Control of gene expression by oligodeoxynucleotides covalently linked to intercalating agents and nucleic acid-cleaving reagents, in *Oligodeoxynucleotides: Antisense Inhibitors of Gene Expression*, Cohen, J. S., Ed., CRC Press, Boca Raton, FL, 1989, 137.

35. Herschlag, D. Implications of ribozyme kinetics for targeting the cleavage of specific RNA molecules in vivo: more isn't always better. *Proc. Natl. Acad. Sci. U.S.A.*, 88, 6921, 1991.

36. Chang, E. H. and Miller, P. S. *Ras*, an inner membrane transducer of growth stimuli, in *Prospects for Antisense Nucleic Acid Therapeutics for Cancer and AIDS*, Wickstrom, E., Ed., Wiley-Liss, New York, 1991, 115.

37. Campbell, J. A., Bacon, T. A., and Wickstrom, E. Oligodeoxynucleoside phosphorothioate stability in serum, cerebrospinal fluid, urine, subcellular extracts and culture media. *J. Biochem. Biophys. Methods*, 20, 259, 1990.

38. Marshall, W. S. and Caruthers, M. H. Phosphorodithioate DNA as a potential therapeutic drug. *Science*, 259, 1564, 1993.

39. Ghosh, M. K., Ghosh, K., Dahl, O., and Cohen, J. S. Evaluation of some properties of a phosphorodithioate oligodeoxyribonucleotide for antisense application. *Nucleic Acids Res.*, 21, 5761, 1993.

40. Sood, A., Shaw, B. R., and Spielvogel, B. F. Boron-containing nucleic acids. 2. Synthesis of oligodeoxynucleoside boranophosphates. *J. Am. Chem. Soc.*, 112, 9000, 1990.

41. Letsinger, R. L., Singman, C. N., Histand, G., and Salunkhe, M. J. Cationic oligonucleotides. *Am. Chem. Soc.*, 110, 4470, 1988.

42. Stirchak, E. P., Summerton, J. E., and Weller, D. D. Uncharged stereoregular nucleic acid analogs. 2. Morpholino nucleoside oligomers with carbamate internucleoside linkages. *Nucleic Acids Res.*, 17, 6129, 1989.

43. Matteucci, M., Lin, K.-Y., Butcher, S., and Moulds, C. Deoxyoligonudeotides bearing neutral analogs of phosphodiester linkages recognize duplex DNA via triple helix formation. *J. Am. Chem. Soc.*, 113, 7767, 1991.

44. Nielsen, P. E., Egholm, M., Berg, R. H., and Buchardt, O. Sequence-selective recognition of DNA by strand displacement with a thymine-substituted polyamide. *Science*, 254, 1497, 1991.

45. Hanvey, J. C., Peffer, N. J., Bisi, J. E., Thomson, S. A., Cadilla, R., Josey, J. A., Ricca, D. J., Hassman, C. F., Bonham, M. A., Au, K. G., Carter, S. G., Bruckenstein, D. A., Boyd, A. L., Noble, S. A., and Babiss, L. E. Antisense and antigene properties of peptide nucleic acids. *Science*, 258, 1481, 1992.

46. Bacon, T. A., Morvan, F., Rayner, B., Imbach, J.-L., and Wickstrom, E. α-Oligodeoxynucleotide stability in serum, subcellular extracts and culture media. *J. Biochem. Biophys. Methods*, 16, 311, 1988.

47. Monia, B. P., Lesnik, E. A., Gonzalez, C., Lima, W. F., McGee, D., Guinosso, C. J., Kawasaki, A. M., Cook, P. D., and Freier, S. M. Evaluation of 2'-modified oligonucleotides containing 2'-deoxy gaps as antisense inhibitors of gene expression. *J. Biol. Chem.*, 268, 14514, 1993.

48. de Smidt, P. C., Doan, T. L., de Falco, S., and van Berkel, T. J. C. Association of antisense oligonucleotides with lipoproteins prolongs the plasma half-life and modifies the tissue distribution. *Nucleic Acids Res.*, 19, 4695, 1991.

49. Eder, P. S., DeVine, R. J., Dagle, J. M., and Walder, J. A. Substrate specificity and kinetics of degradation of antisense oligonucleotides by a 3' exonuclease in plasma. *Antisense Res. Dev.*, 1, 141, 1991.

50. Tang, J. Y., Temsamani, J., and Agrawal, S. Self-stabilized antisense oligodeoxynucleotide phosphorothioates: properties and anti-HIV activity. *Nucleic Acids Res.*, 21, 2729, 1993.

51. Rubin, E., McKec, T. L., and Kool, E. T. Binding of two different DNA sequences by conformational switching. *J. Am. Chem. Soc.*, 115, 360, 1993.

52. Ratajczak, M. Z., Kant, J. A., Luger, S. M., Hijiya, N., Zhang, J., Zon, G., and Gewirtz, A. M. *In vivo* treatment of human leukemia in a *scid* mouse model with c-myb antisense oligodeoxynucleotides. *Proc. Natl. Acad. Sci. U.S.A.*, 89, 11823, 1992.

53. Higgins, K. A., Perez, J. R., Coleman, T. A., Dorshkind, K., McComas, W. A., Sarmiento, U. M., Rosen, C. A., and Narayanan, R. Antisense inhibition of the p65 subunit of NF-B blocks tumorigenicity and causes tumor regression. *Proc. Natl. Acad. Sci. U.S.A.*, 90, 9901, 1993.

54. Brand, R. F., Moore, A. L., Mathews, L., McCormick, J. J., and Zon, G. Immune cell stimulation by antisense oligodeoxynucleotide phosphorothioates complementary to the *rev* gene of HIV-1. *Biochem. Pharmacol.*, 45, 2037, 1993.

55. Chen, T.-L., Miller, P. S., Ts'o, P. O. P., and Colvin, O. M. Disposition and metabolism of oligodeoxynucleoside methylphosphonates following a single IV injection in mice. *Drug Metab. Dispos.*, 18, 815, 1990.

56. Wickstrom, E., Bacon, T. A., and Wickstrom, E. L. Down-regulation of c-*myc* antigen expression in lymphocytes of E$_\mu$-c-*myc* transgenic mice treated with anti-c-*myc* DNA methylphosphonate, *Cancer Res.*, 52, 6741, 1992.

57. Agrawal, S. and Goodchild, J. Oligodeoxynucleoside methylphosphonates: synthesis and enzymic degradation. *Tetrahedron Lett.*, 28, 3539, 1987.

58. Wickstrom, E. L., Bacon, T. A., Gonzalez, A., Freeman, D. L., Lyman, G. H., and Wickstrom, E. Human promyelocytic leukemia HL-60 cell proliferation and c-*myc* protein expression are inhibited by an antisense pentadecadeoxynucleotide targeted against c-*myc* mRNA. *Proc. Natl. Acad. Sci. U.S.A.*, 85, 1028, 1988.

59. Wickstrom, E. L., Wickstrom, E., Lyman, G. H., and Freeman, D. L. HL-60 cell proliferation inhibited by an anti-c-*myc* pentadecadeoxynucleotide. *Fed. Proc. Fed. Am. Soc. Exp. Bio.*, 45, 1708, 1986.

60. Heikkila, R., Schwab, G., Wickstrom, E., Loke, S. L., Watt, R., and Neckers, L. M. A c-*myc* antisense oligodeoxynucleotide inhibits entry into S phase but not progress from G0 to G1. *Nature*, 328, 445, 1987.

61. Haral-Bellan, A., Ferris, D. K., Vinocour, M., Holt, J. T., and Farrar, W. L. Specific inhibition of c-*myc* protein biosynthesis using an antisense synthetic deoxyoligonucleotide in human T lymphocytes. *J. Immunol.*, 140, 2431, 1988.

62. Holt, J. T., Redner, R. L., and Nienhuis, A. W. An oligomer complementary to c-*myc* mRNA inhibits proliferation of HL-60 promyelocytic cells and induces differentiation. *Mol. Cell. Biol.*, 8, 963, 1988.

63. Wickstrom, E. L., Bacon, T. A., Gonzalez, A., Lyman, G. H., and Wickstrom, E. Anti-c-*myc* DNA oligomers increase differentiation and decrease colony formation by HL-60 cells. *In Vitro Cell. Dev. Biol.*, 26, 297, 1989.

64. Stanton, L. W., Watt, R., and Marcu, L. B. Translocation, breakage and truncated transcripts of c-*myc* oncogene in murine plasmacytomas. *Nature*, 303, 401, 1983.

65. Goodchild, J., Kim, B., and Zamecnik, P. C. The clearance and degradation of oligodeoxynucleotides following intravenous injection into rabbits. *Antisense Res. Dev.*, 1, 153, 1991.

66. Agrawal, S. Antisense oligonucleotides: a possible approach for chemotherapy of AIDS, in *Prospects for Antisense Nucleic Acid Therapeutics for Cancer and AIDS,* Wickstrom, E., Ed., Wiley-Liss, New York, 1991, 143.

67. Whitesell, L., Rosolen, A., and Neckers, L. M. *In vivo* modulation of N-*myc* expression by continuous perfusion with an antisense oligonucleotide. *Antisense Res. Dev.*, 1, 343, 350, 1991.

68. Huang, C. H., Kimura, R., Bawarshi-Nassar, R., and Hussain, A. Mechanism of nasal absorption of drugs. II. Absorption of L-tyrosine and the effect of structural modification on its absorption. *J. Pharm. Sci.*, 74, 1298, 1985.

69. Vlassov, V. V., Karamyshev, V. N., and Yakubov, L. A. Penetration of oligonucleotides into mouse organism through mucosa and skin. *FEBS Lett.*, 327, 271, 1993.

70. Blanchard, J., Mohammadi, J. D., and Conrad, K. A. Improved liquid-chromatographic determination of caffeine in plasma. *Clin. Chem.*, 26, 1351, 1980.

71. Miyamoto, C., Chizzonite, R., Crowl, R., Rupprecht, K., Kramer, R., Schaber, M., Kumar, G., Poonian, M., and Ju, G. Molecular cloning and regulated expression of the human c-*myc* gene in *Escherichia coli* and *Saccharomyces cerevisiae*: comparison of the protein products. *Proc. Natl. Acad. Sci. U.S.A.*, 82, 7232, 1985.

72. Siegel, S. *Nonparametric Statistics for the Behaviorial Sciences*, McGraw-Hill, New York, 1956.

73. Young, I. T. Proof without prejudice: use of the Kolmogorov-Smirnov test for the analysis of histograms from flow systems and other sources. *J. Histochem. Cytochem.*, 25, 935, 1977.

74. Durnam, D. M. and Palmiter, R. D. A practical approach for quantitating specific mRNAs by solution hybridization. *Anal. Biochem.*, 131, 385, 1983.

75. Miller, P. S. Non-ionic antisense oligonucleotides, in *Oligodeoxynucleotides: Antisense Inhibitors of Gene Expression,* Cohen, J. S., Ed., CRC Press, Boca Raton, FL, 1989, 79.

76. Lesnikowski, Z. J., Jaworska, M., and Stec, W. J. Octa(thymidine methanephosphonates) of partially defined stereochemistry: synthesis and effect of chirality at phosphorus on binding to pentadecadeoxyriboadenylic acid. *Nucleic Acids Res.*, 18, 2109, 1990.

77. Wickstrom, E. Antisense DNA therapeutics: neutral analogues and their stereochemistry, in *Gene Regulation by Antisense RNA and DNA*, Erickson, R. P. and Izant, J. G., Eds., Raven Press, New York, 1992, 119.

78. Vyazovkina, E. V., Savchenko, E. V., Lokhov, S. G., Engels, J. W., Wickstrom, E., and Lebedev, A. V. Synthesis of specific diastereomers of a DNA methylphosphonate heptamer, d(CpCpApApApCpA), and stability of base pairing with the normal DNA octamer d(TpGpTpTpTpGpGpC). *Nucleic Acids Res.,* submitted.

79. Flotte, T. R., Afione, S. A., Conrad, C., McGrath, S. A., Solow, R., Oka, H., Zeitlin, P. L., Guggino, W. B., and Carter, B. J. Stable *in vivo* expression of the cystic fibrosis transmembrane conductance regulator with an adeno-associated virus vector. *Proc. Natl. Acad. Sci. U.S.A.*, 90, 10613, 1993.

Chapter 7

Self-Stabilized Oligonucleotides as Novel Antisense Agents

Sudhir Agrawal, Jamal Temsamani, and Jinyan Tang

CONTENTS

I. INTRODUCTION

Antisense oligonucleotides and their several modified analogs have been studied in order to regulate the expression of various genes *in vitro* as well as *in vivo*.[1-5] Several antisense oligonucleotides are presently being studied for their efficacy in human clinical trials.[6-8]

A combination of three major properties is necessary for the oligonucleotide to be of potential therapeutic value:

1. They should hybridize selectively to the target nucleic acid and have strong affinity.
2. They should be taken up by the cells.
3. They should be stable to nucleases in cells and body fluids.

Oligonucleotide phosphodiesters hybridize selectively to target nucleic acids and have strong affinity and are taken up very efficiently by various cells; however, they have limited stability against nucleases.[9] Various modified oligonucleotides have been studied to increase resistance toward nucleases. Phosphorothioate analogs, in which one of the nonbridged oxygens of internucleotide phosphodiester linkage is replaced with sulfur, have been extensively studied.[3] Phosphorothioate analogs have increased resistance towards nucleases compared to their phosphodiester counterparts; however, they are also digested by nucleases.[10] In our previous studies, we have shown that degradation of the oligonucleotides containing phosphodiester or phosphorothioate backbone *in vivo* is primarily from the 3′ end.[11,12] Various modifications of oligonucleotides at the 3′ end circumvents this problem. The incorporation of few nuclease-resistant internucleotide linkages at the 3′ end, incorporation of chemical substitutes at the 3′-hydroxyl,[13-17] and circularization of the oligonucleotides by joining the 3′ and 5′ ends[18] have been shown to increase the *in vitro* and *in vivo* persistence of the oligonucleotide.

0-8493-4778-5/95/$0.00+$.50
© 1995 by CRC Press Inc.

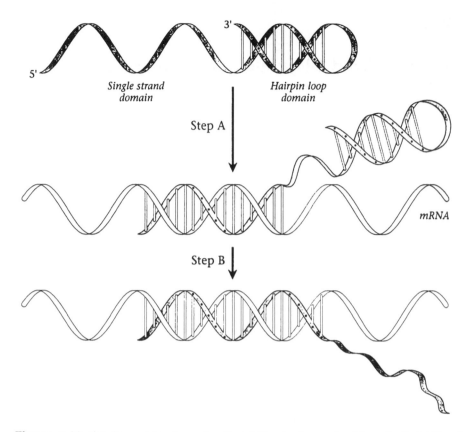

Figure 1 Mechanism of binding of self-stabilized oligonucleotides. Self-stabilized oligonucleotides have two domains. One domain is a single-stranded antisense sequence, and the second domain is a hairpin loop at the 3′ end. In the presence of complementary target nucleic acid, the single-stranded domain of self-stabilized oligonucleotide hybridizes with the target (Step A), and the stability of the duplex formed then destablizes the hairpin loop domain and allows the whole antisense sequence to hybridize to the target nucleic acid (Step B).

In the present study, we have designed oligonucleotides called "self-stabilized oligonucleotides" in which oligonucleotides have been stabilized towards nucleases by structural modification rather than the chemical modification. These oligonucleotides have two domains, one domain is a single-stranded antisense sequence and the second domain is a hairpin loop at the 3′ end (Figure 1). In the presence of a complementary target nucleic acid, the single-stranded domain of the self-stabilized oligonucleotide hybridizes with the target, and the stability of the duplex formed then destabilizes the hairpin loop domain and allows the whole antisense sequence to hybridize to the target nucleic acid. Once the self-stabilized oligonucleotide is released from the target, it should form the hairpin loop domain immediately because of intermolecular hydrogen bonding. Self-stabilized oligonucleotides have increased nuclease resistance, hybridize selectively with target nucleic acid, and are taken up by cells effectively.[19]

II. SYNTHESIS AND PURIFICATION OF OLIGONUCLEOTIDES

Oligonucleotides, both unmodified (PO) and phosphorothioate (PS), were synthesized and purified by the method reported earlier.[20] ^{35}S-Labeled PS oligonucleotides were

Table 1 **Oligonucleotides and their self-stabilized analogs**

Oligonucleotide No.	Sequence	Chain Length
1.	CTCTCGCACCCATCTCTCTCCTTCT	25
2.	CTCTCGCACCCATCTCTCTCC T T T GAGG T C	29
3.	CTCTCGCACCCATCTCTCTCC T T T GAGAGG T C	31
4.	CTCTCGCACCCATCTCTCTCC T T T GAGAGAGG T C	33
5.	CTCTCGCACCCATCTCTCTCC T T T TAGAGAGAGG T C	35
6.	CTCGCACCCATCTCTCTCCT	20
7.	CTCGCACCCATCTCTC T C C TAGAGAG A T C	28
8.	CAGAGCAAAATCATCAGAAGA	21
9.	CAGAGCAAAATCATCAGAAGA T T G GTCTTCT A A	33
10.	GCCGTCTTGGGCTTTGTCTCCATG	24
11.	GCCGTCTTGGGCTTTGTC T C C A AAACAG G A G T	32

synthesized using H-phosphonate chemistry.[20] The specific activities obtained for oligonucleotides 1 (PS), 2 (PS), 3 (PS), 4 (PS), and 5 (PS) were 1.4×10^6, 9.5×10^5, 7.6×10^5, 2.8×10^6, and 1×10^6 cpm/µg, respectively.

Oligonucleotide 1 is a 25-mer oligonucleotide targeted to the gag gene of the human immunodeficiency virus type 1 (HIV-1).[21,22] Oligonucleotides 2, 3, 4, and 5 are self-stabilized analogs of 1, which carry 4, 6, 8, and 10 base pair hairpin loop domain at the

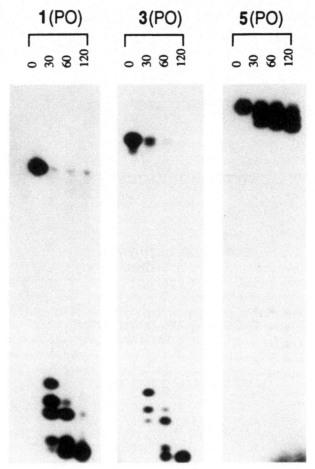

Figure 2 Digestion of (A) oligonucleotides 1 (PO), 3 (PO), and 5 (PO) by DNA polymerase 1 and (B) oligonucleotides 1 (PS) and 5 (PS). ^{32}P-Labeled oligonucleotides (40 pmole) were dissolved in 20 μl buffer (50 mM Tris, pH 7.2, 10 mM Mg SO$_4$, 0.1 mM DTT, 0.5 mg/ml BSA) and incubated with *Escherichia coli* Pol 1 (0.5 μl, 5 units) at 37°C. Aliquots were removed at 0, 30, 60, and 120 min and analyzed by PAGE (20% polyacrylamide containing 8.3 M urea) followed by autoradiography. (From Tang, J-Y., Temsamani, J., and Agrawal, S., *Nucleic Acids Res.*, 21, 2729, 1993. With Permission.)

3′ end (Table 1). Oligonucleotide 6 is a 20-mer oligonucleotide targeted to the gag gene of HIV-1 and oligonucleotide 7 is self-stabilized analog of oligonucleotide 6.

Oligonucleotide 8 is a 24-mer oligonucleotide, complementary to the PB1 gene of influenza A virus,[24,25] and oligonucleotide 9 is a self-stabilized analog of oligonucleotide 8.

Oligonucleotide 10 is complementary to the initiation codon of the ICP 8 gene of herpes simplex virus type 1 and oligonucleotide 11 is a self-stabilized analog of oligonucleotide 10.

Several other control oligonucleotides were synthesized and have been mentioned in respective experiments in the text.

III. NUCLEASE RESISTANCE

The sensitivity of oligonucleotides and their self-stabilized analogs against 3′-exonuclease degradation was studied by (a) digestion with 3′-exonucleolytic activity of DNA

Figure 2 (continued)

polymerase 1 (Pol 1) and (b) incubation with fetal bovine serum. Study of the sensitivity of oligonucleotide 1 (PO), 3 (PO), and 5 (PO) against Pol 1 showed that oligonucleotide 1 (PO) and 3 (PO) were completely digested by the enzyme within 30 min, whereas oligonucleotide 5 (PO) was slowly degraded from 35 mer to 34 mer and 33 mer only (Figure 2A). Similar experiments carried out with oligonucleotides 1 (PS) and 5 (PS) showed that self-stabilized 5 (PS) was more resistant to nucleolytic degradation (Figure 2B).

The sensitivity of the self-stabilized oligonucleotides against nucleases depended on the duplex stability of the hairpin loop domain. The duplex stability of the hairpin loop domain was enhanced with an increased number of base pairs in the hairpin loop domain. The melting temperature (T_m) of the hairpin loop domain of self-stabilized oligonucleotides 3 (PO), 4 (PO), and 5 (PO) containing 6, 8, and 10 base pairs, respectively, was 55, 63, and 66.6°C (Figure 3A). However, the T_m was lower for the hairpin loop domain containing phosphorothioate internucleotide linkages. Oligonucleotides 3 (PS), 4 (PS), and 5 (PS) had T_m of 50, 53, and 57°C, respectively (Figure 3B).

Sensitivity studies of oligonucleotides 1 (PS), 2 (PS), 3 (PS), 4 (PS), and 5 (PS) using cell culture media containing 10% fetal bovine serum showed that oligonucleotides 1 (PS) and 2 (PS) were extensively digested, whereas oligonucleotides 3 (PS), 4 (PS), and particularly 5 (PS) were quite resistant, and most of the oligonucleotides remained intact (Figure 4). Oligonucleotide 2 (PS) has a 4-base pair hairpin domain which is not stable

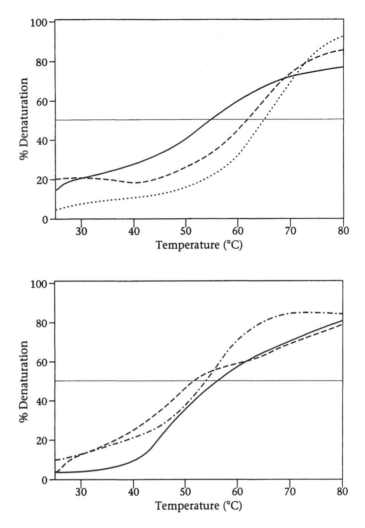

Figure 3 Melting temperatures of self-stabilized oligonucleotides. 0.2 units of A_{260} oligonucleotide were taken up in 500 µl of buffer (10 mM Na$_2$ HPO$_4$, pH 7.4, 10 mM NaCl), heated up to 80°C, and then cooled down to room temperature. The samples were then reheated at a rate of 0.5°C/min and A_{260} was recorded. (A) Oligonucleotides 3 (PO) [——], 4 (PO) [- - - -], and 5 (PO) [· · · ·]; (B) oligonucleotides 3 (PS) [- - - -], 4 (PS) [- · - · -], and 5 (PS) [——].

at 37°C, so it behaves like a single-stranded oligonucleotide and therefore is not resistant to nuclease.

IV. CELLULAR UPTAKE

In order to study whether or not various self-stabilized oligonucleotides have a similar cellular uptake compared to their linear counterparts or whether or not the hairpin loop domain of self-stabilized oligonucleotides interferes in uptake, a comparative uptake study was carried out using oligonucleotides 1 (PS), 4 (PS), and 5 (PS). The cellular uptake studies were carried out using 293 cells and ^{35}S-labeled (at each internucleotide linkage) oligonucleotides. Study showed that there was no significant difference in

1(PS) 2(PS) 3(PS) 4(PS) 5(PS)

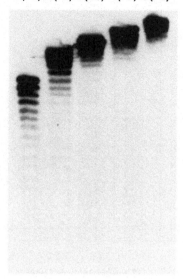

Figure 4 Digestion of oligonucleotide phosphorothioates 1 (PS), 2 (PS), 3 (PS), 4 (PS), and 5 (PS) in the presence of fetal bovine serum. 3.5 μg of each oligonucleotide (^{35}S labeled at each internucleotide linkage, specific activity 6×10^5 cpm/μg) were incubated with 2 ml of culture medium containing 10% fetal bovine serum for 16 h at 37°C. An aliquot (10 μl) was removed, extracted with phenol-chloroform, and precipitated with ethanol. The samples were analyzed by PAGE using 20% polyacrylamide gel containing 8.3 *M* urea. The gel was fixed in acetic acid/methanol/water (10:10:80; v/v/v) solution and dried for 3 h before autoradiography. (From Tang, J-Y., Temsamani, J., and Agrawal, S., *Nucleic Acids Res.*, 21, 2729, 1993. With Permission.)

Table 2 **Cellular uptake**

	Cytoplasmic Fraction (pmoles)	Nuclear Fraction (pmoles)	Total (pmoles)
1 (PS)	2.17	0.73	2.9
2 (PS)	1.96	0.63	2.6
3 (PS)	1.5	0.7	2.2

Note: Cellular uptake of oligonucleotides 1 (PS), 4 (PS), and 5 (PS). 293 cells (10^6 cells) were incubated with 250 n*M* concentration of ^{35}S-labeled oligonucleotides (specific activity 7×10^5 cpm/μg). After 16 h incubation, the cells were fractionated and analyzed by the same procedure as reported earlier.[23]

cellular uptake between self-stabilized oligonucleotides 4 (PS) and 5 (PS) and the linear oligonucleotide 1 (PS) (Table 2). The intracellular distribution was found to be similar for all three oligonucleotides. The concentration in the cytoplasmic fractions was two to three times higher than in the nuclear fraction. The final concentration of oligonucleotides inside the cell, at 16 h of incubation, was in the range of 5 to 7 μ*M*. Oligonucleotides were extracted from the cytoplasmic as well as the nuclear fraction by phenol extraction and analyzed by polyacrylamide gel electrophoresis (PAGE) by the same method as described earlier.[23] The analysis showed extensive degradation of oligonucleotide 1 (PS), whereas self-stabilized oligonucleotides 4 (PS) and 5 (PS) remained intact.

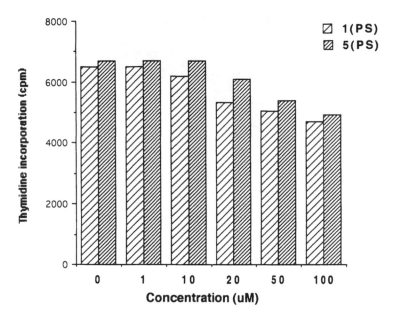

Figure 5 Comparative cytotoxicity of oligonucleotides 1 (PS) and 5 (PS). CEM cells (10⁵ cells) were incubated with oligonucleotides 1 (Ps) and 5 (PS) at the concentration indicated. After 16 h, the cells were washed and pulse labeled with [³H] thymidine (1 μCi/ml) for 2 h. The cells were then lysed and the TCA-precipitated material was counted.

V. CYTOTOXICITY

The study was also carried out to confirm whether or not an increase in the length of the oligonucleotide sequence in self-stabilized oligonucleotides has any additional effect on cytotoxicity. Two oligonucleotides, 1 (PS) and 5 (PS), were studied for their cytotoxicity using CEM cells. The concentrations of oligonucleotides used for the study were 0, 1, 10, 20, 50, and 100 μM and cytotoxicity was followed by thymidine [³H] incorporation into DNA. As it is clear from Figure 5, self-stabilized oligonucleotide 5 (PS) showed no additional toxicity over oligonucleotide 1 (PS), indicating that it is length independent. At 100 μM, only 26% reduction in thymidine incorporation was observed.

VI. AFFINITY AND SELECTIVITY FOR TARGET NUCLEIC ACID

The hairpin loop domain of self-stabilized oligonucleotide does not interfere in its selectivity and affinity for the target nucleic acid. T_m studies of oligonucleotides 1 (PO) and 1 (PS) and their self-stabilized analogs 5 (PO) and 5 (PS) with complementary RNA showed that all four oligonucleotides gave cooperative melting. Similar results were obtained for other self-stabilized oligonucleotides listed in Table 1. In the case of self-stabilized oligonucleotides, Tms were not significantly lower compared to their linear counterparts (Figure 6).

Affinity and selectivity of self-stabilized oligonucleotides were also confirmed by cleavage of a complementary RNA target at specific sites in the presence of RNase H. When target RNA was incubated with oligonucleotide 1 (PS) and its self-stabilized counterpart oligonucleotide 5 (PS) in the presence of RNase H, the cleavage pattern was found to be the same with both oligonucleotides (Figure 7). If the hairpin loop domain of oligonucleotide 5 (PS) remained intact, the cleavage of target RNA would have

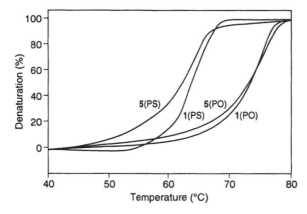

Figure 6 Melting temperature profile of oligonucleotides 1 (PO) and 5 (PO) to their 1 (PS) and 5 (PS) complementary 39-mer gag RNA. Each oligonucleotide (0.2 A_{260} unit) and its complementary 39-mer RNA were annealed in 1 ml buffer (10 mM Na$_2$ HPO$_4$, pH 7.4, 10 mM NaCl) by heating to 80°C and then cooling to 40°C at 2°C/min. The mixture was then reheated to 80°C at a rate of 1°C/min, and the A_{260} was recorded. (From Tang, J-Y., Temsamani, J., and Agrawal, S., *Nucleic Acids Res.*, 21, 2729, 1993. With Permission.)

stopped at single stranded domain of the oligonucleotide. Similar results were obtained with oligonucleotide 6 (PS) and its self-stabilized counterpart oligonucleotide 7 (PS) (Figure 7).

Self-stabilized oligonucleotides should be more selective for their targets over their linear counterparts. As discussed earlier, self-stabilized oligonucleotides have two domains: one single-stranded domain and a hairpin loop domain. The single-stranded domain hybridizes to the target nucleic acid and the stability of the duplex formed then destabilizes the hairpin loop domain of the self-stabilized oligonucleotide. If the duplex formed between the single-strand domain and the target nucleic acid has mismatches, the duplex formed will not be stable enough to destabilize the hairpin loop domain, thereby lowering the chances of interaction with nontargeted nucleic acid. In a model experiment oligonucleotide 5 (PS) was incubated with target RNA in the presence of RNase H, and the expected cleavage pattern of RNA was observed. However, when two mismatches were introduced into the single-strand domain of oligonucleotide 5 (PS) and incubated with target RNA and RNase H under the same condition, the cleavage pattern was quite different (Figure 8). The hairpin loop domain of oligonucleotide 5 (PS) containing two mismatches in the single-strand domain remained intact as it was not destabilized by the duplex formed. However, the same selectivity was not observed with oligonucleotide 1 (PS) and oligonucleotide 1 (PS) containing two mismatches (data not shown).

VII. ANTIVIRAL ACTIVITY

The antiviral activity of self-stabilized oligonucleotide phosphorothioates has been studied against three viruses, HIV-1, influenza virus type A, and herpes simplex virus type 1 (HSV-1), in tissue culture assays.

A. ACTIVITY AGAINST HIV-1

The study of anti-HIV activity of oligonucleotide 1 (PS) and its self-stabilized analogs oligonucleotides 2 (PS), 3 (PS), 4 (PS), and 5 (PS), showed that self-stabilized analogs are more active inhibitors of HIV-1 replication in tissue culture. Oligonucleotide 1 (PS) had IC$_{50}$ of 3.5 × 10^{-7} M, whereas oligonucleotide 5 (PS) had IC$_{50}$ of 0.25 × 10^{-7} M

114

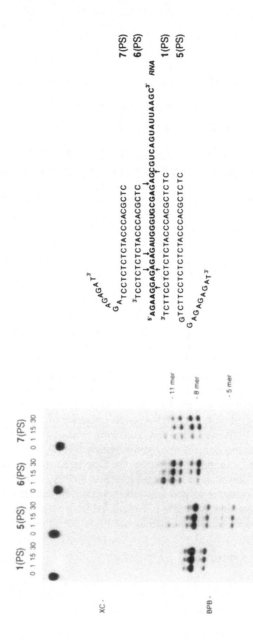

Figure 7 Specific cleavage of 39-mer gag RNA in the presence of oligonucleotides 1 (PS), 5 (PS), 6 (PS), and 7 (PS) by RNase H. Oligonucleotides (10 pmole) were mixed with ^{32}P-labeled 39-mer gag RNA (1 pmole) in 30 µl of 20 mM Tris-HCl (pH 7.5), 10 mM Mg Cl₂, 100 mM KCl, 0.1 mM DTT, and 5% sucrose (w/v) in the presence of RNAsin (40 units). 7 µl of the mixture was removed as control and to the remaining mixture 0.5 µl (0.4 units) of *E. coli* RNase H was added. The mixture was incubated at 37°C and aliquots of 7 µl were removed at 1, 15, and 30 min and mixed with 10 µl formamide. The reaction products were then analyzed by PAGE (20% polyacrylamide gel containing 8.3 M urea) and autoradiographed.

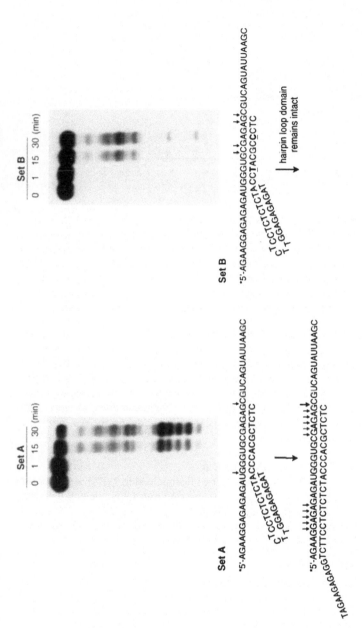

Figure 8 Sequence selectivity and specificity of self-stabilized oligonucleotides. ³²P-labeled gag RNA (1 pmole) was mixed with oligonucleotides (10 pmole) in 10 μl of buffer containing 20 m*M* Tris, pH 7.5, 10 m*M* Mg Cl₂, 100 m*M* KCl, 0.1 m*M* DTT, and 5% glycerol. The mixture was incubated at room temperature for 15 min. 40 units of RNAsin was added and the volume of the reaction was brought up to 30 μl with the above buffer. The samples were then incubated with 0.375 units of RNAse H at 37°C. Aliquots of 7 μl were removed at 0, 1, 15, and 30 min and mixed with 10 μl formamide and then analyzed by PAGE (20% polyacrylamide containing 8.3 *M* urea) and autoradiographed. The underlined nucleotides in Set B represent mutations.

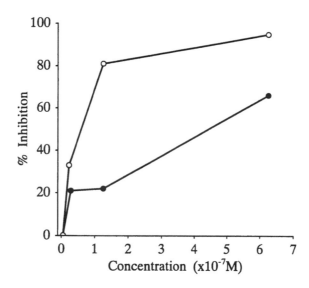

Figure 9 Anti-HIV activity of oligonucleotides 1 (PS) [●] and 5 (PS) [○]. H9 cells were infected with HIV-1 IIIB (~0.1 $TCID_{50}$/cell) for 1 h at 37°C. Unabsorbed virion were then removed by washing and infected cells were divided among wells of 24 well plates in fresh medium. Oligonucleotides at various concentrations were added and the cells were cultured for 3 d. Supernatants from infected cultures were collected and the levels of HIV-1 p24 expression were measured by monoclonal antibody-based p24 antigen capture assay. Both oligonucleotides showed dose-dependent inhibition.

(Figure 9). Both oligonucleotides showed dose-dependent anti-HIV activity. Oligonucleotides 2 (PS), 3 (PS), and 4 (PS) showed intermediate activity between oligonucleotides 1 (PS) and 5 (PS).[19] In our previous studies, in the case of HIV-1, the inhibition of HIV-1 replication was found to be length dependent.[21] The observed increase in activity of self-stabilized oligonucleotides here could be because of their increased length. If this is the case, it still has an advantage over other linear oligonucleotides, as only self-stabilized oligonucleotides were found to be resistant to degradation *in vivo*, not their linear counterpart, as discussed later.

B. ACTIVITY AGAINST INFLUENZA TYPE A

Oligonucleotide 8 (PS) and its self-stabilized analog oligonucleotide 9 (PS) were studied for their comparative inhibitory activity against influenza A virus. Oligonucleotide 8 (PS) had IC_{50} of 65 μM whereas oligonucleotide 9 (PS) had IC_{50} of 2.3 μM (Figure 10). The inhibition of influenza virus A replication was found to be dose dependent. Oligonucleotides were not found to be cytotoxic at these concentrations.

C. ACTIVITY AGAINST HSV-1

Oligonucleotide 10 (PS) and its self-stabilized analog oligonucleotide 11 (PS) were studied for their comparative inhibitory activity against HSV-1 replication. Both oligonucleotides showed dose-dependent inhibition of HSV-1 replication. The IC_{50} of oligonucleotides 10 (PS) and 11 (PS) were >8 μM and 5 μM, respectively (Figure 11).

VIII. PHARMACOKINETICS AND *IN VITRO* STABILITY

The pharmacokinetic and *in vivo* stability of oligonucleotides phosphorothioate was carried out after intravenous administration in mice. Biodistribution of oligonucleotides

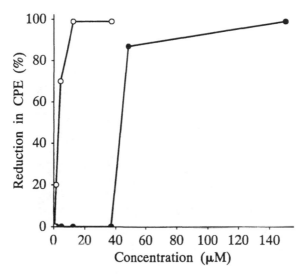

Figure 10 Inhibition of influenza virus replication by oligonucleotide 8 (PS) [●] and 9 (PS) [○]. MDCK cells were seeded in 96-well, flat-bottom tissue culture plates (0.2 ml/well) and incubated overnight in order to establish monolayers of cells. Growth medium was decanted from the plates and oligonucleotide concentrations were added to wells. Influenza A/WSN/33 (HINI) was added to all wells of the plate. The plates were incubated until virus control wells had adequate cytopathic effect (CPE) readings. Cells in test and virus control wells were then examined microscopically.

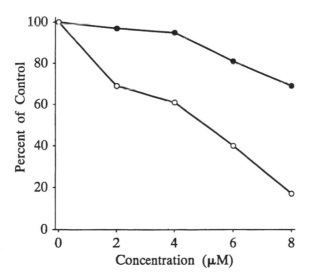

Figure 11 Anti-HSV-1 activity of oligonucleotides 10 (PS) [●] and 11 (PS) [○]. Approximately 100 plaque-forming units of HSV-1 KOS strain was plated on each 3.5-cm well containing Vero cells. After a 1-h absorption period, the cells were overlaid with 2 ml of medium 199 containing 1% calf serum and 0.2% immune human immunoglobin and an appropriate concentration of oligonucleotides. After 3 d, the cells were fixed with methanol and stained with Giemsa stain. The plaques were then counted.

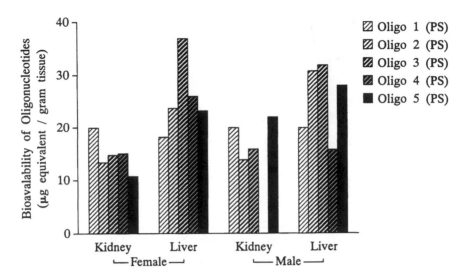

Figure 12 Biodistribution of oligonucleotides in kidney and liver tissues. 5 mg/kg of
^{35}S-labeled oligonucleotides 1 (PS), 2 (PS), 3 (PS), 4 (PS), and 5 (PS) (15 µCi/200
µg) were administered intravenously in one male and one female mouse for each
oligonucleotide. At 26 h post-dosing, animals were sacrificed, and the kidneys and
liver of each animal were removed and homogenized. The oligonucleotides were
extracted from tissue homogenate by the same procedure as reported earlier.[11] The
sample 4 (PS) in kidney (male) was lost in handling.

1 (PS), 2 (PS), 3 (PS), 4 (PS), and 5 (PS) in liver and kidney showed no significant
differences (Figure 12). The excretion rate in urine was similar; an overall 30% of the
administered dose of oligonucleotides was excreted at 24 h post-dosing. However, when
oligonucleotides were analyzed for their stability by gel electrophoresis, self-stabilized
oligonucleotides showed increased resistance towards nucleases and remained intact after
24 h. Oligonucleotides 1 (PS) and 2 (PS) were found to be degraded extensively in kidney
and liver tissues, whereas in the case of oligonucleotides 3 (PS), 4 (PS), and 5 (PS) the
majority of the extracted oligonucleotide was found to be intact in these organs (Figure 13).

These results of *in vivo* stability also suggest that the resistance towards nucleases
increases with stability of the hairpin domain. These results also corroborate with the *in
vitro* study described in Figure 4.

IX. CONCLUSION

The approach to self-stabilized oligonucleotide described here is applicable to both
oligodeoxyribonucleotides and to oligoribonucleotides containing phosphorodiester,
phosphorothioate, or other nuclease-susceptible internucleotide linkages. The approach is
applicable to both synthetic oligonucleotides and endogenously expressed RNA. Based
on the studies described here, it is clearly evident that self-stabilized oligonucleotides are
more active in regulating gene expression in tissue culture assays, and are very stable *in
vivo* compared to their linear counterparts. Taking these two parameters together, the self-
stabilized oligonucleotide should prove to be a better therapeutic agent.

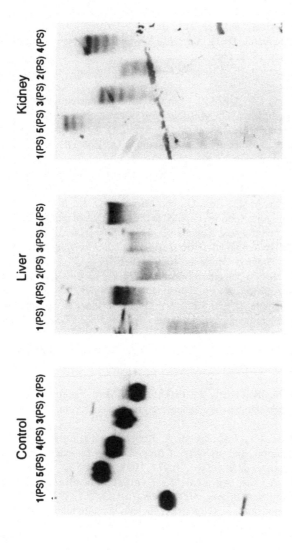

Figure 13 Stability of oligonucleotides 1 (PS), 2 (PS), 3 (PS), 4 (PS), and 5 (PS) in kidney and liver after intravenous administration in mice. The left hand side (control) is an autoradiograph of ^{35}S-labeled oligonucleotides before administration, and the right hand side shows the oligonucleotide extracted from liver and kidney tissues at 24 h post-dosing. (From Tang, J-Y., Temsamani, J., and Agrawal, S., *Nucleic Acids Res.*, 21, 2729, 1993. With Permission.)

REFERENCES

1. Zamecnik, P. C., Introduction: oligonucleotide base hybridization as a modulator of genetic message readout, in *Prospects for Antisense Nucleic Acid Therapy for Cancer and AIDS*, Wickstrom, E., Ed., Wiley-Liss, New York, 1991, 1.

2. Agrawal, S., Antisense oligonucleotides: a possible approach for chemotherapy of AIDS, *Prospects for Antisense Nucleic Acid Therapy for Cancer and AIDS*, Wiley-Liss, New York, 1991, 143.

3. Stein, C. A. and Cheng, Y. C., Antisense oligonucleotides as therapeutic agents — is the bullet really magical?, *Science*, 261, 1004, 1993.

4. Baserga, R. and Denhardt, D. T., *Antisense Strategies*, The New York Academy of Sciences, New York, 1992, 660.

5. Crook, S. T. and Lebleu, B., *Antisense Research & Applications*, CRC Press, Boca Raton, FL, 1993.

6. Agrawal, S. and Tang, J-Y., GEM 91 — an antisense oligonucleotide phosphorothioate as a therapeutic agent for AIDS, *Antisense Res. Dev.*, 2, 261, 1992.

7. Bayeuer, E., Iversen, P. L., Bishop, M. R., Sharp, G. J., Tewary, H. K., Arneson, M. A., Pirruccello, S. J., Ruddon, R. W., Kessinger, A., Zon, G., and Armitage, J. O., Systematic administration of a phosphorothioate oligonucleotide with a sequence complementary to p53 for acute myelogenous leukemia and myelodysplastic syndrome: initial results of a Phase I trial, *Antisense Res. Dev.*, 3, 383, 1993.

8. Cowsert, L. M., Fox, M. C., Zon, G., and Mirabelli, C. K., *In vitro* evaluation of phosphorothioate oligonucleotides targeted to the E2mRNA of papillomavirus: potential treatment for genital warts, *Antimicrob. Agents Chemother.*, 37, 171, 1993.

9. Wickstrom, E., Oligodeoxynucleotide stability in subcellular extracts and culture media, *J. Biochem. Biophys. Methods*, 13, 97, 1986.

10. Zon, G., Oligonucleoside phosphorothioates, in *Protocols for Oligonucleotides and Analogs*, Vol. 20, Agrawal, S., Ed., Humana Press, Totowa, NJ, 1993, chap. 8.

11. Agrawal, S., Temsamani, J., and Tang, J-Y., Pharmacokinetics, biodistribution, and stability of oligonucleotide phosphorothioates in mice, *Proc. Natl. Acad. Sci. U.S.A.*, 88, 7595, 1991.

12. Temsamani, J., Tang, J-Y., and Agrawal, S., Capped oligodeoxynucleotide phosphorothioates; pharmacokinetics and stability in mice, *Ann. N.Y. Acad. Sci.,* 660, 318, 1992.

13. Shaw, J. P., Kent, K., Bird, J., Fishback, J., and Froehler, B., Modified oligodeoxynucleotide stable to exonuclease degradation in serum, *Nucleic Acids Res.*, 19, 747, 1991.

14. Temsamani, J., Tang, J-Y., Padmapriya, A., Kubert, M., and Agrawal, S., Pharmacokinetics, biodistribution, and stability of capped oligodeoxynucleotide phosphorothioates in mice, *Antisense Res. Dev.*, 3, 277, 1993.

15. Padmapriya, A. A. and Agrawal, S., Synthesis of oligodeoxynucleoside methylphosphonothioates, *Bioorg. Med. Chem. Lett.*, 3, 761, 1993.

16. Agrawal, S. and Goodchild, J., Oligodeoxynucleotide methylphosphonate: synthesis and enzyme degradation, *Tetrahedron Lett.*, 28, 3539, 1987.

17. Zhao, Q. Y., Matson, S., Herrera, C. J., Fisher, E., Yu, H., and Krieg, A. M., Comparison of cellular binding and uptake of antisense phosphodiester, phosphorothioate, and mixed phosphorothioate and methylphosphonate oligonucleotides, *Antisense Res. Dev.*, 3, 53, 1993.

18. Prakash, G. and Kool, E. T., Molecular recognition by circular oligonucleotides: strong binding of single-stranded DNA and RNA, *J. Chem. Soc. Chem. Commun.*, 1161, 1991.

19. Tang, J-Y., Temsamani, J., and Agrawal, S., Self-stabilized antisense oligodeoxynucleotide phosphorothioates: properties and anti-HIV activity, *Nucleic Acids Res.*, 21, 2729, 1993.
20. Agrawal, S., *Methods in Molecular Biology: Protocols for Oligonucleotides and Analogs*, Vol. 20, Humana Press, Totowa, NJ, 1993.
21. Agrawal, S., Sarin, P. S., Zamecnik, M., and Zamecnik, P. C., Cellular uptake and anti-HIV activity of oligonucleotides and their analogs, in *Gene Regulation: Biology of Antisense RNA and DNA*, Erickson, R. P. and Izant, J. G., Eds., Raven Press, New York, 1992, 273.
22. Lisziewicz, J., Sun, D., Metelev, V., Zamecnik, P., Gallo, R. C., and Agrawal, S., Long-term treatment of human immunodeficiency virus-infected cells with antisense oligonucleotide phosphorothioates, *Proc. Natl. Acad. Sci. U.S.A.*, 90, 3860, 1993.
23. Temsamani, J., Kubert, M., Tang, J-Y., Padmapriya, A., and Agrawal, S., Cellular uptake of oligodeoxynucleotide phosphorothioates and their analogs, *Antisense Res. Dev.*, 4, 35, 1994.
24. Leiter, J. M. E., Agrawal, S., Palese, P., and Zamecnik, P. C., Inhibition of influenza virus replication by phosphorothioate oligodeoxynucleotides, *Proc. Natl. Acad. Sci. U.S.A.*, 87, 3430, 1990.
25. Agrawal, S. and Leiter, J. M. E., Alternative antiviral approaches to influenza virus: antisense RNA and DNA, in *Antisense RNA and DNA*, Vol. 11, Murray, J. A. H., Ed., Wiley-Liss, New York, 1992, 305.

Chapter 8

Circular Oligonucleotides as Potential Modulators of Gene Expression

Eric T. Kool

CONTENTS

I. INTRODUCTION

A considerable fraction of the DNA in the world is in circular form. This can be stated with confidence because the genomes of most prokaryotic organisms are circular.[1] Clearly, circular nucleic acids exhibit important differences in their properties as compared to linear nucleic acids, or this topological form would not be selected for by nature. For example, circular duplex DNAs undergo supercoiling, which plays an important role in control of gene expression and in replication.[2,3]

It is interesting to speculate why the circular topological form has evolved. It seems likely that one reason for this is that polynucleotide-degrading enzymes, and in particular exonucleases, are common in living organisms. Since bacteria, viruses, and viroids must often operate in close contact with other life forms, they may have evolved the circular structure in part to avoid enzymatic digestion. This would seem especially true for viruses, which must expose their genome to a host cell during their replication cycle.

A second probable reason for evolution of circularity is the ease and simplicity of genomic replication made possible by this structure. The complex problem of completing the ends of a chromosome is simplified with a circular structure.[4,5] In addition, polymerase enzymes processing multiple times around a circle allows for amplified copying of sequences without complete dissociation of the product duplex or the enzyme. This strategy may well have evolved during prebiotic times as a way to avoid the problem of product inhibition during strand copying.[6,7] In any case, the circular genomes in many lower organisms are replicated by this "rolling circle" mechanism to take advantage of these features.

While circular DNAs and RNAs are abundant in nature, synthetically constructed circular oligonucleotides are only recently beginning to receive attention.[8-10] Since oligonucleotides and related analogs have been proposed as potential agents for modulation of gene expression, it would seem prudent to study the properties of circular oligonucleotides in this light as well. The development of efficient synthetic methods for covalent closure of linear oligonucleotides into circular form is now making such studies possible.

As is true in nature, synthetically constructed circular oligonucleotides have a number of unusual and potentially advantageous properties. For example, we have found that single-stranded circular oligonucleotides can display nuclease resistance, binding affinity, and sequence selectivity which are superior to those of linear structures. Such characteristics may eventually prove to be advantageous in the sequence-specific inhibition of gene expression. In this report I will discuss the design and synthesis of cyclic oligonucleotides, their hybridization, and possible strategies for their eventual biological use.

II. DESIGN OF SYNTHETIC CIRCULAR OLIGONUCLEOTIDES FOR BINDING A SPECIFIC TARGET SEQUENCE

In this account I will discuss circular oligonucleotides which are not internally self-complementary.[11-15] Avoidance of self-complementarity ensures that a circle will remain open under most conditions when alone in solution, and will thus have sufficient nucleotides available for Watson-Crick hybridization with a target strand.

A. HYBRIDIZATION BY SIMPLE WATSON-CRICK COMPLEXATION

Because of the right-handed twist of duplex DNA, a circular strand of DNA is topologically limited in its ability to hybridize with another nucleic acid strand, if only Watson-Crick bonds are involved. Such a duplex may be difficult to form beyond one turn of the helix or so, unless the target strand is threaded through the circle, or unless some left-handed turns are added to compensate. This same constraint is seen in pseudoknot structures in RNA.[16] It has been proposed, however, that as few as 7 to 14 bases of recognition may have therapeutic potential with the antisense or antigene strategies,[17] and it is likely that this length of hybridization could be achieved with a circle ~20 to 30 nucleotide (nt) in size, since one full turn involves about 11 to 12 base pairs (see Figure 1).

An example of this kind of hybridization is shown below. A 34-nt cyclic DNA oligomer was tested for its ability to bind a 12-nt Watson-Crick complement. Optical melting studies (100 mM Na[+], 10 mM Mg^{2+}, pH 7.0) show that the circle binds the

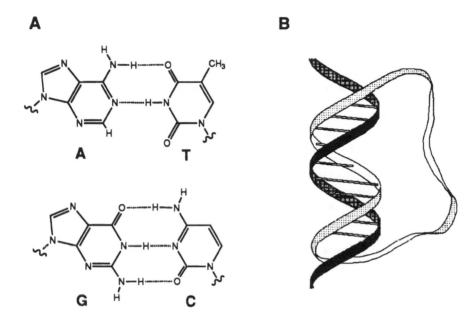

Figure 1 (A) The Watson-Crick base pairs involved in a circle-single-strand duplex. (B) Illustration of a circular oligonucleotide bound to a single-stranded target by Watson-Crick hybridization. The right-handed twist brings the target strand through the circle once per turn.

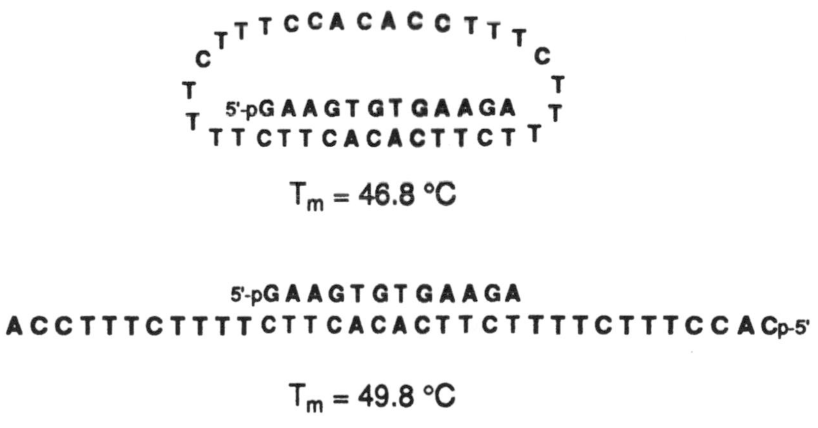

$$C^{T^{T \ T \ CCA \ CA \ CC}T_{T}_{T}}_{C}$$

```
            T T CCA CA CC T T
         C T                   T C
     T                             T
       T   5'-pG AAGT GT GAAGA   T
         T T CTT CACACTT CT T T
```

$$T_m = 46.8 \ °C$$

```
            5'-pG AAGT GT GAAGA
ACCTTTCTTTTCTT CACACTT CTTTTCTTTCCACp-5'
```

$$T_m = 49.8 \ °C$$

Figure 2 Binding of a 34-nt circular oligodeoxynucleotide to a 12-nt complement by duplex formation, with comparison to the uncircularized sequence.

complement with a T_m of 46.8°C, which is slightly lower than the T_m of a linear version of the 34 mer bound to the same sequence (Figure 2). Thus, a 34-nt circle can form a Watson-Crick complex about 12 base pairs in length, although there is no significant advantage in binding affinity from the circularity. If used simply as a Watson-Crick hybridizing agent, a circular DNA or analog may have some advantages over a linear analog, however. For example, it would be resistant to exonuclease degradation, and at the same time would likely be able to induce RNase H cleavage[18] of a targeted RNA strand.

B. HYBRIDIZATION WITH A SINGLE STRAND BY TRIPLEX FORMATION[8,9,19,20]

Much of our work has been aimed at formation of a different type of complex. A circular single-stranded oligomer with two opposite pyrimidine-rich domains can form a triple-helical complex with a complementary purine-rich target strand of nucleic acid.[21-24] This is possible because the pyrimidine-purine-pyrimidine triple helix has pyrimidine strands which are antiparallel in orientation,[25] and this occurs on opposite sides of a circle which contains natural 5' to 3' connectivity. Because such a circle contains mostly pyrimidines, it is not internally self-complementary, and so contains no strong secondary structure which would have to be broken in order to form the desired complex.

The design of such a circle would therefore include an antiparallel Watson-Crick domain, which utilizes standard T-A and C-G pairing rules. It would also include a parallel Hoogsteen[26] domain, in which a C is used for recognition of G and a T is used for recognition of A. These two domains are connected by short loops, the design of which is discussed below (Figure 3).

Such a circular oligomer is, on average, open in solution (in the absence of complement), and a 34-nt circle (designed to bind to a 12-base purine sequence) has a circumference of ~110 Å and a cross-ring diameter of ~33 Å. When presented with a complementary sequence, hybridization occurs, and the sides of the circle close down, sandwiching the target strand in between. In this fashion, recognition of a sequence involves formation of four H-bonds to an adenine and five H-bonds to a guanine, considerably more than with simple Watson-Crick recognition alone. The binding involves a significant change in shape, with the two opposed pyrimidine domains moving from ~33 Å to as close as ~5 Å in the triple-helical complex.

Interestingly, and importantly, such a complex is not topologically constrained in the length of hybrid which can be formed. This is true because for every turn of the helix formed between the Watson-Crick domain and the target, there is a compensating turn in the Hoogsteen domain which reverses the winding. No threading of the target through the circle is therefore required. For example, we have formed strong complexes between a circular DNA and a single-stranded target DNA sequence as long as 32 nt, or about two and one-half turns of the helix, in length.[27]

As mentioned above, loops are required for bridging the two binding domains in such a complex. Points of interest in the design are what length of loop is optimum for complexation, whether there are any interactions between the target strand and the loop as it passes across, and whether there is some sequence preference in loops as a result. We have investigated these questions, and they will be discussed in the "Structural Variations and Effects" section below.

III. SYNTHESIS

There are a number of published studies aimed at nontemplated (random) cyclization of short synthetic oligonucleotides.[28-31] Both solid-phase and solution-phase methods have been investigated. In practice, cyclization is entropically difficult to achieve, with yields dropping off rapidly with increasing length. To date, no one has been successful in cyclizing oligomers longer than about 14 nt using random approaches.[31]

To overcome the entropic barrier to cyclization, we developed a simple template-directed strategy for the nonenzymatic construction of circular oligonucleotides from linear precursors which are phosphorylated on one end. Reagents such as cyanogen bromide (BrCN)/imidazole[32,33] or a water-soluble carbodiimide, EDC,[14,33] can activate a phosphate for attack by a hydroxyl group. Addition of these reagents to an aqueous solution of a linear phosphorylated precursor of ~30 nt does not yield a cyclic product, however. It is well known that formation of very large macrocycles can be quite

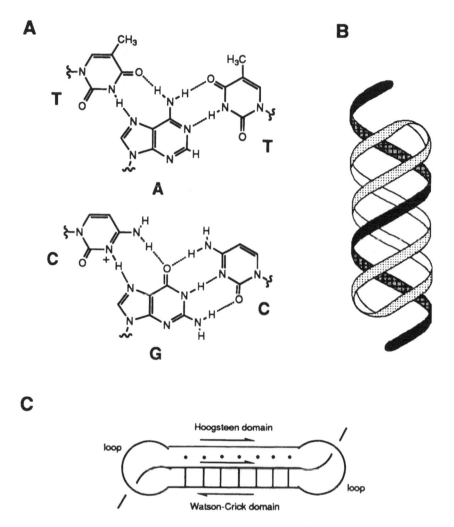

Figure 3 (A) The pyrimidine-purine-pyrimidine base triads formed in a triple-helical complex. (B) Illustration of a circular oligonucleotide bound to a single-stranded target by triplex formation. The twist of the duplex is reversed by the third strand of the complex, so that no topological linking results. (C) Schematic drawing of the design of a triplex-forming circle.

difficult,[34] since the ends must find each other for reaction to occur. To overcome this entropic problem, we use a short oligonucleotide as a template which aligns the ends directly adjacent to each other (Figure 4).[10,35]

A reaction takes about 12 h to complete and can be carried out using crude unpurified oligomers for circle precursor and template. Conversions from linear to circular forms are on the order of 50 to 95%,[35] depending on sequence, and the circular product is easily separated from the mixture by preparative denaturing gel electrophoresis or by ion-exchange, high-performance liquid chromatography.

Characterization of a circular product requires different methods than for a linear sequence. For example, standard sequencing would require that the circle first be nicked at a specific site. One common characterization of circles involves the comparison of

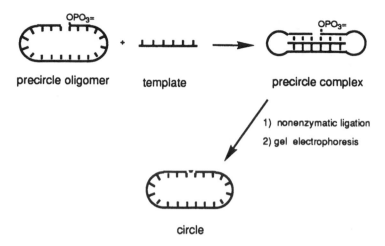

Figure 4 The strategy used for template-directed nonenzymatic cyclization of linear oligonucleotides to form triplex-forming circular oligonucleotides.

relative electrophoretic mobility in gels of different percentages of polyacrylamide.[37] For example, a 34-nt circular sequence travels at 0.9 times the rate of a linear 34 mer in a 20% denaturing PAGE gel.[10]

Perhaps the most reliable confirmation of circularity comes from exposure to nuclease enzymes. With a circular oligonucleotide, the initial cleavage by a single-strand-specific nuclease (such as S1) creates a distinctive product: instead of producing two fragments, as occurs with a linear sequence, it produces one linear product, with a faster mobility than the circular structure. Another test for circularity of synthetic oligonucleotides results from exposing them to exonuclease-type activities, since circles are, by definition, completely resistant. For example, circles are uncleaved by T4 DNA polymerase, which carries a 3' exonuclease activity, under conditions in which a linear 5'-phosphate precursor is completely cleaved to mononucleotides.[38] If a 3'-phosphate precursor is used in the cyclization, even the linear oligomer is resistant to cleavage by this enzyme, and so we substitute a two-enzyme system for cleavage: T4 polynucleotide kinase (in the absence of ATP it removes a 3'-phosphate)[38] and T4 DNA polymerase. We have also radiolabeled a linear precursor on the 5' end with ^{32}P-phosphate and then cyclized it; the product then is resistant to removal of the label by calf alkaline phosphatase, while the uncyclized precursor has its label completely removed.[39]

IV. BINDING PROPERTIES WITH DNA

A. STRENGTH OF BINDING

Comparison of binding affinities of circular and linear oligodeoxynucleotides (ODN) for a complementary strand of DNA shows that circular structures can have a very large binding advantage. This is apparently the result of at least three effects: first, a circular triplex-forming oligomer forms four or five hydrogen bonds with each base in a target strand, as compared to two or three for standard Watson-Crick recognition. Second, even long hairpin-type or nicked-circle oligomers which form the same number of hydrogen bonds do not bind as tightly as the circular case. This advantage of circularity is likely due to the entropic effect of limiting bond rotations in the oligomer prior to binding (the "preorganization" effect).[40] Third, base stacking in the bridging loops[41] in such a complex also adds to the overall affinity. Thus, the increased binding affinity is due both to entropic and enthalpic effects.[42]

Table 1   **Melting temperatures and free energies of complexation for the binding of purine oligodeoxynucleotides by linear, nicked circular, and closed circular oligodeoxynucleotides at pH 7.0**

Complex	T_m, °C	$-\Delta G°_{37}$(kcal/mol)
5'-AAAAAAAAAAAA 3'-TTTTTTTTTTTT	37.1	8.1
5'-AAGAAAAGAAAG 3'-TTCTTTTCTTTC	43.8	10.3
$_A{}^C$TTTTTTTTTTTT$^C{}_A$ C 5'-AAAAAAAAAAAA C $^A{}_C$TTTTTT TTTTTT$_C{}^A$ $\quad\quad$OPO$_3^=$	44.7	10.5
$_A{}^C$TTCTTTTCTTTC$^C{}_A$ C 5'-AAGAAAAGAAAG C $^A{}_C$TTCTTT TCTTTC$_C{}^A$ $\quad\quad$OPO$_3^=$	47.0	10.8
$_A{}^C$TTTTTTTTTTTT$^C{}_A$ C 5'-AAAAAAAAAAAA C $^A{}_C$TTTTTTTTTTTT$_C{}^A$	57.4	16.7
$_A{}^C$TTCTTTTCTTTC$^C{}_A$ C 5'-AAGAAAAGAAAG C $^A{}_C$TTCTTTTCTTTC$_C{}^A$	62.3	16.4

Reprinted with permission from *J. Am. Chem. Soc.*, 1992, 114, 3523. Copyright 1992, American Chemical Society.

For example, we used thermal melting studies to compare the ability of three ODNs to bind to the sequence dAAGAAAAGAAAG (Table 1).[10] The 12-nt Watson-Crick complement binds (pH 7.0, 100 Na$^+$, 10 mM Mg^{2+}) with a free energy of −10.3 kcal·mol^{-1}. A nicked version of the circle binds only slightly more tightly, with a free energy of −10.8 kcal·mol^{-1}. The covalently closed compound, however, binds the most tightly, with a free energy of −16.4 kcal·mol^{-1}. Interestingly, for nicked circles, the position of the nick makes a difference in the binding strength.[10] Clearly, the final covalent closure of the compound makes an important difference. In a separate study, we showed that there is a second way to close a circle: namely, the use of a Watson-Crick stem to close a loop with noncovalent bonds (see Table 2).[43] This also results in stronger binding (although not as strong as the covalently closed case).

The contributions of the loops to the binding is currently under investigation. To keep the binding analysis simple, we often first test binding of a circle to the minimal triplex recognition sequence, so that interaction of the target sequence with loops does not come into play. In practice, the question of how loops interact with the target sequence as it passes beyond may be an important one. To examine the general magnitude of such an effect, we compared the ability of circles to bind to minimal 12-nt recognition oligomers

Table 2 **The effect of closing a triplex-forming DNA loop with a Watson-Crick stem. Melting temperatures (T_m) are for the first (low temperature) transition, measured at pH 7.0.[43]**

Complex	T_m, °C
AAAAAAAA TTTTTTTT	~11
5'-C ^{C} TTTTTTTT ^{C} T AAAAAAAA C ^{C} $_C$ TTTTTTTT $_C$ ^{T}	32.2
5'-T C G $_C$^{C} TTTTTTTT ^{C} T A G C $_C$$_C$ TTTTTTTT $_C$ ^{T} AAAAAAAA C	34.0
5'-G C A T C G $_C$^{C} TTTTTTTT ^{C} T C G T A G C $_C$$_C$ TTTTTTTT $_C$ ^{T} AAAAAAAA C	40.4
5'-G C A G C A T C G $_C$^{C} TTTTTTTT ^{C} T C G T C G T A G C $_C$$_C$ TTTTTTTT $_C$ ^{T} AAAAAAAA C	39.6
5'-G C A G C A G C A T C G $_C$^{C} TTTTTTTT ^{C} T C G T C G T C G T A G C $_C$$_C$ TTTTTTTT $_C$ ^{T} AAAAAAAA C	38.7

and to a central site in longer 36-nt DNA oligomers.[10] We found that a circle can bind even more strongly to a longer strand than to the minimal sequence (Figure 5).

It should be noted that in this experiment the first and last bases in the loops of the circle are complementary to the first flanking nucleotide on either side of the recognition site. We now know that circles may bind either more strongly or less strongly to longer sequences, depending on the nature of these specific nucleotides.[44] We believe that some base-pairing interactions may occur with one of these loop bases with the flanking base, and we are currently studying this in detail using model systems.

B. EFFECT OF CONDITIONS

The strength of binding of a circular oligonucleotide is, as with all DNA complexes, dependent on the conditions in solution.[45] Pyrimidine-purine-pyrimidine triple helices involving C-G-C triads are sensitive to pH changes, since the pK_a of the C in the Hoogsteen strand of a triplex is approximately 6.5.[46] A circular DNA complex is more stable at pH 6 than at neutral pH, and is less stable at pH 8. Because Watson-Crick interactions are also involved in the association, a circular DNA complex is atypical of standard py·pu·py triplexes, however. We have measured circle complexes with reasonable stability at pH values greater than 9.[47]

Circle complexes are also dependent on the ionic strength of solution and require >5 mM Mg^{2+} or ~1 M Na$^+$ or ~50 μM spermine for full stabilization.[47] Again, the circle complex is atypically strong: for example, we have measured a T_m of 43°C for such a

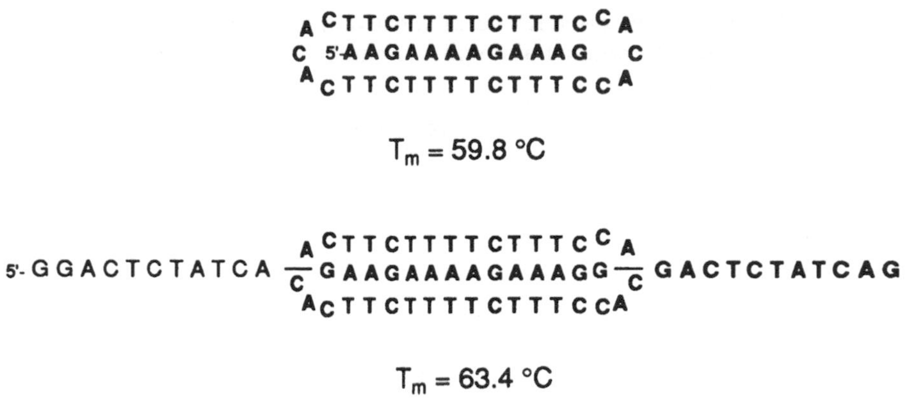

$$_A{}^{C\ T\ T\ C\ T\ T\ T\ T\ C\ T\ T\ C}{}^C{}_A$$
$$_C\ _{5'}A\ A\ G\ A\ A\ A\ A\ G\ A\ A\ A\ G\quad C$$
$$^A{}_{C\ T\ T\ C\ T\ T\ T\ T\ C\ T\ T\ T\ C\ C}{}^A$$

$$T_m = 59.8\ °C$$

$$_{5'-}G\ G\ A\ C\ T\ C\ T\ A\ T\ C\ A\ \overline{_C^{\ A\ C\ T\ T\ C\ T\ T\ T\ T\ C\ T\ T\ C^C}_A}\ G\ A\ C\ T\ C\ T\ A\ T\ C\ A\ G$$

$$T_m = 63.4\ °C$$

Figure 5 The relative binding strengths for a circular ODN hybridized with a short minimal purine target sequence and with the same sequence embedded within a longer target. (Reprinted with permission from *J. Am. Chem. Soc.*, 114, 3523, 1992.)

complex in the presence of only 10 mM tris buffer (and no other added cations) for stabilization.

C. KINETICS OF BINDING

Since circular ODNs bind very tightly, it is of interest to ask whether this is the result of an especially fast hybridization rate, or an especially slow dissociation rate (or a combination of both). Hybridization of short Watson-Crick complementary oligonucleotides occurs at a rate almost independent of the sequence involved.[48] Several studies have reported a second-order rate constant of 10^6 to 10^7 $M^{-1}s^{-1}$ for duplex formation.[49] For triplexes, studies of rate constants for the reaction (single strand + duplex = triplex) in DNA have shown that this hybridization is considerably slower than for duplex formation, with rate constants commonly found in the range 10^2 to 10^4 $M^{-1}s^{-1}$.[50] Preliminary studies in our laboratory followed by stopped-flow UV[51] or by quenching of fluorescence in a fluorescein-tagged target sequence[52] have shown that the circle complex is again atypical of standard triplexes. We have measured rate constants for association of ~2 × 10^6 $M^{-1}s^{-1}$, which makes this hybridization as fast as that for duplex formation.

This rapid hybridization has some interesting implications. First, rate constants for third-strand hybridization to form triplexes are apparently slowed by reorganization of the duplex major groove and/or the surrounding metal ions and water molecules.[50] The circle, by contrast, has no strong secondary structure prior to complexation, and so this kind of reorganization is probably not as important. The second inference to be drawn from this data is that the dissociation of triplex-forming circles from their complements is exceedingly slow. A complex having a free energy (37°C) of −16 kcal/mol (T_m = 56°C) will thus have a half-life for dissociation on the order of ~40 years under physiological conditions.

D. SELECTIVITY OF BINDING

The circular triplex-forming structure has another distinct advantage in binding: very high sequence selectivity in recognition. We constructed a series of DNA target strands which varied by a single base and measured the resulting differences in free energy of binding by a circular oligomer.[8] We found that while a Watson-Crick complement showed 3 to 4 kcal/mol of selectivity for the correct sequence (Table 3), the circle showed 6 to 7 kcal/mol of selectivity. This is among the highest selectivities yet seen for DNA sequence recognition. This degree of discrimination means that the circle binds to its correct sequence with a binding constant four to six orders of magnitude higher than to a sequence different by only one base.

Table 3 Comparison of selectivities of linear and circular oligodeoxynucleotides against a single mismatch in a 12-nt complementary DNA sequence at pH 7.0

Complex (X = A,T,G,C)		ΔT_m for Mismatch	$\Delta\Delta G^{\circ}_{37}$ for Mismatch
5'-AAGAXAAGAAAG	X = A	—	—
3'-TTCTTTTCTTTC	T	12.7°C	3.9 kcal·mol⁻¹
	G	10.0	3.2
	C	15.5	4.4
$_A{}^C$TTCTTTTCTTTC$^C{}_A$ $_C$ 5'-AAGAXAAGAAAG ^{C} $^A{}_C$TTCTTTTCTTTC$_C{}^A$	X = A	—	—
	T	21.5°C	7.3 kcal·mol⁻¹
	G	18.1	6.2
	C	22.5	7.6

Reprinted with permission from *J. Am. Chem. Soc.,* 1991, 113, 6265. Copyright 1991, American Chemical Society.

Balancing tight binding and selectivity is extremely important if oligonucleotides and analogs are to be used as therapeutic agents. It is often found that tighter binding gives better inhibition of translation or transcription,[53] especially for analogs which are not active by a catalytic mechanism such as induction of RNase H cleavage. The translation and transcription machineries have evolved to be able to move beyond weak secondary structure,[54-56] and so efficient blocking can only occur if the blocking group has a very high local free energy of binding and a slow dissociation rate. At the same time, selectivity is important to limit the blocking effect at undesired sequences that are closely related to the actual target.[57] There are now several strategies available for increasing binding strength of an oligonucleotide strand to a target.[58,59] Nearly all of these strategies raise binding in a nonsequence-specific way, which means that binding of undesired sequences is also increased. In the future, it seems likely that the most successful oligonucleotide analogs will have not only increased binding affinity, but also increased sequence selectivity as well.

E. BINDING OF MIXED PURINE-PYRIMIDINE SEQUENCES
A standard py·pu·py triplex requires a purine-rich sequence in the central strand. Third-strand binding of duplex DNA is weaker than Watson-Crick binding,[60,61] and so the presence of even one or two pyrimidines in this strand can prevent binding.[62-64] Since triplex-forming circles recognize a target sequence with both Hoogsteen and Watson-Crick bonds, however, it is possible to form complexes even in cases where some pyrimidine bases occur. It is even possible to form a complex with a mostly pyrimidine sequence using the Watson-Crick domain of a circle (see Figure 2 above), although no Hoogsteen interactions (and no added binding advantage) are then seen.

As a test of binding of sequences containing pyrimidines, we constructed a circle containing 5-methyl-dC residues which is designed to bind to a 12-nt sequence which contains two pyrimidines out of the twelve bases.[65] At pH 6.0 we find (Figure 6) that the circle binds significantly more strongly than is possible for simple Watson-Crick binding. Further studies will be necessary to fully explore the limitations of binding using the naturally occurring bases. In addition, synthetic studies are underway in several laboratories,[66] including our own, to develop nonnatural bases which will selectively recognize pyrimidines within the Hoogsteen motif, and to explore new strategies for binding pyrimidine-rich sequences (see *J. Am. Chem. Soc.,* 116, 8857, 1994).

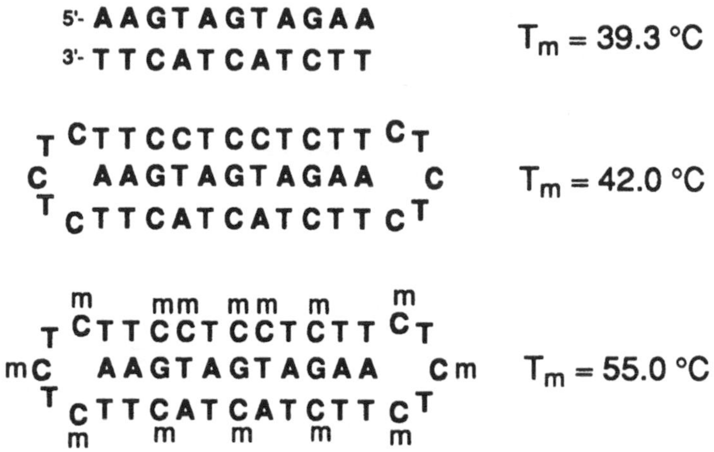

$$\begin{array}{l} \text{5'- A A G T A G T A G A A} \\ \text{3'- T T C A T C A T C T T} \end{array} \qquad T_m = 39.3\ ^\circ C$$

$$\begin{array}{l} {}_{T}\text{C T T C C T C C T C T T}\,{}^{C}_{T} \\ \text{C}\quad\text{A A G T A G T A G A A}\quad\text{C} \\ {}^{T}\text{C T T C A T C A T C T T}\,\text{C}^{T} \end{array} \qquad T_m = 42.0\ ^\circ C$$

$$\begin{array}{l} {}^{m}\quad{}^{mm}\ {}^{mm}\ {}^{m}\qquad{}^{m} \\ {}_{T}\text{C T T C C T C C T C T T}\,{}^{C}_{T} \\ {}^{m}\text{C}\quad\text{A A G T A G T A G A A}\quad\text{C m} \\ {}^{T}\text{C T T C A T C A T C T T}\,\text{C}^{T} \\ {}_{m}\qquad{}_{m}\quad{}_{m}\qquad{}_{m}\qquad{}_{m} \end{array} \qquad T_m = 55.0\ ^\circ C$$

Figure 6 The binding of a 12-nt sequence containing purines interrupted by two pyrimidines.

V. STRUCTURAL VARIATIONS AND EFFECTS

We have investigated the effects of a number of structural variations on the strength of binding of triplex-forming circles.[10] Among these are varying circle size and binding site size, varying loop size, and substitution of cytosine (dC) by 5-methyl-dC residues.

A. SIZE OF CIRCLE

Given a binding site length of n residues, the design of a triplex-forming circle would contain at least 2n + x residues, where x is the total number of nucleotide units in the bridging loops. To date, we have successfully constructed DNA circles ranging from 24 to 74 nt in length, and we have formed complexes with DNA sequences ranging from 4 to 32 nt (or from one third to two and one half turns of the helix) in length. In all these cases the circle binds with considerably greater affinity than does a standard Watson-Crick complement. Circles smaller than 24 to 28 residues have so far proved difficult to synthesize by our method because the reaction requires binding of the template strand, which is inefficient for short sequences. On the long end, the only limitation would appear to be the length which can be constructed on a DNA synthesizer. In fact, one might imagine methods in which two oligomers are dimerized to form a single circle,[67] so that longer circular sequences might be constructed with reasonable efficiency.

B. SIZE OF BRIDGING LOOPS

The size of bridging loops is also important in optimizing binding strength. If a loop is too short to bridge the distance between the pyrimidine binding domains, it will pull nucleotides from the binding domain to bridge the gap, and this distortion will lessen binding. If a loop is very large, the binding domains will begin to act independently, and the cooperative entropic effect on binding will be lost. These two effects are illustrated by the experiments shown below (Figure 7), in which we constructed circles having a constant binding domain, but varied loop size.[10] We find that, at least for the sequences studied, there is a clear preference for 5-nt loops. This preference is seen whether the target strand is short or long. Watson-Crick duplex hairpins prefer loops of four nucleotides;[68,69] this difference is likely due to the somewhat longer distance between the pyrimidine strands in a triplex[70] relative to the two strands of a duplex.

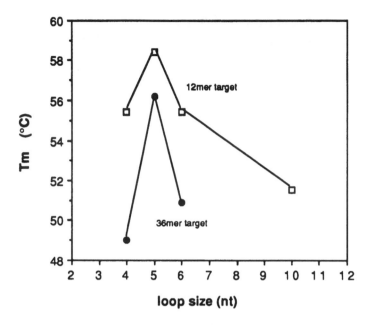

Figure 7 The effect of loop size on strength of binding of circles to dA_{12} and to a 36 mer with dA_{12} embedded. Circles contain binding domains of dT_{12} interrupted by the loop sequences (5') -CACA-, -CACAC-, -CACACA-, and -CACACACACA-. (Reprinted with permission from *J. Am. Chem. Soc.*, 114, 3523, 1992.)

C. EFFECT OF CYTOSINE METHYLATION

Methylation of cytosines has been commonly used in triple helices to increase the strength of third-strand binding.[71-73] The stabilization effect is due in part to an increase in basicity[72] and also to an increase in base-stacking ability[74] of the substituted base. We examined the magnitude of the effect in a circular oligomer which contains 12 cytosines, 6 each in binding domains and in loops.[75] We found (Figure 8) that methylation of the six Cs in the binding domains adds another –6 kcal/mol of binding free energy at pH 7. Thus, with cytosine methylation, a 34-nt circle can bind its complement with a free energy of –22 kcal/mol at pH 7.0. Interestingly, methylation of the Cs in the loops also has a positive effect, although considerably smaller.

VI. RESISTANCE TO DEGRADATION

In biological applications it is desirable that an oligonucleotide or analog be stable for a reasonable period of time in the medium of interest. Many studies have shown that linear ODNs are rapidly degraded in human and other mammalian sera.[53,76] As mentioned above, circular oligonucleotides are by definition resistant to exonuclease enzymes. Since the circles are single stranded, only single-strand-specific endonucleases can degrade them. To check the effect of circularity on stability, we incubated linear and circular oligonucleotides in undiluted human serum and checked for any degradation by gel electrophoresis.[77] We found that a linear 34-nt sequence remained intact with a half-life of only 20 to 30 min. By contrast, a circular oligomer of the same sequence remained completely intact for at least 2 days (Table 4). This confirms both the stability of circles and the fact that the primary DNA-degrading activity in serum is of the exonuclease type. In cellular extracts it has been reported that even linear oligomers are not as rapidly cleaved,[53] indicating that nuclease activity is not as prevalent inside cells. Studies of the stability of circles in cellular extracts is currently under way.

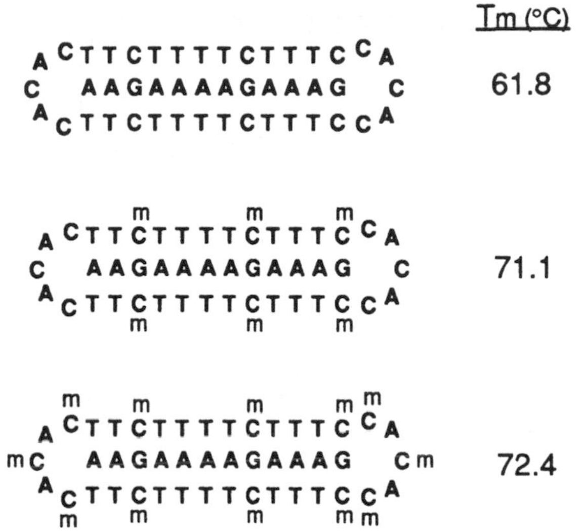

Tm (°C)

$_A$C T T C T T T T C T T T C ^{C}C$_A$
C A A G A A A A G A A A G C   61.8
^{A}C T T C T T T T C T T T C C ^{A}

```
    m       m       m
 A CTTCTTTTCTTTC C A
 C  AAGAAAAGAAAG  C          71.1
 A CTTCTTTTCTTTCC A
    m       m   m
```

```
   m    m       m   m m
 A CTTCTTTTCTTTC C A
mC  AAGAAAAGAAAG  Cm        72.4
 A CTTCTTTTCTTTCC A
   m    m       m mm
```

Figure 8 The effect of cytosine C-5 methylation on the strength of binding (as measured by T_m) by a circular oligodeoxynucleotide.

Table 4 Rates of degradation of linear and circular forms of the same oligodeoxynucleotide in human serum at 37°C.[77] Half-life ($t_{1/2}$) is that for the first cleavage event.

Sequence	$t_{1/2}$
$_C$^{A} ^{C}TTCTTTTCTTTC ^{C}$_A$$_C$ ^{A}$_C$TTCTTT TCTTTC$_C$^{A} OPO$_3^=$ 5′	30 min
$_C$^{A} ^{C}TTCTTTTCTTTC ^{C}$_A$$_C$ ^{A}$_C$ TTCTTTTCTTTC$_C$ ^{A}	>2 d

It is interesting that the issue of stability of an oligonucleotide against degradation when bound to a target strand has not often been approached experimentally. This may be important, since an RNA strand is translated multiple times by the ribosome: an oligonucleotide or analog which relies on steric blockage for activity may be most effective if it stays bound for a considerable amount of time. Linear ODNs which form duplexes with their targets are likely to be substrates for cleavage both when free in solution and when bound, since many nuclease enzymes recognize both duplexes and single strands. A triplex-forming circle, by contrast, is resistant even to single-strand-specific endonucleases when it is bound,[77] because the enzyme cannot recognize the triplex structure. There is, to date, no known nuclease enzyme that can cleave a triple helix.

Figure 9 The replacement of nucleotide loops in a circular oligodeoxynucleotide by hexaethylene glycol linkers. The sequence is synthesized in eight fewer steps than the all-DNA case.[77]

VII. SYNTHETIC MODIFICATIONS

One can envision many types of synthetic modifications which could be made to a circular oligonucleotide in order to increase binding strength, increase resistance to degradation, increase cellular uptake, or to label the circle for various biophysical studies.[81] Modifications might be made to the backbone or to the bases themselves. Described below are experiments and strategies for carrying out both kinds of modifications.

A. NONNUCLEOTIDE LOOPS

One backbone modification which we have been pursuing is the replacement of the 5-nt loops with a single nonnucleotide bridging group.[77] This replacement has several potential advantages: it shortens the circle synthesis by eight steps (two linkers replace ten nucleotides); it increases resistance to degradation when bound to a target strand; and it is also possible that certain linkers might increase the rate of cellular uptake. In addition, if a linker is more rigid than the original five nucleotides, it has the potential for increasing binding strength even further.

To explore the length requirements for such a linker group, we attempted synthesis of circles with linkers made from tetra-, penta-, and hexaethylene glycol (Figure 9).[77] We found that the precircle oligomer with tetraethylene glycol linkers did not cyclize (presumably because the linker was too short). After cyclization, comparison of the penta- and hexaethylene glycol cases showed that hexaethylene glycol is the optimum linker of these cases, and that binding to a short sequence is nearly as good as for a circle with five-nucleotide loops. Subsequent studies showed that, when binding a circle to longer strands, the hexaethylene glycol loop is not as proficient as the nucleotide loop.[80] This may be due to either a lack of positive interactions of the loop with the flanking target strand bases, or to an unfavorable steric interaction of the loop with this strand (for example, the loop may not be long enough to curve around the target strand). Interestingly, studies of the circle with hexaethylene glycol loops did show that, when bound to a complementary DNA strand, the circle was not cleaved by S1 nuclease, a single-strand-specific endonuclease.[77] Further studies on other nonnucleotide linker groups are currently under way.

B. ATTACHMENT OF LABELS AND OTHER MOIETIES

Many methods have been developed recently for attaching other functionalities, such as fluorescent groups or intercalators, to oligonucleotides.[81] Perhaps the simplest approach synthetically is the attachment of amino or thiol groups, which are easily functionalized after oligonucleotide synthesis, at either the 5′ or 3′ terminus of an oligonucleotide. Circular oligonucleotides require unfunctionalized ends, however, and so different methods must be used in this case. When using a circle containing only natural nucleotides, we have employed C-5-functionalized deoxyuridine bases. For example, an acrylic acid ester side chain at the C-5 position can be derivatized as an amide with ethylenediamine,[82,83] and the resulting free amine can be used to attach a fluorescent group to the base prior to DNA synthesis. Alternatively, this amine can be protected as the trifluoroacetamide derivative,[84] and the amine can then be functionalized after DNA synthesis. We have prepared fluorescent-labeled circular DNAs using both these methods.

Another method of C-5 functionalization which has found use in our laboratory is the palladium (Pd)-catalyzed coupling of alkynes to iododeoxyuridine.[84-86] This is a simple reaction which is widely applicable to many C-5 derivatives,[86] which can then be incorporated as the phosphoramidite into a DNA strand as readily as the natural thymidine bases.

For circles with synthetically modified loops, there are in principle many ways that functionalization could be achieved in the loops themselves without disrupting binding of the circle. For example, a linker which is modified with a suitably protected amine or thiol could be incorporated into the circle during synthesis.

C. CYCLIC RNAS AND BACKBONE-MODIFIED DNAS

Because RNA can be easily ligated using BrCN,[32,33] it is anticipated that one could synthesize cyclic RNAs as readily as DNA. One slight complication is that, if a 5'-phosphorylated linear precursor to the circle is used, two products may be formed: the 2'-5' phosphodiester and the 3'-5' phosphodiester. This might be avoided by cyclizing the 3' phosphate rather than the 5'-phosphate, or by using one 2'-deoxynucleotide unit at the 3' terminus to avoid the possibility of 2'-5' attachment.[35]

Several backbone modifications of DNA have been the subjects of study for antisense applications.[53] In principle, if a backbone modification allows py·pu·py triplex formation, then incorporation into a circle may increase binding strength and selectivity. One can envision, for example, circular phosphorothioate[87] DNA oligomers, or circular 2'-O-methyl RNA[88] oligomers as strong-binding analogs of natural DNA.

VIII. BINDING OF RNA

To date we have concentrated our study on the binding of circular ODNs to single-stranded DNA oligomers. The conclusions of these studies would, of course, be most directly applicable to antigene applications (see below). For antisense-type application, one must be concerned with the binding of RNA sequences as well. In one early experiment, we reported[7] that a cyclic DNA would bind to an RNA strand, implying the formation of a "DNA-RNA-DNA"-type triplex.[89,90] We now know,[35] however, that while a circular DNA does bind to an RNA strand, it appears to bind only about as strongly as a DNA-RNA duplex, with the Hoogsteen domain probably remaining largely unbound, giving a complex probably resembling that in Figure 1B. Work is presently continuing on this topic.[35]

IX. BINDING OF DUPLEX DNA

Recent studies in our laboratory (K. Ryan, in progress) have established that circular oglionucleotides can bind duplex DNA. There are at least two potential modes of binding for a circular DNA to a targeted sequence in duplex DNA. One is the simple binding of the Hoogsteen domain to the purine-rich strand in the major groove of the duplex, which we term the "outside" mode (see Figure 10). This is analogous to standard third-strand triplex formation of an oligonucleotide with a duplex. The second mode is one in which both sides of the circle bind to the purine-rich strand of the duplex, displacing the other strand to form a D-looped structure. We term this invasive mode of binding the "inside" mode.

The "outside" mode of binding involves parallel-stranded Hoogsteen-type bonding, and the physical properties of this binding are likely to be similar to those of a linear

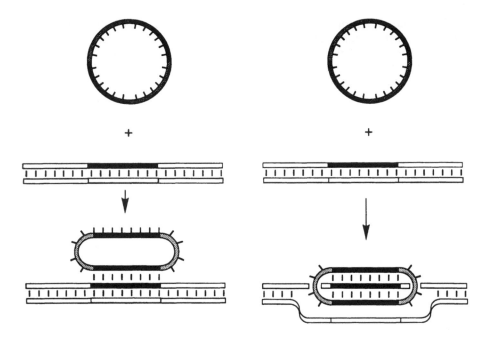

Figure 10 Two possible modes of binding of a circular oligonucleotide to duplex DNA. The "outside" mode involves only Hoogsteen H-bonds between the circle and the target, while the "inside" mode involves both Hoogsteen and Watson-Crick bonds with the circle, and displacement of one strand of the duplex into a D-loop.

oligonucleotide binding to a site in a duplex by triple-helix formation. The binding strength is expected to be pH dependent, and the rate constant for binding should be on the order of 10^2 to 10^4 $M^{-1}s^{-1}$.[50]

It is less clear what properties the "inside" mode of association should have, since there are very few precedents for this kind of binding. To bind in this fashion, there are two major physical requirements for an oligonucleotide analog: first, thermodynamically very strong binding. This is necessary because the displacement of the duplex structure requires a significant amount of energy. The second requirement is a kinetic pathway to the displaced duplex, which is necessary because long duplex DNAs are kinetically very stable under physiological conditions.

The total energy of binding in the strand-displaced case is the combination of the energy gained in binding of the purine strand by an oligonucleotide analog, minus the energy lost in opening the necessary base pairs of the duplex[91] for access by the analog. The first quantity is simple to determine by thermal melting or other calorimetric methods. The second quantity — the energy required to open a base-paired region within a longer duplex — is less straightforward to determine.

A. PRECEDENTS FOR INVASIVE BINDING

Recent work by Nielsen et al.,[92] Egholm et al.,[93] and by workers at Glaxo[94] has resulted in studies of a backbone-altered DNA analog called PNA, which contains a linear peptide-like backbone appended with nucleic acid bases. PNA has sufficient binding energy to displace sections of duplex DNA at equilibrium under physiological conditions. This is an important precedent, and it is very interesting to note that this binding was sufficiently strong to terminate transcription downstream from the promoter[94] — a difficult or impossible task for most noncovalently bound DNA analogs.[95,96] It is also

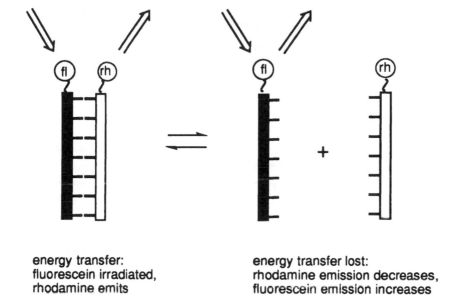

energy transfer:
fluorescein irradiated,
rhodamine emits

energy transfer lost:
rhodamine emission decreases,
fluorescein emission increases

Figure 11 Strategy used for following kinetics of duplex dissociation by fluorescence energy transfer. Fluorescein fluorescence increases over time as the rhodamine-tagged strand dissociates and diffuses away.

important to note that the kinetic barrier to binding is evident in this case: at physiological salt conditions, the PNA oligonucleotide is apparently extremely slow kinetically to invade the duplex.[93,97] In order to bypass this problem, these workers lower the ionic strength of solution to near zero (the equivalent to melting the duplex), where the duplex is kinetically unstable. The PNA then binds invasively, and the ionic strength can then be raised to physiological levels. This work represents, to our knowledge, the first case of an analog which can bind stably in this fashion, but it also illustrates the difficulty of surmounting the kinetic barrier to this mode of binding.

B. CIRCULAR OLIGONUCLEOTIDES AND DUPLEX BINDING

We believe that circular oligonucleotides have properties which may help to overcome both the thermodynamic and kinetic barriers to strand displacement. First, we know that these circles can bind a strand of DNA very strongly, with free energies approaching two to three times those of a linear complement binding the same sequence.[10,75] We know that simple C-5 modifications can raise binding strength by several kilocalories more if needed.[86,99] Thus, it seems likely that thermodynamically it should be possible for a circular ODN or analog to bind in invasive or "inside" fashion.

In addition, we now know that, at least in some cases, such a circle also has a kinetic advantage in binding,[100] and may be able to overcome this problematic kinetic barrier. We studied the rates of dissociation of short model duplex DNAs by attaching two fluorescent labels — fluorescein and teramethylrhodamine — to the opposite strands.[100,101] These two labels can be used to measure kinetically the presence of duplex relative to single strands by following energy transfer (Figure 11) between the chromophores (which is distance dependent) as a function of time. We found that, as expected, the duplex was kinetically quite stable at temperatures below its T_m, with a half-life of several hours.

When we added a circular ODN which was complementary to one of the strands, however, the half-life of the duplex was about 30 s. The circle apparently actively separates the strands of the substrate duplex, and then remains bound to its complementary

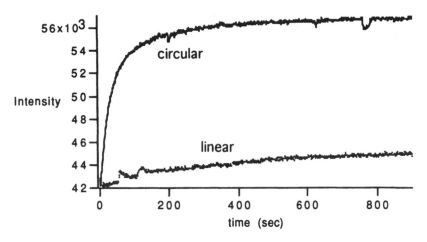

Figure 12 Kinetics runs for duplex dissociation in the presence of trapping linear or circular oligonucleotides. Fluorescein fluorescence increases as the quenching rhodamine-tagged strand is lost. (Reprinted with permission from *J. Chem. Soc., Chem. Commun.*, 215, 1993.)

half (Figure 12). Studies showed that the reaction is first order in circle concentration. We fit this reaction to Michaelis-Menten kinetic parameters, and found that the rate constant for the rate-limiting step is $7.3 \times 10^4 M^{-1}s^{-1}$. Other evidence leads us to believe that this step is the binding of the circle to the substrate duplex in "outside" fashion. The subsequent step, strand separation, is faster than the rate-limiting step, and thus is not kinetically measurable. This means that the circle is increasing the rate of duplex dissociation by at least three orders of magnitude. Based on the experimental evidence, we proposed a mechanism for the reaction, shown in Figure 13. The first step is the "outside" binding of the duplex, forming a Hoogsteen complex. The second, more rapid, step is the strand separation and Watson-Crick bond formation with the circle.

How does this second step proceed? It is probable that the duplex partially opens by fraying, breathing, or slipping on a timescale much faster than full dissociation. When this happens, we propose, the circle is already bound, and can reach across with its unbound Watson-Crick domain and trap a few base pairs in this strong complex. The rest of the duplex is then easily separated, and the final complex is formed.

Studies are under way to more fully understand the mechanism involved in this strand separation. It is clear that such a circle can have a strong kinetic advantage in binding duplex DNA invasively. It is also true, however, that this initial study involves only a simple model system in which the duplex is short and is fully separated at the end. In most practical cases, the duplex is much longer, and the displaced strand would remain attached, which is clearly a different case. Studies are currently in progress in our laboratory to investigate the scope of this phenomenon in general.

It is interesting to speculate that this kind of kinetic strand-separating activity may be useful in the binding of RNA as well. While many potential target RNAs, such as mRNA and viral RNA, are ostensibly single stranded, in reality they exist in folded structures containing regions of duplex.[102] This secondary structure is known to be able to inhibit binding by oligonucleotides.[103] In the case of a circular oligonucleotide analog, however, one can imagine that a cyclic molecule could diffuse and bind to a purine-rich region of homology in a duplex of RNA by a Hoogsteen complex, and then unwind or unfold the secondary structure to form the intended strong triplex.

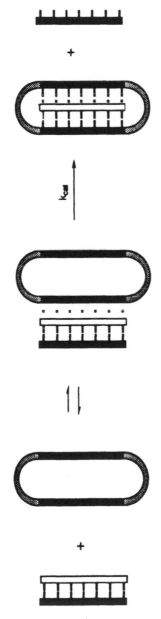

Figure 13 The proposed two-step mechanism for the acceleration of dissociation of duplex substrate by a circular oligodeoxynucleotide. (Reprinted with permission from *J. Chem. Soc., Chem. Commun.*, 215, 1993.)

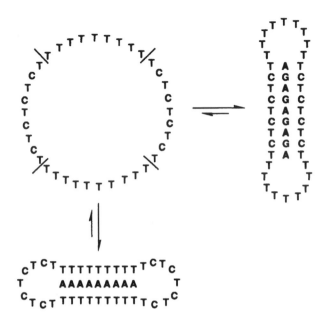

Figure 14 The binding of two different sequences by conformational switching. Loops and binding domains are interchanged to give bifunctional behavior. (Reprinted with permission from *J. Am. Chem. Soc.*, 115, 360, 1993.)

X. MULTIPLE SITE BINDING AND CONFORMATION SWITCHING

It has recently been shown in antisense studies that the use of more than one oligonucle-otide targeting multiple sites can be advantageous over single-target binding.[104,105] Two oligonucleotides can have synergistic effects in inhibiting gene expression,[105] and the use of multiple oligomers can have advantages in inhibition of viral replication by limiting the ability of the viruses to avoid inhibition by mutating the targeted sequence.[105]

We recently took advantage of the design of a circular triplex-forming oligonucleotide to synthesize a structure which can bind to more than one sequence in DNA by switching conformation.[106] This new design (Figure 14) takes advantage of nucleotides in the bridging loops, turning loops into a second pair of binding domains for a second comple-mentary sequence. Thus, the loops which bridge the binding domains in one complex will reverse roles and become the binding domains for a second targeted sequence. Above is the sequence of a circular compound we synthesized which can bind in this bifunctional mode.

This molecule is a cyclic 36 mer which contains four alternating nine-base binding domains. One pair of domains is complementary to the sequence dA$_9$, while the other is complementary to the sequence dAGAGAGAGA. The binding to each of these two sequences is 1:1 in stoichiometry and is strong, as seen for other triplex-forming circles. In this case, the two sets of binding domains are completely separate. The act of binding induces a conformational change, and nine-base loops are formed on both ends of the triplex. These loops cannot bind to the second sequence at 25°C because they are in an unproductive conformation.

One advantage of this double targeting is that fewer nucleotides are used than if one synthesized two separate circles. One can also imagine designing two-site-binding circles in which the binding domains are overlapping, such that some of the nucleotides are used for binding both sequences.[107,108] In that case, even fewer nucleotides are used in the synthesis. It would also be possible to design cyclic compounds which could recognize even more than two sequences, by designing in more than two pairs of binding domains.[107,108] It remains to be seen whether or not this kind of multiple-site binding will find use in biological systems.

XI. PROSPECTS FOR THE FUTURE

It is clear that circular oligonucleotides possess properties different from their linear relatives, and it is equally clear that further studies of circular oligonucleotides are needed to understand more of their basic physical properties. It will certainly be helpful to learn more about the specific structures of complexes involving circular DNAs and RNAs. Indeed, questions remain about the detailed structure of triple helices,[109,110] and we are only beginning to understand the structural and energetic properties of the bridging loops in our system. Further synthetic improvements, both to enhance binding properties and to increase biological activity, can be imagined and are being actively pursued. In addition, it would be attractive if oligonucleotides, linear or circular, were available on much larger scales;[111,112] to this end, we are pursuing enzymatic methods for oligonucleotide synthesis which involve the use of circular DNAs as catalysts.[113]

Clearly, there are many questions to be answered about the physical, biochemical, and pharmacological properties of both linear and circular oligonucleotides before an application can be made to therapeutic use. However, the properties of strong, highly selective binding, resistance to degradation, and favorable kinetics of binding lead us to believe that circular oligonucleotides are worth studying for their ability to inhibit gene expression. We have begun to focus on two primary strategies: targeting of single-stranded RNAs and targeting of duplex DNA.

A. TARGETING SINGLE-STRANDED RNA

When single-stranded RNA is the target, primary sites of interest are messenger RNAs (or pre-mRNAs) and viral RNAs. Inhibition of gene expression in these cases would involve inhibiting splicing, ribosome binding, or ribosome progression (in the mRNA case), or inhibiting binding or progression of reverse transcriptase in the retroviral case. One can imagine two approaches to binding and inhibition: first, the use of circular DNAs which will hybridize by Watson-Crick bonding alone; in this case binding is not as strong, but the possibility of inducing RNAse H activity[114] may make this approach attractive. The second approach is the use of analogs which can form triplexes with RNA,[35] in which case the primary mode of activity would probably be inhibition by formation of strong secondary structure. We are currently examining these approaches both in our own laboratory and with collaborators.[115]

B. TARGETING DUPLEX DNA

Targeting duplex DNA is attractive because of the smaller number of targets within a cell.[116] In this case, inhibiting the binding of transcription factors or progression of the RNA polymerase would be primary aims. It may be possible to do this in more than one way. If a circle is bound in "outside" (noninvasive) fashion, it may inhibit binding of proteins at that site by occupying the DNA major groove. If bound in this fashion downstream of a transcription start site, it is also possible that the RNA polymerase complex, which opens up the duplex ahead of itself, will allow the circle to form an invasively bound triplex, and that this complex will stall the polymerase. Finally, direct invasive binding may inhibit gene expression by stopping polymerase progression as well.

ACKNOWLEDGMENTS

I thank my co-workers, whose names are listed in the references, for their enthusiasm and dedication. This work is supported by the National Institutes of Health. I also gratefully acknowledge awards from the Young Investigator Programs of the Office of Naval Research, the Arnold and Mabel Beckman Foundation, and the Army Research Office, and a Teacher-Scholar Award from the Camille and Henry Dreyfus Foundation.

REFERENCES

1. Watson, J. D., Hopkins, N. H., Roberts, J. W., Steitz, J. A., Weiner, A. M., *Molecular Biology of the Gene*, 4th ed., Benjamin/Cummings, Menlo Park, CA, 1987, 193–195.
2. ibid., pp. 257–265.
3. Saenger, W., *Principles of Nucleic Acid Structure*, Springer-Verlag, New York, 1984, 450–457.
4. Watson, J. D., Hopkins, N. H., Roberts, J. W., Steitz, J. A., Weiner, A. M., *Molecular Biology of the Gene*, 4th ed., Benjamin/Cummings, Menlo Park, CA, 1987, 301–303.
5. Branch, A. D., Robertson, H. D., A replication cycle for viroids and other small infectious RNAs, *Science*, 223, 450–455, 1984.
6. Orgel, L. E., RNA catalysis and the origins of life, *J. Theor. Biol.*, 123, 127–149, 1986.
7. Usher, D. A., Early chemical evolution of nucleic acids: a theoretical model, *Science*, 196, 311–313, 1977.
8. Kool, E. T., Molecular recognition by circular oligonucleotides. Increasing the selectivity of DNA binding, *J. Am. Chem. Soc.*, 113, 6265–6266, 1991.
9. Prakash, G., Kool, E. T., Molecular recognition by circular oligonucleotides: strong binding of single-stranded DNA and RNA, *J. Chem. Soc., Chem. Commun.*, 1161–1162, 1991.
10. Prakash, G., Kool, E. T., Structural effects in the recognition of DNA by circular oligonucleotides, *J. Am. Chem. Soc.*, 114, 3523–3528, 1992.
11. Other researchers have studied circular DNAs and RNAs which are internally self-complementary.[12-14] These are essentially duplexes closed by hairpin loops on both ends, and are sometimes referred to as "dumbbells" because of their shape. Such molecules may find use as decoys for double-stranded DNA- or RNA-binding proteins.[15]
12. Wemmer, D. E., Benight, A. S., Preparation and melting of single strand circular DNA loops, *Nucleic Acids Res.*, 13, 8611–8621, 1985.
13. Erie, D. A., Jones, R. A., Olson, W. K., Sinha, N. K., Breslauer, K. J., Melting behavior of a covalently closed, single-stranded, circular DNA, *Biochemistry*, 28, 268–273, 1989.
14. Ashley, G. W., Kushlan, D. M., Chemical synthesis of oligodeoxynucleotide dumbbells, *Biochemistry*, 30, 2927–2933, 1991.
15. Ma, M. Y.-X., Reid, L. S., Climie, S. C., et al., Design and synthesis of RNA miniduplexes via a synthetic linker approach, *Biochemistry*, 32, 1751–1758, 1993.
16. Pleij, C. W. A., Rietveld, K., Bosch, L., A new principle of RNA folding based on pseudoknotting, *Nucleic Acids Res.*, 13, 1717–1731, 1985.
17. Thuong, N. T., Hélène, C., Sequence-specific recognition and modification of double-helical DNA by oligonucleotides, *Angew. Chem. Int. Ed. Engl.*, 32, 666–690, 1993.
18. Minshull, J., Hunt, J., The use of single-stranded DNA and RNase H to promote quantitative hybrid arrest of translation of mRNA/DNA hybrids in reticulocyte lysate cell-free translations, *Nucleic Acids Res.*, 14, 6433–6451, 1986.
19. Xodo, L. E., Manzini, G., Quadrifoglio, F., Spectroscopic and calorimetric investigation on the DNA triplex formed by d(CTCTTCTTTCTTTTCTTTCTTCTC) and d(GAGAAGAAAGA) at acidic pH, *Nucleic Acids Res.*, 18, 3557–3564, 1990.

20. Giovannangeli, C., Montenay-Garestier, T., Rougée, M., Chassignol, M., Thuong, N. T., Hélène, C., Single-stranded DNA as a target for triple-helix formation, *J. Am. Chem. Soc.*, 113, 7775–7776, 1991.

21. Felsenfeld, G., Davies, D. R., Rich, A., Formation of a three-stranded polynucleotide molecule, *J. Am. Chem. Soc.*, 79, 2023–2024, 1957.

22. Riley, M., Maling, B., Chamberlin, M. J., Physical and chemical characterization of two- and three-stranded adenine-thymine and adenine-uracil homopolymer complexes, *J. Mol. Biol.*, 20, 359–389, 1966.

23. Morgan, A. R., Wells, R. D., Specificity of the three-stranded complex formation between double-stranded DNA and single-stranded RNA containing repeated nucleotide sequences, *J. Mol. Biol.*, 37, 63–80, 1968.

24. Arnott, S., Selsing, E., Structures for the polynucleotide complexes poly(dA)·poly(dT) and poly(dT)·poly(dA)·poly(dT), *J. Mol. Biol.*, 88, 509–521, 1974.

25. Moser, H. E., Dervan, P. B., Sequence-specific cleavage of double-helical DNA by triple helix formation, *Science*, 238, 645–650, 1987.

26. Hoogsteen, K., The structure of crystals containing a hydrogen-bonded complex of 1-methylthymine and 9-methyladenine, *Acta Crystallogr.*, 12, 822–823, 1959.

27. Rubin, E., Rumney, S., Kool, E., manuscript submitted.

28. Barbato, S., DeNapoli, L., Mayol, L., Piccialli, G., Santacroce, C., Solid-phase synthesis of cyclic oligodeoxyribonucleotides, *Tetrahedron Lett.*, 28, 5727–5728, 1987.

29. de Vroom, E., Broxterman, H. J. G., Sliedregt, L. A. J. M., van der Marel, G., A., van Boom, J. H., Synthesis of cyclic oligonucleotides by a modified phosphotriester approach, *Nucleic Acids Res.*, 16, 4607–4620, 1988.

30. Capobianco, M. L., Carcuro, A., Tondelli, L., Garbesi, A., Bonora, G. M., One pot solution synthesis of cyclic oligodeoxyribonucleotides, *Nucleic Acids Res.*, 18, 2661–2669, 1990.

31. DeNapoli, L., Messere, A., Montesarchio, D., Piccialli, G., Santacroce, C., PEG-supported synthesis of cyclic oligodeoxyribonucleotides, *Nucleosides Nucleotides*, 12, 21–30, 1993.

32. Kanaya, E., Yanagawa, H., Template-directed polymerization of oligoadenylates using cyanogen bromide, *Biochemistry*, 25, 7423–7430, 1986.

33. Dolinnaya, N. G., Sokolova, N. I., Asherbikova, D. T., Shabarova, Z. A., The use of BrCN for assembling modified DNA duplexes and DNA-RNA hybrids; comparison with water-soluble carbodiimide, *Nucleic Acids Res.*, 19, 3067–3080, 1991.

34. Illuminati, G., Mandolini, L., Masci, B., Ring-closure reactions. 9. A comparison with related cyclization series, *J. Am. Chem. Soc.*, 99, 6308–6312, 1977.

35. Wang, S., Kool, E. T., Circular RNA oligonucleotides. Synthesis, nucleic acid binding properties, and a comparison with circular DNAs, *Nucleic Acids Res.*, 22, 2326–2333, 1994. (Note: extinction coefficients for cyclic oligonucleotides are calculated by the nearest neighbor method[36]).

36. Borer, P. N., in *Handbook of Biochemistry and Molecular Biology*, Vol. I, 3rd ed., Fasman, G. D., Ed., CRC Press, Cleveland, 1985, 589.

37. Serwer, P., Hayes, S. J., Atypical seiving of open circular DNA during pulsed field agarose gel electrophoresis, *Biochemistry*, 28, 5827–5832, 1989.

38. Sambrook, J., Fritsch, E. F., Maniatis, T., in *Molecular Cloning*, Nolan, C., Ed., Cold Spring Harbor Laboratory, Cold Spring Harbor, NY, 1989, 5.44–5.47.

39. ibid., p. 5.72.

40. Cram, D. J., Molecular cells, their guests, portals, and behavior, *Chemtech*, 17, 120–125, 1987.

41. Heus, H. A., Pardi, A., Structural features that give rise to the unusual stability of RNA hairpins containing GNRA loops, *Science*, 253, 191–194, 1991.

42. Searle, M. S., Williams, D. H., On the stability of nucleic acid structures in solution: enthalpy-entropy compensations, internal rotations and reversibility, *Nucleic Acids Res.*, 21, 2051–2056, 1993.

43. D'Souza, D. J., Kool, E. T., Strong binding of single-stranded DNA by stem-loop oligonucleotides, *J. Biomol. Struct. Dyn.*, 10, 141–152, 1992.

44. Wang, S., Booher, M. A., Kool, E. T., Stabilities of nucleotide loops bridging the pyrimidine strands in DNA pyrimidine-purine-pyrimidine triplexes, *Biochemistry*, 33, 4639–4644, 1994.

45. Porschke, D., Eigen, M., Kinetics of the helix-coil transition of the oligoribo-uridylic·oligoriboadenylic acid system and of oligoriboadenylic acid alone at acidic pH, *J. Mol. Biol.*, 62, 361–381, 1971.

46. Xodo, L. E., Manzini, G., Quadrifoglio, F., van der Marel, G. A., van Boom, J. H., Effect of 5-methylcytosine on the stability of triple-stranded DNA — a thermodynamic study, *Nucleic Acids Res.*, 19, 5625–5631, 1992.

47. D'Souza, D. J., Kool, E. T., Solvent, pH, and ionic effects on the binding of single-stranded DNA by circular oligodeoxynucleotides, *Biomed. Chem. Lett.*, 4, 965–970, 1994.

48. Williams, A. P., Longfellow, C. E., Freier, S. M., Kierzek, R., Turner, D. H., Laser temperature-jump, spectroscopic, and thermodynamic study of salt effects on duplex formation by dGCATGC, *Biochemistry*, 28, 4283–4291, 1989.

49. Turner, D. H., Sugimoto, N., Freier, S. M., in *Nucleic Acids* (subvolume C), Saenger, W., Ed., Springer-Verlag, Berlin, 1990, 201–227.

50. Rougée, M., Faucon, B., Mergny, J. L., Barcelo, F., Giovannangeli, C., Garestier, T., Hélène, C., Kinetics and thermodynamics of triple-helix formation: effects of ionic strength and mismatches, *Biochemistry*, 31, 9269–9278, 1992.

51. Wang, S., Friedman, A., Kool, E., manuscript submitted.

52. Perkins, T., Goodman, J. L., Kool, E. T., unpublished data, 1992.

53. Uhlmann, E., Peyman, A., Antisense oligonucleotides: a new therapeutic principle, *Chem. Rev.*, 90, 543–584, 1990.

54. Nussinov, R., Tinoco, I., Jr., Sequential folding of a messenger RNA molecule, *J. Mol. Biol.*, 151, 519–533, 1981.

55. Kozak, M., Influences of mRNA secondary structure on initiation by eukaryotic ribosomes, *Proc. Natl. Acad. Sci. U.S.A.*, 83, 2850–2854, 1986.

56. Kramer, F. R., Mills, D. R., Secondary structure formation during RNA synthesis, *Nucleic Acids Res.*, 9, 5109–5124, 1981.

57. Woolf, T. M., Melton, D. A., Jennings, C. G. B., Specificity of antisense oligonucle-otides in vivo, *Proc. Natl. Acad. Sci. U.S.A.*, 89, 7305–7309, 1992.

58. Asseline, U., Delarue, M., Lancelot, G., Toulmé, F., Thuong, N. T., Montenay-Garestier, T., Hélène, C., Nucleic acid binding molecules with high affinity and base sequence specificity: intercalating agents covalently linked to oligodeoxynucleotides, *Proc. Natl. Acad. Sci. U.S.A.*, 81, 3297–3301, 1984.

59. Mergny, J. L., Duval-Valentin, G., Nguyen, C. H., Perrouault, L., Faucon, B., Rougée, M., Montenay-Garestier, T., Bisagni, E., Hélène, C., Triple helix-specific ligands, *Science*, 256, 1681–1684, 1992.

60. Plum, G. E., Park, Y.-W., Singleton, S. F., Dervan, P. B., Breslauer, K. J., Thermodynamic characterization of the stability and the melting behavior of a DNA triplex: a spectroscopic and calorimetric study, *Proc. Natl. Acad. Sci. U.S.A.*, 87, 9436–9440, 1990.

61. Pilch, D. S., Brousseau, R., Shafer, R. H., Thermodynamics of triple helix formation: spectrophotometric studies on the $dA_{10} \cdot 2dT_{10}$ and $dC_3T_4C_3 \cdot dG_3A_4G_3 \cdot dC_3T_4C_3$ triple helices, *Nucleic Acids Res.*, 18, 5743–5750, 1990.

62. Griffin, L. C., Dervan, P. B., Recognition of thymine·adenine base pairs by guanine in a pyrimidine triple helix motif, *Science*, 245, 967–971, 1989.

63. Roberts, R. W., Crothers, D. M., Specificity and stringency in DNA triplex formation, *Proc. Natl. Acad. Sci. U.S.A.*, 88, 9397–9401, 1991.

64. Mergny, J.-L., Sun, J.-S., Rougée, M., Montenay-Garestier, T., Barcelo, F., Chomilier, J., Hélène, C., Sequence specificity in triple-helix formation: experimental and theoretical studies of the effect of mismatches on triplex stability, *Biochemistry*, 30, 9791–9798, 1991.

65. Wang, S., Kool, E., unpublished studies.

66. Griffin, L. C., Kiessling, L. L., Beal, P. A., Gillespie, P., Dervan, P. B., Recognition of all four base pairs of double-helical DNA by triple-helix formation: design of nonnatural deoxyribonucleosides for pyrimidine·purine base pair binding, *J. Am. Chem. Soc.*, 114, 7976–7982, 1992.

67. Rubin, E., Rumney, S., Kool, E., manuscript submitted.

68. Xodo, L. E., Manzini, G., Quadrifoglio, F., van der Marel, G., van Boom, J. H., Hairpin structures in synthetic oligodeoxynucleotides: sequence effects on the duplex-to hairpin transition, *Biochimie*, 71, 793–803, 1989.

69. Hirao, I., Nishimura, Y., Tagawa, Y., Watanabe, K., Miura, K., Extraordinarily stable mini-hairpins: electrophoretical and thermal properties of the various sequence variants of d(GCGAAAGC) and their effect on DNA sequencing, *Nucleic Acids Res.*, 20, 3891–3896, 1992.

70. Harvey, S. C., Luo, J., Lavery, R., DNA stem-loop structures in oligopurine-oligopyrimidine triplexes, *Nucleic Acids Res.*, 16, 11795–11809, 1988.

71. Povsic, T. J., Dervan, P. B., Triple helix formation by oligonucleotides on DNA extended to the physiological pH range, *J. Am. Chem. Soc.*, 111, 3059–3061, 1990.

72. Lee, J. S., Woodsworth, M. L., Latimer, L. J. P., Morgan, A. R., Poly-pyrimidine•polypurine synthetic DNAs containing 5-methylcytosine form stable triplexes at neutral pH, *Nucleic Acids Res.*, 12, 6603–6614, 1984.

73. Xodo, L. E., Manzini, G., Quadrifoglio, F., van der Marel, G., van Boom, J. H., Effect of 5-methylcytosine on the stability of triple- stranded DNA — a thermodynamic study, *Nucleic Acids Res.*, 19, 5625–5631, 1991.

74. Sowers, L. C., Shaw, B. R., Sedwick, W. D., Base stacking and molecular polarizability: effect of a methyl group in the 5-position of pyrimidines, *Biochem. Biophys. Res. Commun.*, 148, 790–794, 1987.

75. D'Souza, D. J., Kool, E. T., unpublished studies.

76. Wickstrom, E., Oligodeoxynucleotide stability in subcellular extracts and culture media, *J. Biochem. Biophys. Methods*, 13, 97–102, 1986.

77. Rumney S., IV, Kool, E. T., DNA recognition by hybrid oligoether-oligodeoxy-nucleotide macrocycles, *Angew. Chem. Int. Ed. Engl.*, 31, 1617–1619, 1992.

78. Harel-Bellan, A., Ferris, D. K., Vinocour, M., Holt, J. T., Farrar, W. L., Specific inhibition of c-myc protein biosynthesis using an antisense synthetic deoxy-oligo-nucleotide in human T lymphocytes, *J. Immunol.*, 140, 2431–2435, 1988.

79. Beaucage, S. L., Iyer, R. P., Advances in synthesis of oligonucleotides by the phosphoramidite approach, *Tetrahedron*, 48, 2223–2311, 1992.

80. Rumney, S., Kool, E. T., manuscript submitted.

81. Beaucage, S. L., Iyer, R. P., The functionalization of oligonucleotides via phosphoramidite derivatives, *Tetrahedron*, 49, 1925–1963, 1993.

82. Bergstrom, D. E., Ogawa, M. K., C-5 Substituted pyrimidine nucleosides. 2. Synthesis via olefin coupling to organopalladium intermediates derived from uridine and 2'-deoxyuridine, *J. Am. Chem. Soc.*, 100, 8106–8112, 1978.

83. Dreyer, G. B., Dervan, P. B., Sequence specific cleavage of single-stranded DNA: oligodeoxynucleotide-EDTA-Fe(II), *Proc. Natl. Acad. Sci. U.S.A.*, 82, 968–972, 1985.

84. Hobbs, F., Jr., Palladium-catalyzed synthesis of alkynylamino nucleosides. A universal linker for nucleic acids, *J. Org. Chem.*, 54, 3420–3422, 1989.

85. Robins, M. J., Barr, P. J., Nucleic acid related compounds. 39. Efficient conversion of 5-iodo to 5-alkynyl and derived 5-substituted uracil bases and nucleosides, *J. Org. Chem.*, 48, 1854–1862, 1983.

86. Sági, J., Szemzö, A., Ébinger, K., Szabolcs, A., Sági, G., Ruff, E., Ötvös, L., Base-modified oligodeoxynucleotides. I. Effect of 5-alkyl, 5-(1-alkenyl), 5-(1-alkynyl) substitutions of the pyrimidines on duplex stability and hydrophobicity, *Tetrahedron Lett.*, 34, 2191–2194, 1993.

87. Connolly, B. A., Potter, B. V. L., Eckstein, F., Pingoud, A., Grotjahn, L., Synthesis and characterization of an octanucleotide containing the EcoRI recognition sequence with a phosphorothioate group at the cleavage site, *Biochemistry*, 23, 3443–3453, 1984.

88. Inoue, H., Hayase, Y., Imura, A., Iwai, S., Miura, K., Ohtsuka, E., Synthesis and hybridization studies on two complementary nona(2'-O-methyl)ribonucleotides, *Nucleic Acids Res.*, 15, 6131–6148, 1987.

89. Roberts, R. W., Crothers, D. M., Stability and properties of double and triple helices: dramatic effects of RNA or DNA backbone composition, *Science*, 258, 1463–1466, 1992.

90. Han, H., Dervan, P. B., Sequence-specific recognition of double helical RNA and RNA·DNA by triple helix formation, *Proc. Natl. Acad. Sci. U.S.A.*, 90, 3806–3810, 1993.

91. Breslauer, K. J., Frank, R., Blocker, H., Marky, L. A., Predicting DNA duplex stability from the base sequence, *Proc. Natl. Acad. Sci. U.S.A.*, 83, 3746–3750, 1986.

92. Nielsen, P. E., Egholm, M., Berg, R. H., Buchardt, O., Sequence selective recognition of DNA by strand displacement with a thymine- substituted polyamide, *Science*, 254, 1497–1500, 1991.

93. Egholm, M., Behrens, C., Christensen, L., Berg, R. H., Nielsen, P. E., Buchardt, O., Peptide nucleic acids containing adenine or guanine recognize thymine and cytosine in complementary DNA sequences, *J. Chem. Soc., Chem. Commun.*, 800–801, 1993.

94. Hanvey, J. C., Peffer, N. J., Bisi, J. E., et al., Antisense and antigene properties of peptide nucleic acids, *Science*, 258, 1481–1485, 1992.

95. Young, S. L., Krawscyk, S. H., Matteucci, M. D., Toole, J. J., Triple helix formation inhibits transcription enlongation in vitro, *Proc. Natl. Acad. Sci. U.S.A.*, 88, 10023–10026, 1991.

96. Maher III, L. J., Dervan, P. B., Wold, B., Analysis of promoter-specific repression by triple-helical DNA complexes in a eukaryotic cell-free transcription system, *Biochemistry*, 31, 70–81, 1992.

97. Cherny, D. Y., Beloserkovskii, B. P., Frank-Kamenetskii, M. D., Egholm, M., Buchardt, O., Berg, R. H., Nielsen, P. A., DNA unwinding upon strand-displacement binding of a thymine-substituted polyamide to a double-stranded DNA, *Proc. Natl. Acad. Sci. U.S.A.*, 90, 1667–1670, 1993.

98. Froehler, B. C., Wadwani, S., Terhorst, T. J., Gerrard, S. R., Oligodeoxynucleotides containing C-5 propyne analogs of 2'-deoxyuridine and 2'-deoxycytidine, *Tetrahedron Lett.*, 33, 5307–5310, 1992.

99. Wagner, R. W., Matteucci, M. D., Lewis, J. G., Gutierrez, A. J., Moulds, C., Froehler, B. C., Antisense gene inhibition by oligonucleotides containing C-5 propyne pyrimidines, *Science*, 260, 1510–1513, 1993.

100. Perkins, T. A., Goodman, J. L., Kool, E. T., Accelerated displacement of duplex DNA strands by a synthetic circular oligodeoxynucleotide, *J. Chem. Soc., Chem. Commun.*, 215–217, 1993.

101. Cardullo, R. A., Agrawal, S., Flores, C., Zamecnik, P., Wolf, D. E., Detection of nucleic acid hybridization by nonradiative fluorescence resonance energy transfer, *Proc. Natl. Acad. Sci. U.S.A.*, 85, 8790–8794, 1988.

102. Turner, D. H., Sugimoto, N., RNA structure prediction, *Annu. Rev. Biophys. Biophys. Chem.*, 17, 167–202, 1988.
103. Lima, W. F., Monia, B. P., Ecker, D. J., Freier, S. M., Implication of RNA structure on antisense oligonucleotide hybridization kinetics, *Biochemistry*, 31, 12055–12061, 1992.
104. Maher L. J., III, Dolnick, B. J., Specific hybridization arrest of dihydrofolate reductase mRNA in vitro using anti-sense RNA or anti-sense oligonucleotides, *Arch. Biochem. Biophys.*, 253, 214–220, 1987.
105. Lisziewicz, J., Sun, D., Klotman, M., Agrawal, S., Zamecnik, P., Gallo, R., Long-term treatment of human immunodeficiency virus-infected cells with antisense oligonucleotide phosphorothioates, *Proc. Natl. Acad. Sci. U.S.A.*, 89, 11209–11213, 1992.
106. Rubin, E., McKee, T. L., Kool, E. T., Binding of two different DNA sequences by conformational switching, *J. Am. Chem. Soc.*, 115, 360–361, 1993.
107. Rubin, E., Kool, E. T., Strong, specific binding of six different DNA sequences by a single conformation-switching DNA macrocycle, *Angew. Chem., Int. Ed. Engl.*, 33, 1004–1006, 1994.
108. Rubin, E., Kool, E. T., unpublished data.
109. Rajagopal, P., Feigon, J., Triple-strand formation in the homopurine-homopyrimidine DNA oligonucleotides d(G-A)$_4$ and d(T-C)$_4$, *Nature*, 339, 637–640, 1989.
110. Howard, F. B., Miles, H. T., Liu, K., Frazier, J., Raghunathan, G., Sasisekharan, V., Structure of d(T)$_n$·d(A)$_n$·d(T)$_n$: the DNA triple helix has B-form geometry with C2'-endo sugar pucker, *Biochemistry*, 31, 10671–10677, 1992.
111. Bonora, G. M., Biancotto, G., Maffini, M., Scremin, C. L., Large scale, liquid phase synthesis of oligonucleotides by the phosphoramidite approach, *Nucleic Acids Res.*, 21, 1213–1217, 1993.
112. Wright, P., Lloyd, D., Rapp, W., Andrus, A., Large-scale synthesis of oligonucleotides via phosphoramidite nucleosides and a high-loaded polystyrene support, *Tetrahedron Lett.*, 34, 3373–3376, 1993.
113. Zillmann, M. A., Daubendiek, S. L., Ryan, K., Kool, E. T., unpublished data.
114. Mizrahi, V., Analysis of the ribonuclease H activity of HIV-1 reverse transcriptase using RNA·DNA hybrid substrates derived from the gag region of HIV-1, *Biochemistry*, 28, 9088–9094, 1989.
115. Rowley, P., Thomas, M., Kosciolek, B., Kool, E., Circular antisense oligonucleotides inhibit proliferation of chronic myeloid leukemia cells in sequence-specific manner, *Blood*, 82, 330a, 1993.
116. Toulmé, J.-J., Hélène, C., Antimessenger oligodeoxyribonucleotides: an alternative to antisense RNA for artificial regulation of gene expression — a review, *Gene*, 72, 51–58, 1988.

Chapter 9

Peptide Nucleic Acids as Antisense Therapeutic Agents

Jeffery C. Hanvey and Lee E. Babiss

CONTENTS

I. INTRODUCTION

Peptide nucleic acids (PNAs) are oligomers in which the typical deoxyribose-phosphodiester backbone is absent, and the purine and pyrimidine bases are instead linked via N-ethylaminoglycine monomer units (Figure 1A). PNA oligomers bind to complementary sequences in either RNA or DNA with remarkably high affinity. For example, a complex consisting of a PNA T_{10} and dA_{10} has a melting temperature (T_m) of 73°C, while the DNA·DNA duplex would have a T_m of only 23°C.[1] The reason for the dramatic increase in stability of a PNA·DNA complex has not been determined, although the lack of charge repulsion between the two strands may partially explain the phenomena. Initially, all PNAs were synthesized with a lysine group at the carboxyl terminus; however, the lysine group is not necessary for binding to either DNA or RNA.

A unique property of PNAs is their ability to bind duplex DNA via a strand-invasion reaction (Figure 1B). This property was originally identified for a homothymidine PNA (T_{10}),[1] but other polypyrimidine (or pyrimidine-rich) sequences have similar behavior.[2,3] Unlike homopyrimidine oligodeoxynucleotides (ODNs), which can bind in the major groove of DNA to form a triple helix, homopyrimidine PNAs displace the pyrimidine strand of DNA and bind to the polypurine strand via Watson-Crick hydrogen bonds. The structure which is formed, a D-loop, was identified by the sensitivity of the PNA-DNA complex to single-strand-specific reagents such as S1 nuclease.[1] A differential sensitivity to S1 nuclease was seen between the two DNA strands, which indicated that only one of the two DNA strands was single stranded. The complex actually consists of two PNA strands associated with the polypurine DNA strand (Figure 1B). The second PNA strand most likely lies in the major groove of the PNA·DNA duplex, with this PNA binding via Hoogsteen hydrogen bonds, as evidenced by the pH dependence of this interaction.[4] The PNA·DNA triplex which forms is a right-handed helix, based on circular dichroism spectroscopy of PNA-ODN interactions.[5] Formation of the D-loop within a closed circular plasmid results in unwinding of the double helix; one turn of helix is unwound per 10 bases of PNA bound.[6]

D-Loop formation using a mixed sequence PNA has not yet been observed. While we believe that G·C-rich target sequences will have reduced DNA breathing and restrict D-loop formation, a PNA sequence with 33% cytidines (such as $(TC)_5T_5$) is capable of strand-invasion. Also, mixed-sequence PNAs with a low G·C content do not strand

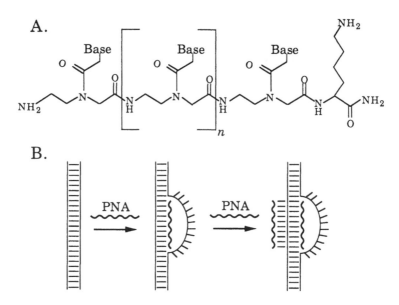

Figure 1 (A) Structure of a peptide nucleic acid. The repeating unit, an *N*-ethylaminoglycine, is shown in brackets. A lysine amide is at the carboxyl terminus (right end) of the molecule. Base represents either A, T, G, or C. (B) Schematic of PNA strand invasion of duplex DNA. Strand invasion may occur in a multistep process. After localized breathing (melting) of the DNA duplex, one PNA molecule binds to form a D-loop. This structure is stabilized by binding of a second PNA molecule to the same DNA strand via Hoogsteen hydrogen bonds.

invade, which implies that the strand-invasion potential must not be due solely to the G·C content of the sequence. It is likely that addition of the second PNA strand to the complex is necessary for D-loop stability, and that the mixed-sequence PNA cannot form the 2:1 complex because of failure to correctly Hoogsteen hydrogen bond (as only polypyrimidine ODN sequences can form a triple helix). Further support for this idea comes from analysis of PNA binding to mismatched sites within DNA. Although a single mismatch within a PNA 10 mer should be thermodynamically stable at 37°C (based on T_m data), its binding to duplex DNA was greatly reduced.[6] Therefore, the authors suggested a kinetic model in which the stability of the complex is controlled by binding of the second PNA molecule.

II. PNA ANTIGENE POTENTIAL

The ability of PNAs to strand invade DNA and form a D-loop indicated that PNAs may be useful as antigene agents. Inhibition of gene expression at the transcription level (antigene) as opposed to translation arrest (antisense) may represent the most fundamental means of ablating gene function. If PNAs could form sequence-specific D-loops *in vivo*, they could inhibit pol II transcription, either at the level of transcriptional initiation or elongation. Recent studies suggest that PNAs could inhibit transcription initiation *in vitro*, by demonstrating that a PNA-dependent D-loop blocks the binding of proteins to their DNA target site.[2,6] Although this is a reasonable approach to inhibit gene expression, more sites in the DNA and greater gene specificity could be achieved by blocking transcription elongation.

We examined the ability of a PNA to inhibit eukaryotic pol II transcription elongation by forming a PNA D-loop on DNA downstream of a transcription start site and then

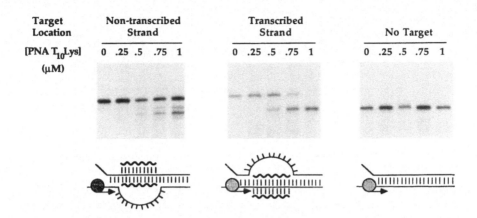

Figure 2 Site-specific inhibition of transcription elongation. When a dA_{10} binding site for a PNA is on the transcribed strand of the DNA, complete and site-specific inhibition of transcription (lower band, transcribed strand lanes) was seen with increasing concentrations of PNA T_{10}Lys. Complete inhibition of transcription was seen only when the PNA and RNA polymerase were bound to the same strand of DNA (schematics).

initiating *in vitro* transcription. Site-specific inhibition of transcription elongation was observed at the location of the PNA binding site (Figure 2). No inhibition of transcription was observed in DNA without the PNA target sequence, demonstrating the specificity of the PNA·DNA interaction.

By forming a D-loop, a PNA is hydrogen bonded to only one strand of DNA, while the other DNA strand is unpaired. Therefore it was not surprising to discover that the inhibition of transcription elongation was strand specific. When the PNA was bound to the transcribed strand (the strand with which RNA polymerase is interacting), complete and dose-dependent inhibition of transcription was observed. However, only partial inhibition of transcription elongation was seen when the same PNA target sequence was placed on the nontranscribed strand of the DNA (Figure 2). It is likely that some RNA polymerase molecules bypass the D-loop when the PNA is bound to the nontranscribed strand of DNA.

The two strands of DNA associate in an antiparallel orientation due to the configurational and conformational restraints inherent in each nucleotide monomer. PNAs, which have an achiral backbone without such configurational forces, have the potential to bind DNA or RNA in an orientation-independent manner. The two possible binding motifs are for the PNA to bind NH_2 to 3' (antiparallel orientation) or NH_2 to 5' (parallel orientation) (Figure 3). The nomenclature for parallel and antiparallel is derived from the standard convention of writing nucleic acids and peptides.

To examine the binding orientation of PNAs to DNA and RNA, we prepared an asymmetrical polypyrimidine PNA, $(TC)_5T_5$Lys, containing a lysine amide at the COOH end of the PNA, and examined its binding properties against DNA and RNA containing a complementary binding site in either orientation (5' $(GA)_5A_5$ or 5' $A_5(GA)_5$). The ability of this PNA to bind duplex DNA was examined by inhibition of pol II transcription elongation. Similarly, we examined the binding of this PNA to RNA using inhibition of reverse transcriptase elongation. In both assays the PNA $(TC)_5T_5$Lys gave a dose-dependent inhibition with the target sequence in both orientations.[3] The extent of inhibition was similar for both orientations, indicating that the PNA could bind to both duplex DNA and RNA regardless of orientation.

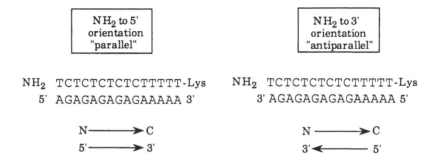

Figure 3 PNAs bind to DNA and RNA in both orientations. A PNA such as $(TC)_5T_5$ can bind to its complementary sequence $(AG)_5A_5$ in DNA or RNA with the amino terminus of the PNA aligned to either the 5' or 3' ends of the nucleic acid strand. Amino to 5' is called binding in a parallel orientation, and amino to 3' is antiparallel binding.

The binding orientation of PNAs to single-stranded DNA has been examined by T_m analysis. In a T_m analysis of either DNA or RNA and a 15-mer PNA containing all four bases, the antiparallel binding orientation was preferred, although the PNAs could readily bind in both orientations.[7]

In our binding orientation studies with the PNA sequence $(TC)_5T_5$, complete binding of two PNA strands to the target would occur only if the two PNA strands bind parallel to one another. Based on molecular mechanics calculations, it was proposed that the PNA T_{10} binds to a dA_{10} target with the two PNA strands antiparallel to one another.[8,9] Further work is necessary to completely describe the binding properties of PNA molecules to DNA and RNA. The indication that PNAs bind independent of orientation dictates that each PNA would have two possible target sequences, which would decrease their specificity when used as antisense or antigene agents.

The antigene potential of PNAs *in vivo* may be limited due to the solution conditions necessary for D-loop formation. Formation of the D-loop on DNA is highly dependent on ionic strength. D-Loop formation occurs readily in a low-ionic strength solution (such as TE buffer), but the extent of strand invasion decreases rapidly as NaCl is added to the solution.[3,6] In Figure 4A, the extent of D-loop formation was measured by a gel mobility shift assay after preincubation of the DNA and PNA for 1 h. D-Loop formation is essentially completely inhibited by 100 m*M* NaCl. As the melting of DNA decreases (T_m increases) with increasing salt, the observed salt dependence likely reflects the importance of DNA breathing (localized melting) as a prerequisite of D-loop formation. Further evidence for the role of DNA breathing is our observation that, given a 20-h incubation in 100 m*M* NaCl, D-loop formation increased from 10 to 50%. The salt effect was not restricted to NaCl, as inhibition of D-loop formation was also seen with solutions of KCl or $MgCl_2$.

Once the PNA·DNA complex is formed, it is extremely stable and the complex is not dissociated by increasing the ionic strength. For example, when a PNA D-loop is formed in a low-salt solution and then shifted to transcription buffer before initiation of transcription, greater than 50% inhibition of transcription elongation was seen even after a 5-h incubation in the buffer (Figure 4B). In this experiment, control reactions (not shown) showed that the PNA was not dissociating and then reassociating during the incubation. The slow off rate of the PNA from the DNA further demonstrates that the transcription inhibition we observed was due to termination of transcription and not pausing of the polymerase at the D-loop. In contrast, when a triplex-forming ODN was used in similar experiments, the inhibition of transcription seen was transient, likely caused by pausing of the polymerase; complete inhibition could not be achieved due to the dissociation of the ODN from the DNA.[10] The stability of the PNA·DNA D-loop complex can be

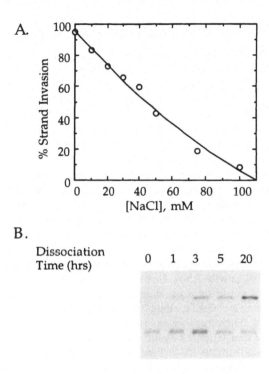

Figure 4 (A) Salt dependence of PNA strand invasion. The graph shows the percent strand invasion observed as a function of NaCl concentration in a gel mobility shift assay after a 1-h preincubation of T_{10}Lys PNA with a plasmid containing a dA_{10} target sequence. (B) Dissociation of PNA from duplex DNA. The PNA·DNA complex was formed in low salt, shifted to transcription buffer, and then at various times a transcription reaction was performed. The persistence of the termination product (lower band) indicates the stability of the PNA·DNA complex.

attributed to several factors such as those proposed by Almarsson et al.:[8] (i) the van der Waals attractions between the PNA strands (as two PNA strands bind one strand of DNA), (ii) the stabilizing electrostatic effect caused by the local separation of the phosphodiester groups in the two DNA strands, and (iii) the increased solvation of the phosphates and bases on the displaced strand.

Binding of PNAs to duplex DNA causes a change in the migration of the DNA in gel electrophoresis.[2] As a PNA such as $(TC)_5T_5$Lys has a positive charge at neutral pH (+2 charge due to the terminal NH_2 and the NH_2 of the Lys moiety), clearly a PNA bound to DNA or RNA increases the overall mass of the DNA and changes the mass to charge ratio. However, we have observed that the position of a PNA D-loop with respect to the ends of the DNA have an effect on migration of the DNA in polyacrylamide gels.[3] This behavior is similar to that of alterations which cause a bend in DNA. A larger gel retardation is seen when a bend is in the middle of the DNA molecule than when the bend is near one end (as migration of DNA through a polyacrylamide gel is dependent on the end-to-end distance of the DNA).[11-13] To test PNA-induced DNA bending, we used a plasmid in which a PNA binding site could be positioned at various distances from the ends of restriction fragments which are precisely the same length. A larger gel mobility shift (slower migration) was seen when the PNA binding site was positioned in the center of the restriction fragment, with faster migration seen as the site progressively moved toward the end of the fragment.[3] Therefore, the PNA-induced D-loop may bend or kink DNA.

III. PNA ANTISENSE POTENTIAL

The interactions of PNAs with RNA and single-stranded DNA do not have the sequence and ionic strength limitations mentioned above for duplex DNA. PNAs bind to RNA in physiological salt solutions, and bind to complementary sequences regardless of the base composition. However, it is likely that polypyrimidine PNAs bind polypurine RNA sequences in a 2:1 ratio to form a triple helix. Therefore, PNAs could potentially be used as an antisense agent to target both hnRNA and mRNA.

We initially measured the interactions of PNAs with RNA using an inhibition of reverse transcriptase assay. This assay provides a simplified method to evaluate the binding capabilities of PNAs to RNA independent of T_m measurements. After preincubating a PNA with an RNA transcript containing the target sequence, we observed inhibition of reverse transcriptase elongation. Reverse transcription terminated specifically at the target site, and no inhibition was seen in the absence of the target.[2] The inhibition was dependent on the relative concentration of the PNA to the RNA target sequence, and dose-response curves could be generated.

The inhibition of reverse transcriptase assay allowed us to evaluate the *in vitro* binding properties of many PNAs with RNA. However, it did not address that actual question of whether or not a PNA could serve as an antisense agent by binding to the coding region of mRNA and blocking translation elongation, which is probably the most stringent test of an antisense agent. Therefore, we examined the antisense potential for PNAs by the inhibition of *in vitro* translation elongation. In one study, a mixed-sequence target site ($A_3TGA_2GA_2$) derived from the SV40 T-antigen mRNA was cloned and RNA containing this sequence was generated. Both complementary and scrambled PNA sequences were generated and hybridized to the RNA prior to initiation of translation using a rabbit reticulocyte lysate. Analysis of the peptide fragments (^{35}S-labeled) by PAGE analysis revealed that the complementary PNA inhibited translation precisely at the position of the SV40 T-antigen site.[2] This PNA had no effect on translation in the absence of its complementary site in the RNA. Furthermore, the scrambled sequence PNA did not inhibit *in vitro* translation elongation. Figure 5 shows the results of a similar experiment, except that the target sequence for PNA binding in the RNA was A_{10} and the PNA used was a T_{10}. Again, site-specific inhibition of translation elongation was observed at the PNA binding site.

We have found that a PNA·RNA heteroduplex is not a substrate for RNase H (using HeLa nuclear extract as a source of the enzyme). This was expected as all ODN analogs whose backbones are not negatively charged (for example, methylphosphonate oligomers) or which contain a ribose sugar (2'-O-methyl oligomers, for example) do not function via an RNase H mechanism. Thus the inhibition of translation elongation we observed was likely the result of ribosomes stalling or being displaced from the RNA.

The reverse transcription and translation studies demonstrated the superior RNA-binding capabilities of PNAs over ODNs. Although ODNs can inhibit both reverse transcription and translation elongation *in vitro*, this inhibition is only a result of RNA cleavage at the site of the RNA·ODN heteroduplex by RNase H. The binding of ODNs to RNA does not inhibit translation elongation by a steric blocking mechanism. Increased binding strength as seen with PNAs, however, may not be sufficient for a therapeutic effect by an antisense compound, as the compound may need to function catalytically through RNase H or inactivate the RNA through some other mechanism.

IV. PNA CELL-BASED STUDIES

The ability of eukaryotic cells to internalize PNAs has been examined using a PNA with a fluorescent tag. The pattern and rate of PNA uptake was compared to a fluorescently tagged phosphorothioate-ODN (*S*-ODN) as a control. *S*-ODNs are taken up most likely

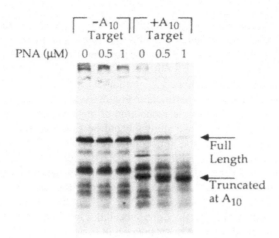

Figure 5 Site-specific inhibition of *in vitro* translation. RNA with or without an A_{10} target site was incubated with PNA T_{10}Lys followed by *in vitro* translation with a rabbit reticulocyte lysate. The position of both the full-length protein and the peptide formed by termination at the A_{10} site on the RNA are shown by arrows. No inhibition of translation by T_{10}Lys was seen in the absence of the A_{10} target.

by an adsorptive endocytosis process, which results in the sequestering of the ODN in intracellular vesicles. A characteristic punctate fluorescent pattern of S-ODNs is observed in many cell types, with little or no fluorescence observed in the nucleus. When a cloned rat embryo fibroblast cell line was treated with 20 μ*M* of a fluorescein-tagged 15-mer PNA, we observed predominantly a cytoplasmic/perinuclear punctate fluorescent pattern (Figure 6A). It should be noted that the fluorescent vesicles seen in the nucleus are either above or below the plane of the nucleus itself. Often the cytoplasm of these cells has a weak uniform fluorescence which may indicate some freely diffusible PNA. However, it is clear that the majority of the PNA is contained in vesicles, and as such is segregated from the rest of the cytoplasm. Also, this pattern of PNA uptake was not unique to these cells, as we have observed similar punctate staining patterns with a variety of cell types.

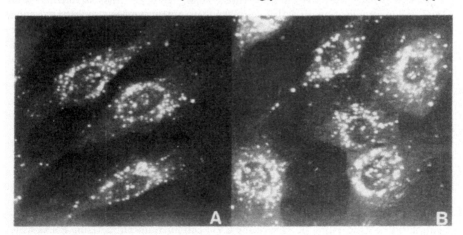

Figure 6 Fluorescent microscopy of cloned rat embryo fibroblast cells incubated for 4 h with either: (A) 20 μ*M* fluorescein-tagged 15-mer PNA or (B) 5 μ*M* fluorescein-tagged 15-mer S-ODN.

In general, the cellular fluorescent pattern of PNAs and *S*-ODNs were identical (Figure 6 A and B). However, the efficiency of the uptake of fluorescently labeled PNAs by several human and rodent cell types was typically no better, and often not as efficient when compared to the *S*-ODNs. To observe a similar intensity of fluorescent staining shown in Figure 6, a fourfold-higher concentration of the PNA relative to the *S*-ODN was required. Although the feeding experiment in Figure 6 was monitored for only 4 h, similar results were obtained with overnight incubation of the PNA or *S*-ODN with the cells.

One potential criticism of these studies could be that monitoring only the fluorochrome was not indicative of the subcellular localization of the ODN or PNA. To determine if the fluorescein-tagged PNAs or *S*-ODNs were metabolized, resulting in the dissociation of the fluorochrome and PNA/ODN, a laser-based capillary zone electrophoresis (CZE) assay was developed. We determined that both fluorochrome-tagged oligomers were stable during the course of the cell feeding studies (at least 24 h). In another series of studies, we examined the subcellular localization of fluorescein-tagged PNAs and ODNs when they were microinjected in the cytoplasm of a variety of human and rodent cell types. In these studies, we found a very rapid movement of both the ODN and PNA from the cytoplasm to the nucleus. Again the stability of the agent was confirmed by the CZE method. Of greater significance, this suggested that if either compound could leak out of the cytoplasmic endosome, it would rapidly translocate to the cell nucleus. Since we never observed nuclear staining following any cell-feeding study, we believe that the leak rate of PNAs or ODNs form endosomes is not significant.

To circumvent the problems associated with uptake of PNAs, we have bypassed the barrier of the cell membrane by performing cell microinjection experiments. By microinjection we can introduce the PNA into the nucleus and monitor its effects on expression of specific genes. We have analyzed the effects of PNAs on both an endogenous gene and one introduced simultaneously with the PNA. For all of these experiments, the expression of a control gene, β-galactosidase, which was cointroduced with the PNA, was monitored. This latter gene allowed us to monitor the gene specificity of the compound and allowed us to identify successfully injected and still metabolically active cells. Based on ease of detection and uniqueness to all cell types, we chose T-antigen as our model gene. For endogenous studies we used TSA8 cells, which express a temperature-sensitive T-antigen. In experiments using TSA8 cells, the cells were maintained at the restrictive temperature (38.8°C) prior to microinjection, and shifted to the permissive temperature (33°C) following injection, such that new T-antigen protein synthesis could be examined by immunohistochemistry. In these experiments, PNAs were synthesized complementary to a target sequence in the coding region of the T-antigen mRNA known to be permissive for antisense intervention.[14] We have seen inhibition of T-antigen expression following microinjection of approximately 1 μ*M* antisense PNA (estimated intracellular concentration).[2] Injection of control PNAs which were not complementary to T-antigen RNA did not inhibit T-antigen expression. Because nonspecific inhibition of our control gene, β-*galactosidase*, was seen at levels of PNA only fivefold higher, it was not possible to obtain complete sequence-specific inhibition of T-antigen in all cells.

While catalytically acting ODNs have been shown to be capable of inhibiting both RNA translation initiation and elongation *in vivo*, we wondered whether or not the steric-acting nature of the PNA might limit its use to certain RNA regions. A known PNA binding site was cloned into the 5′ untranslated region of the T-antigen gene to evaluate the ability of a PNA to block gene expression in this region of the RNA. As before, we monitored the antisense effects of our PNAs by measuring both T-antigen and β-galactosidase gene expression following CV-1 cell microinjection. Using PNAs which were antisense to this target sequence, we observed 100% inhibition of T-antigen expression at 1 μ*M* PNA, and decreasing inhibition at lower PNA concentrations. Therefore, the 5′ untranslated region of the RNA, where a PNA could block assembly or initiation of the translation machinery, may be a more effective target region for PNAs than in an exonic sequence. In a similar experiment, when the 5' splice-donor sequence of the large T-antigen hnRNA was targeted for binding by an antisense PNA, inhibition of T-antigen expression was similarly observed. In this case a slightly higher dose (2 μM) was required to obtain complete inhibition.

Unlike all of the steric blocking agents that have been described in the literature and tested in our laboratory, PNAs are unique in that they can sterically inhibit both RNA translation initiation and elongation *in vivo*. This difference once again points to the greater avidity of the PNA for its target sequence, when compared to an ODN. In addition, this likely represents a feature that is novel to the PNA; the ability to bind to a single-stranded RNA as a 2:1 complex. While the potent binding nature might be cause for concern, the sequence dependence of our antisense compound suggests that promiscuity is not a problem with the PNAs. Of greater concern, however, is the high intracellular concentrations that are required to observe gene-specific inhibition using PNAs. The steric-blocking mechanism by which PNAs operate may require that PNAs contain additional functionalization to evoke a catalytic mechanism. For this type of compound one could envision a significant decrease in dose requirements and still maintain high potency.

The ideal antisense compound for a therapeutic agent, or even one used to block gene expression in cell culture or animal model systems, would be freely cell permeable, stable to degradation, have high affinity and specificity for the target site, and would work catalytically. While PNAs have been shown to be serum stable and have high affinity and gene specificity, their poor bioavailability precludes their current use for therapeutic intervention for human diseases. Considerable effort is underway to chemically modify the physicochemical properties of our PNAs to render them more bioavailable. These studies will aim to alter charge and compound volume, and to increase the lipophilicity of the PNAs. As an alternate approach various types of liposomes will be used to formulate PNAs to allow better penetration when exposed to human cells. In either experimental scheme, an increase in the bioavailability of PNAs will allow these compounds to be tested as a therapeutic agent in the near future.

REFERENCES

1. Nielsen, P. E., Egholm, M., Berg, R. H., and Buchardt, O., Sequence-selective recognition of DNA by strand displacement with a thymine-substituted polyamide, *Science*, 254, 1497, 1991.
2. Hanvey, J. C., Peffer, N. J., Bisi, J. E., Thomson, S. A., Cadilla, R., Josey, J. A., Ricca, D. J., Hassman, C. F., Bonham, M. A., Au, K. G., Carter, S. G., Bruckenstein, D. A., Boyd, A. L., Noble, S. A., and Babiss, L. E., Antisense and antigene properties of peptide nucleic acids, *Science*, 258, 1481, 1992.
3. Peffer, N. J., Hanvey, J. C., Bisi, J. E., Thomson, S. A., Hassman, C. F., Noble, S. A., and Babiss, L. E., Strand-invasion of duplex DNA by peptide nucleic acid oligomers, *Proc. Natl. Acad. Sci. U.S.A.*, 90, 10648, 1993.
4. Egholm, M., Nielsen, P. E., Buchardt, O., and Berg, R. H., Recognition of guanine and adenine in DNA by cytosine and thymine containing peptide nucleic acid (PNA), *J. Am. Chem. Soc.*, 114, 9677, 1992.
5. Kim, S. K., Nielsen, P. E., Egholm, M., Buchardt, O., Berg, R. H., and Norden, B., Right-handed triplex formed between peptide nucleic acid PNA-T_8 and poly (A) shown by linear and circular dichroism spectroscopy, *J. Am. Chem. Soc.*, 115, 6477, 1993.
6. Nielsen, P. E., Egholm, M., Berg, R. H., and Buchardt, O., Sequence specific inhibition of DNA restriction enzyme cleavage by PNA, *Nucleic Acids Res.*, 21, 197, 1993.

7. Egholm, M., Buchardt, O., Christensen, L., Behrens, C., Frier, S. M., Driver, D. A., Berg, R. H., Kim, S. K., Norden, B., and Nielsen, P. E., PNA hybridizes to complementary oligonucleotides obeying the Watson-Crick hydrogen-bonding rules, *Nature*, 365, 566, 1993.

8. Almarsson, O., Bruice, T. C., Kerr, J., and Zuckermann, R. N., Molecular dynamics calculations of the structures of polyamide nucleic acid DNA duplexes and triple helical hybrids, *Proc. Natl. Acad. Sci. U.S.A.*, 90, 7518, 1993.

9. Almarsson, O. and Bruice, T. C., Peptide nucleic acid (PNA) conformation and polymorphism in PNA-DNA and PNA-RNA hybrids, *Proc. Natl. Acad. Sci. U.S.A.*, 90, 9542, 1993.

10. Young, S. L., Krawczyk, S. H., Matteucci, M. D., and Toole, J. J., Triple helix formation inhibits transcription elongation *in vitro*, *Proc. Natl. Acad. Sci. U.S.A.*, 88, 10023, 1991.

11. Hagerman, P. J., Sequence-directed curvature of DNA, *Annu. Rev. Biochem.*, 59, 755, 1990.

12. Crothers, D. M., Gartenberg, M. R., and Shrader, T. E., DNA bending in protein-DNA complexes, *Methods Enzymol.*, 208, 118, 1991.

13. Lane, D., Prentki, P., and Chandler, M., Use of gel retardation to analyze protein-nucleic acid interactions, *Microbiol. Rev.*, 56, 509, 1992.

14. Wagner, R. W., Matteucci, M. D., Lewis, J. G., Gutierrez, A. J., Moulds, C., and Froehler, B. C., Antisense gene inhibition by oligonucleotides containing C-5 propyne pyrimidines, *Science*, 260, 1510, 1993.

Chapter 10

Stabilized RNA Analogs for Antisense and Ribozyme Applications

Helle Aurup, Olaf Heidenreich, and Fritz Eckstein

CONTENTS

I. INTRODUCTION

Antisense oligonucleotides have become of increasing interest as tools for the modulation of gene expression. The emphasis has been so far on the exploitation of oligo*deoxy*nucleotides, as methods for their chemical synthesis have been available for many years.[1,2] However, oligo*ribo*nucleotides have attracted more and more attention, not only because methods of synthesis have now been developed, but also because this class of compounds is endowed in the form of ribozymes, with the catalytic power to hydrolyze RNA sequence-specifically.[3] Thus oligoribonucleotides can be used either in the form of strict antisense-oligonucleotides, to block translation by formation of RNA:RNA duplexes, or as ribozymes, to hydrolyze the target RNA in a RNA:RNA duplex. As partial complementarity is required for the formation of the ribozyme/target RNA complex, a certain degree of antisense character is also contained in the action of the ribozyme. These two modes of action are sometimes not easy to differentiate; however, a prerequisite of both is the formation of RNA:RNA, and possibly sometimes of RNA:DNA complexes, is a prerequisite. It follows that the stability of such duplexes is of the utmost importance. This aspect is dealt with in the first part of this chapter. Moreover, if application of the oligoribonucleotides is envisaged for *ex vivo* (in cell culture) or *in vivo* (in animals) experiments, their stability towards the action of nucleases is another important consideration. This will be dealt with in the second part of this review.

0-8493-4778-5/95/$0.00+$.50

The review is restricted to the description of sugar and phosphate modifications which stabilize oligoribonucleotides. The description of alternative approaches to this end such as the use of peptide nucleic acid (PNA) oligomers or of self-stabilized oligomers is dealt with in other chapters of this book. More general reviews on the chemical modification of RNA can be found elsewhere.[4,5] Details on the chemical synthesis of modified as well as unmodified oligoribonucleotides are also available.[6,7]

II. THERMAL STABILITY OF OLIGORIBONUCLEOTIDES

Studies on the thermal stability of RNA:RNA and RNA:DNA duplexes have a long history. Polynucleotide phosphorylase was then used to prepare homo- or heteropolymers including those with 2'-modifications of the ribose moiety.[8,9] The modified polymers provided at that time a new series of compounds to explore the influence of the sugar-phosphate backbone on polynucleotide structure and duplex stability.

As early as 1965 it was shown with enzymatically synthesized polymers containing ribo- and 2'-deoxynucleotides that the nature of the sugar has a strong influence on the structure and thermal stability of the polynucleotide duplexes, with RNA:RNA duplexes being more thermodynamically stable than DNA:DNA duplexes.[9] It is now well established that the sugar conformation is a major determinant for the structure and the thermostability of oligo- and polynucleotides.[10,11] A RNA:RNA duplex adopts an A- or A'-type helix, where the ribose is predominantly in the 3'-endo conformation; DNA:DNA duplexes, on the other hand, exist in several helical conformations of which the B form is most prevalent under physiological conditions, where the ribose adopts mainly the 2'-endo conformation. It is now accepted that the purine-rich strand determines duplex stability with a purine-rich RNA strand conferring the highest stabilization. Stability follows the general rule **RNA**:RNA > **RNA**:DNA > **DNA**:DNA > **DNA**:RNA, where the bold type indicates the purine-rich strand.[12-14] Thermal stabilities of DNA:DNA and RNA:RNA duplexes can now be predicted using next-neighbor rules to calculate the free energy of formation.[15-17] Most thermostability studies are now performed with oligonucleotides consisting of 11 to 15 nucleotides. In cells, however, the target is not that short, but is typically a RNA of much larger size which also possesses significant secondary structure. This has been shown to dramatically influence binding of the antisense oligonucleotide.[18] The potency of 2'-modified antisense oligonucleotides in inhibiting gene expression *in vivo* has been shown to correlate directly with the affinity of a given 2'-modification for its complementary RNA.[19] Thus the type and stability of the duplex is an important design feature.

A. OLIGORIBONUCLEOTIDES WITH A MODIFIED SUGAR RESIDUE
1. 2'-O-Alkylated Oligonucleotides

The sugar conformation of a nucleoside is determined by the size and by the polarity of the 2'-substituent. A linear relationship exists between the electronegativity of the 2'-ribose modification and the population of 3'-endo conformation in the decreasing order 2'-fluoro > 2'-O-methyl > 2'-hydroxyl > 2'-deoxy > 2'-amino (Figure 1).[20,21] The very first thermal stability studies on 2'-O-alkyl-modified polymers were reported for duplexes containing unmodified or fully 2'-O-methyl-modified homoribopolymers.[22] From these data it was clear that the most stable duplex was obtained when both strands were substituted with the 2'-O-methyl group. However, whereas a poly(rA):poly(rU) duplex containing 2'-O-methyl modifications in the pyrimidine strand was stabilized, compared to the unmodified duplex, the duplex with the 2'-O-methyl modification in the purine-containing strand was destabilized. It was also shown that such modifications did not inhibit formation of the characteristic A-type RNA:RNA helix. The effect of 2'-O-methylation on the thermal stability of oligoribo- or oligodeoxynucleotides of defined

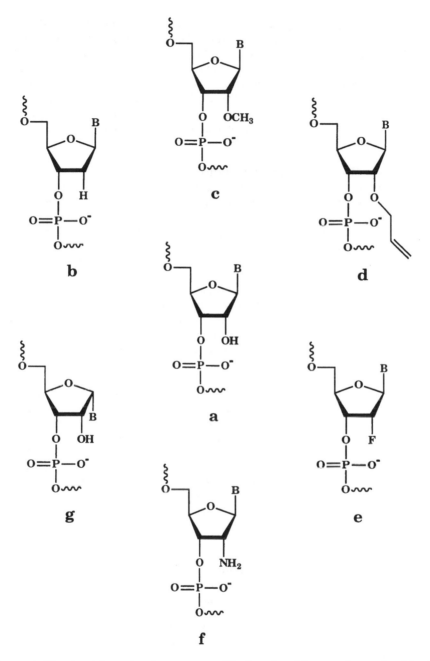

Figure 1 Structural formulae for sugar-modified nucleosides. a, ribonucleoside; b, 2'-deoxynucleoside; c, 2'-O-methyl-2'-deoxynucleoside; d, 2'-O-allyl-2'-deoxynucleoside; e, 2'-fluoro-2'-deoxynucleoside; f, 2'-amino-2'-deoxynucleoside; g, α-ribonucleoside. B-nucleo base.

sequence was also examined.[23,24] This data was consistent with that obtained from the 2'-O-methylated homopolymers, i.e., an increase in stability is obtained when the modification is incorporated into the pyrimidine-rich strand and a slight decrease in stability when the modification is in the purine-rich strand.

The size of the 2'-O-alkyl group also has an influence on the thermal stability. Experiments for various 2'-O-alkyl-modified oligonucleotides with a complementary unmodified oligoribonucleotide indicate the following order of stability: 2'-O-methyl > 2'-O-allyl > 2'-OH > 2'-O-butyl > 2'-O-pentyl > 2'-O-(3,3-dimethylallyl).[25,26] Nonspecific interactions to nucleic acid-binding proteins are probably lower for 2'-O-allyl-containing oligomers than for those bearing a 2'-O-methyl group.[27] In conclusion, the data available at present concerning 2'-O-alkyl-modified oligonucleotides indicate that although the 2'-O-allyl group does not confer as much thermal stability as the 2'-O-methyl group to a duplex, it is considered slightly more nuclease stable.

2. 2'-Fluoro-Modified Oligonucleotides

The synthesis of poly(2'-fluoro-2'-deoxyuridylic acid) was reported back in 1972 by using the 2'-fluoronucleoside diphosphate and polynucleotide phosphorylase.[28] It was reported that poly(rA):poly(2'-fluoro-2'-deoxyuridylic acid) had a thermal stability 17°C higher than poly(rA):poly(rU) and 19°C higher than poly(rA):poly(dU). The geometry of the A-type poly(rI):poly(rC) helix was maintained if one or both strands were fully substituted by 2'-fluoro-ribose.[29,30] In addition, it has also been shown that doubly modified duplexes possess the highest T_m and that 2'-fluoro modification of the purine or the pyrimidine strand has the same stabilizing effect on the duplex. Poly(2'-fluoro-2'-deoxyadenylic acid) was shown to have physical properties, such as UV and CD spectra resembling those obtained with poly(rA), but not poly(dA).[31] The thermal stability of the duplex formed with poly(rU) was slightly increased compared to that of the unmodified duplex. The incorporation of the 2'-fluoro as well as the 2'-amino function into RNA of a size beyond that obtainable by chemical synthesis can nowadays be achieved with the T7 RNA polymerase.[32]

Increasing the content of 2'-fluoro modification in an oligodeoxynucleotide leads to a consistently additive increase in thermal stability of a pentadecamer when annealed to its oligoribonucleotide target.[33] In agreement with this data, duplexes of oligodeoxynucleotides with the corresponding oligoribonucleotides exhibit CD spectra which are increasingly like that of an A-type helix with an increasing number of 2'-fluoronucleosides in the deoxy strand of the duplex. The stabilizing effect of 2'-fluoro modifications on RNA:DNA duplexes is superior to that of 2'-O-methyl substitutions. In order to increase stability towards nucleases, phosphorothioate linkages were introduced in addition. The fully 2'-fluoro-phosphorothioate oligonucleotide showed a higher thermal stability than both the unmodified phosphodiester and the 2'-O-methyl-phosphodiester oligonucleotide.

3. 2'-Amino-Modified Oligonucleotides

There are not much data available on the thermal stability of 2'-amino-modified oligonucleotides. Hobbs et al.[34] showed that the thermostability of the duplex poly(2'-amino-2'-deoxycytidylic acid):poly(rI) was dramatically decreased compared to that of the unmodified duplex. No duplex formation between poly(2'-amino-2'-deoxyuridylic acid) and poly(rA) was observed.[34] Ikehara et al.[35] showed that poly(2'-aminoadenylic acid) formed a similar three-stranded complex, with poly(rU), as did poly(rA):2poly(rU). The thermal stability of the modified complex was only slightly decreased relative to the unmodified complex. We have examined duplex formation of oligoribonucleotides of nine nucleotides in length containing between one and three 2'-aminocytidines.[36] The modification had a considerable destabilizing effect on duplex stability, an effect which has also been described for oligodeoxynucleotides.[37] Although the destabilizing effect on RNA duplexes would be consistent with the predominant 2'-endo conformation of 2'-aminonucleosides, this can not explain the destabilization observed with oligodeoxynucleotides.

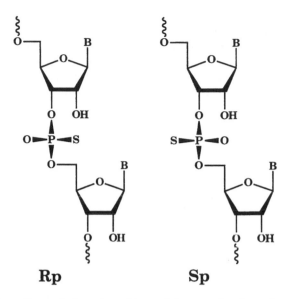

Figure 2 Configurations of phosphorothioate linkages. Configurations are indicated by Rp or Sp.

4. α-Oligoribonucleotides

The α-ribonucleoside analogs differ from the natural β-ribonucleosides in the configuration of C1'. The nucleo base is attached to the sugar in the "up" position in the β configuration and in the "down" configuration in the α-ribonucleoside (Figure 1). Duplex formation was seen between an α-anomeric and the complementary β-anomeric sequence in parallel orientation.[38] No duplex formation was obtained with the β-anomer in the antiparallel orientation. A comparison of the stabilities of complexes between eight modified oligonucleotides and the same 12-mer RNA target revealed that the complex with the α-oligoribonucleotide was by far the weakest, surprisingly much more so than that with the α-oligodeoxynucleotide.[39]

B. PHOSPHOROTHIOATE-MODIFIED OLIGORIBONUCLEOTIDES

The introduction of a modification into the phosphodiester linkage of an oligonucleotide generally serves two purposes, first to render the oligonucleotide resistant to nucleases and second to facilitate the transport by endocytosis across the cell membrane.[2] A methylphosphonate or an alkylphosphotriester linkage adjacent to a 2'-hydroxyl group is hydrolytically unstable under the conditions of the ammonia treatment required in the synthesis of oligoribonucleotides.[7] They are thus in contrast to the oligodeoxynucleotide series, unavailable for the oligoribonucleotide series, although they might be the most desirable analogs because of their neutral character. This leaves essentially the phosphorothioate modification (Figure 2) as the only one of practical value for oligoribonucleotides at present.

When a nonbridging oxygen atom in the phosphodiester linkage is replaced, by chemical synthesis, by sulfur it results in the formation of a Rp and Sp diastereomeric mixture. This is in contrast to RNA and DNA polymerase-mediated synthesis which yields only the Rp diastereomer of the phosphorothioate internucleotidic linkage.[40] In this configuration the sulfur points into the major groove of the helix.[41] The presence of diastereomers leads to problems in elucidating a structure-activity relationship in antisense oligonucleotide applications. Results obtained with phosphorothioate oligodeoxynucleotides and polynucleotides indicate that the complexes with the phosphorothioate of the Rp configuration are less stable than both the unmodified and the oligonucleotide with the phosphorothioate of the Rp configuration.[42-44] However, the influence of the phosphorothioate does not only depend on its configuration, but also on its position whether it is located 5' to a pyrimidine or a purine nucleoside.

Caution has to be exercised in extrapolating from the data obtained from oligodeoxynucleotides to oligoribonucleotides, particularly as duplexes with phosphorothioate-containing homoribopolymers such as poly[r(AS-US)] were as stable as the parent duplexes.[45,46] These polymers contained only the Rp diastereomer of the phosphorothioate as they were prepared enzymatically. A duplex formed of a chemically synthesised 14-nucleotide-long phosphorothioate oligoribocytidine with poly(rI) was less stable (T_m, 37°C) than the corresponding duplex with an unmodified oligoribonucleotide (T_m, 46°C).[47] When a phosphorothioate linkage was introduced in either an oligoribo- or an oligodeoxynucleotide the thermal stability was decreased compared to the corresponding unmodified oligonucleotide, but the decrease was more dramatic when the modification was introduced into the oligodeoxynucleotides. When this modification is combined with the 2'-fluoro modification the stability is increased and is in fact higher than that of the corresponding unmodified oligoribonucleotides.[33] The thermal stability of some fully phosphorothioated 20-mer duplexes decreased in the following order: RNA:RNA > RNA:DNA > RNA:RNA (phosphorothioate) > RNA:DNA (phosphorothioate).[48] As no diastereomerically pure fully phosphorothioated oligoribonucleotides are available yet by chemical synthesis, it is too early to say whether this decrease in stability is due to the mixture of diastereomers. For oligodeoxynucleotides the problem of chirality of the phosphorothioates has been circumvented by the synthesis of phosphorodithioates.[49] These have not been described for oligoribonucleotides yet. In favorable cases, however, oligoribonucleotides can be synthesized with only one diastereomer if the mixture can be separated. This has been successful for two hammerhead ribozymes where separation of the diastereomeric mixture was achieved by reverse-phase chromatography.[50,51]

III. STABILITY OF MODIFIED OLIGORIBONUCLEOTIDES TOWARDS NUCLEASES

One of the hurdles for the exogenous delivery of oligoribonucleotides is their high susceptibility to degradation by extra- and intracellular nucleases and especially by 2'-hydroxyl-dependent RNases such as pancreatic RNaseA. An exogenously delivered oligonucleotide should not only exhibit a considerable half-life time in the extracellular medium, but should also be stable in several cellular compartments like endosomes, cytoplasm, and nucleus. As the nucleases present in serum are different from nucleases present in the cytoplasm or the nucleus of the cell it is important to keep into account that the stability of a given oligoribonuceotide might change, depending on the environment.

The first step in examining the influence of a chemical modification on the stability of an oligonucleotide are degradation assays with specific nucleases like phosphodiesterases or RNases. These systems are well defined and easy to interpret. However, despite their valuable information about enzyme-specific degradation, such systems do not necessarily reflect the *ex vivo* (i.e., cell culture) or *in vivo* (i.e., animals) conditions. More suitable systems to predict the *ex vivo* or *in vivo* stability of a given oligonucleotide are those consisting of cell culture medium, serum, or cell extracts. Such test systems contain sets of different nucleases. Serum, for example, is known to contain not only pyrimidine-specific RNases, but also a 3'-exonuclease activity.[52-54] Cell culture medium usually contains 5 to 20% heat-inactivated fetal calf serum (FCS). In this case FCS is incubated for between 20 min to 2 h at 55°C before addition to the medium. Therefore, the nuclease activities present in serum are not only diluted, but also partially heat inactivated.[54] Extracts from whole cells are prepared, for instance, by opening the cells by sonication.[55] Cytoplasmic or nuclear extracts are obtained by limited destruction of the cell membrane followed by several centrifugation steps. The intracellular stability of oligonucleotides is examined by another approach whereby whole cells are either simply incubated with fluorescently labeled oligonucleotides or the oligonucleotides are applied to cells by injection or by transfection methods.[56] However, only very limited data are available on the intracellular stability of oligoribonucleotides (see below).

A. STABILIZATION OF ANTISENSE OLIGORIBONUCLEOTIDES
1. Phosphorothioate Linkages
Several modifications of the sugar-phosphate backbone are known to interfere with nuclease action, thereby considerably increasing the stability of the corresponding oligo-nucleotide towards degradation. In particular, phosphorothioate linkages are known to protect oligonucleotides against such nuclease degradation.[40,57,58] This is demonstrated by the fact that phosphorothioate analogs of polyribonucleotides, such as poly[r(AS-US)] or poly[r(IS-CS)], are 10 to 100 times less sensitive to ribonucleases and also exhibit an increased stability in different sera compared to the unmodified compounds.[45,59,60] The degree of protection against a particular nuclease is often dependent on the configuration of the phosphorothioate linkage. While the Sp-diastereomer of the phosphorothioate is hydrolyzed by spleen phosphodiesterase 100-fold slower than a phosphate linkage, the Rp-diastereomer is completely resistant to this enzyme. On the other hand, however, snake venom phosphodiesterase cleaves the Rp-diastereomer only tenfold slower than the unmodified phosphate diester, but is almost unable to hydrolyze the Sp-diastereomer.[40] Similar differences exist for hydrolysis by RNaseT1.[61,62] As chemically synthesized phosphorothioate oligonucleotides are mixtures of diastereomers, their susceptibility to degradation by a nuclease will differ from one molecule to the other. However, introduction of phosphorothioate linkages usually increases the stability of a oligoribonucleotide. A 14-nucleotide-long phosphorothioate oligoribocytidine exhibited some tenfold increased stability against degradation by several nucleases compared to the unmodified oligomer.[47]

2. Sugar Modifications
Ribonucleases like RNaseA or RNaseT1 employ the 2'-hydroxyl group of RNA for the hydrolysis of phosphate diesters. These enzymes catalyze a two-step transesterification reaction where the first step involves attack of the 2'-oxygen at the phosphorus resulting in strand cleavage with the formation of a 2',3'-cyclic phosphate and a free 5'-hydroxyl group. In the second step the cyclic phosphate is hydrolyzed to a nucleoside 3'-phosphate. Thus, modification of ribonucleosides at the 2' position should result in complete resistance against this class of nucleases. The simplest 2'-modification is replacement of the 2'-hydroxyl group by a hydrogen, resulting in 2'-deoxyribonucleotides. Oligodeoxy-nucleotides are, of course, not substrates for 2'-OH-specific RNases. This type of modification has been extensively examined for the synthesis of chemically modified ribozymes (see below).

Another 2'-modification, also employed by nature, is the 2'-O-methyl group. Substitution of all nucleosides by their 2'-O-methyl analogs stabilizes the resulting oligonucleotide not only against degradation by 2'-OH-dependent RNases, but also against several other nucleases such as S1 nuclease, mung bean nuclease, exonuclease III, and DNaseI. However, such oligomers are still susceptible to degradation by micrococcal nuclease, nuclease P1, and snake venom phosphodiesterase.[63] 2'-O-Methyloligoribonucleotide-RNA duplexes are not substrates for RNase H.[64] Similar results have been obtained with oligomers consisting of 2'-O-allylnucleotides, which were as stable as their 2'-O-methyl counterparts, but more selective in hybridization to their target sequences.[27] Both 2' modifications were also examined for their intracellular stability.[56] 5'-Fluorescein-labeled thymidine octamers, their 2'-O-methyl, or their 2'-O-allyl analogs were microinjected into Rat2 cells. Each of the oligonucleotides was localized in the nucleus within 5 min. Degradation of the oligonucleotides was determined by the decrease of the fluorescent signal over time. Whereas the 2'-O-methyl-modified oligonucleotide exhibited an eight-fold increase in stability compared to the oligodeoxynucleotide, the 2'-O-allyl oligonucleotide was 60 times more stable than the latter.

The polymerization of 2'-fluorouridine diphosphate by polynucleotide phosphorylase resulted in a polynucleotide being completely resistant against pancreatic ribonuclease A and DNaseI, but not against phosphodiesterases and DNaseII.[28] Similar results were obtained for polynucleotides containing 2'-amino-2'-deoxypyrimidine nucleosides.[34] Like 2'-O-methyloligonucleotides, 2'-fluorooligonucleotides do not support RNase H cleavage.[33,65] Oligonucleotides containing exclusively 2'-fluoronucleosides were degraded after incubation for 5 h in heat-inactivated FCS, whereas an oligonucleotide of the same sequence containing 2'-fluoronucleosides in combination with phosphorothioate linkages remained stable for at least 24 h.

Another class of analogs modified at the sugar moiety are α-anomeric oligonucleotides (see Section II.A.4). Oligoribonucleotides consisting of α-anomeric units exhibited a very low susceptibility to several nucleases, including ribonuclease A.[66]

B. STABILIZATION OF RIBOZYMES

Trans-cleaving ribozymes like the hairpin or the hammerhead ribozymes (Figure 3) are considered to have great potential as anticancer and antiviral therapeutics. Endogenously delivered to cells, they have been successfully applied to the inhibition, e.g., of HIV replication.[68,69] However, as in the case of simple antisense oligoribonucleotides, the exogenous delivery of unmodified ribozymes without any supporting systems like liposomes seems to be almost impossible because of their low stability to nucleases. Thus, it is not surprising that most of the recent efforts to stabilize oligoribonucleotides by chemical modifications have been undertaken with hammerhead ribozymes.

Stabilization of ribozymes by chemical modifications is somewhat more demanding than the stabilization of simple antisense oligonucleotides. The reason for this is the possibly negative effect of the chemical modification on the cleavage activity of the ribozyme. Therefore, any chemical modification has to be carefully examined not only for its benefits in stability, but also for a possible detrimental effect on activity. Unfortunately, most chemical modifications of ribozymes are accompanied by a decrease in cleavage activity.

1. Phosphorothioate Linkages

The influence of phosphorothioate linkages on the activity and stability of ribozymes has been studied by several groups. Partial incorporation of nucleoside α-thiotriphosphates by run-off transcription into hammerhead ribozymes revealed several phosphate linkages where a replacement by a phosphorothioate linkage seriously impairs the cleavage activity.[70,71] Therefore, it is impossible to completely substitute the phosphates by phosphorothioates without abolishing the catalytic activity of this ribozyme. However, the activity of a trans-cleaving hairpin ribozyme was less than twofold decreased by the substitution of all guanidylate and cytidylate residues by their phosphorothioate analogs. This ribozyme exhibited a tenfold increased stability in cytoplasmic and nuclear extracts compared to the unmodified ribozyme.[72] As it had been synthesized by run-off transcription with phage RNA polymerases, all phosphorothioate linkages incorporated into the ribozymes are of the Rp configuration. As the chemical synthesis of phosphorothioate oligoribonucleotides yields a mixture of Rp- and Sp-diastereomers, the results obtained for phosphorothioate ribozymes synthesized by run-off transcription might not necessarily be the same as for those obtained by chemical synthesis.

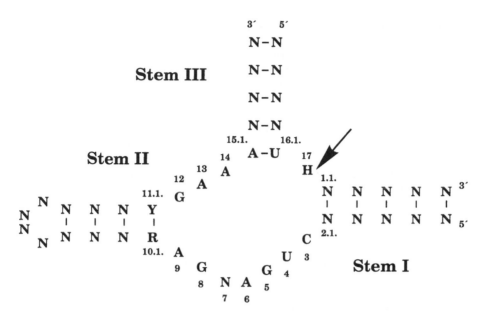

Figure 3 Hammerhead ribozyme. The ribozyme is shown complexed to its substrate. The cleavage site is indicated by an arrow. The nomenclature is according to Reference 67. R, purine nucleoside; Y, pyrimidine nucleoside; H, any nucleoside except a guanosine.

2. 2′ Modifications at the Sugar Moiety

The first 2′-modified nucleosides employed to elucidate the necessity of 2′-hydroxyl groups for the hammerhead ribozyme-catalyzed RNA cleavage were 2′-deoxynucleotides.[55] Extensive substitution of all nucleosides except seven purine nucleosides located in the central core of the ribozyme by their 2′-deoxy analogs caused a 100-fold decrease in catalytic efficiency compared to the unmodified ribozyme. However, the 2′-deoxynucleotide-containing ribozyme was some 1000-fold more stable to RNaseA degradation and some 10,000-fold more stable in yeast cell extract. Chimeric ribozymes containing 2′-deoxynucleotides in the hybridizing arms and ribonucleotides in the core region and in stem-loop II exhibited a fourfold increased catalytic efficiency compared to the unmodified ribozyme.[73] This ribozyme was also examined for its intracellular stability. After transfection of H9 lymphocytes with ribozyme-Lipofectin™ complexes the 2′-deoxymodified ribozyme was about threefold more stable than the unmodified one. In addition, the chimeric ribozyme was also relatively stable in serum-free cell culture medium. Hendry and collaborators[74] reported a 15-fold increased catalytic efficiency for a ribozyme where only the central core contained ribonucleotides. The hybridizing arms and stem-loop II consisted of DNA. However, no remarkable stabilization in serum-containing solutions could be observed. Even in the presence of only 0.1% FCS the half-life time was around 1 to 2 min. Goodchild[75] reported that while after 6 min of incubation in 1% FCS an unmodified ribozyme was completely degraded, replacement of all nucleosides in the hybridizing arms by 2′-O-methylnucleosides resulted in a fourfold increased stability. The cleavage efficiency of this ribozyme was about twofold higher than that of the unmodified one.

Recently, chimeric ribozymes have been synthesized which contain 2′-deoxynucleotides in the hybridizing arms and in stem II.[76] In addition, all linkages in these regions and three linkages in the central core were phosphorothioates. This ribozyme exhibited a 15-fold decreased cleavage specificity compared to the unmodified ribozyme. Such phosphorothioate chimeric ribozymes were still unstable in 10% FCS. However, when this ribozyme was incubated in 20% human serum for 3 h, 50% of the ribozyme remained intact. In contrast, the unmodified ribozyme was already degraded after 5 min to 90%. This difference between human serum and FCS was explained by the smaller amount of endoribonucleases present in the human serum.

Ribozymes containing 2'-O-alkyl residues have also been examined for their catalytic efficiency and stability.[77] On the one hand, replacement of all but six nucleosides located in the central core, by their 2'-O-allyl analogs, resulted in a ribozyme with a fivefold reduced catalytic efficiency compared to its unmodified counterpart. On the other, the 2'-O-allyl-modified ribozyme was only 50% degraded after 2 h of incubation in bovine serum, whereas the unmodified ribozyme was already completely degraded after 1 min.

Substitution of all pyrimidine nucleosides by their 2'-amino or 2'-fluoro analogs yielded hammerhead ribozymes with an at least 10^6 times improved resistance against RNaseA degradation and a more than 1000-fold increased stability in rabbit serum compared to the unmodified ribozyme. The catalytic efficiency of such modified ribozymes was 25- to 50-fold decreased.[78]

Determination of the kinetic parameters of modified ribozymes mentioned so far were performed with short oligoribonucleotides of 12 to 30 nucleotides in length as substrates. In a somewhat different approach, the influence of 2'-modifications and terminal phosphorothioate linkages on the ribozyme-mediated cleavage of a 1000-nucleotide-long transcript of the HIV-LTR was examined.[79] A ribozyme containing 2'-fluoro-2'-deoxypyrimidine nucleosides instead of ribopyrimidine nucleosides and four terminal phosphorothioate linkages cleaved such a substrate some sevenfold less efficiently than its unmodified counterpart. However, while the unmodified ribozyme was almost completely degraded after 10 min incubation in cell culture supernatant containing 10% heat-inactivated FCS, the chemically modified ribozyme was completely stable for at least 1 h. Two uridine residues located in the central core were identified to be responsible for the observed decrease in catalytic efficiency.[80] Replacement of these by 2'-amino-2'-deoxyuridines restored the catalytic efficiency. Whereas the unmodified ribozyme was completely degraded in less than 2 min, such modified ribozymes were stable for at least 24 h in untreated FCS. Ribozymes containing only 2'-fluoropyrimidine nucleosides, but no terminal phosphorothioate linkages, revealed half-life times in FCS between 15 min and 1 h. The degree of stability conferred by the phosphorothioates was more pronounced for a ribozyme where the 3'-end was single stranded than for one where it was double stranded. Determination of half-life times for these ribozymes was limited by the stability of the label attached to the 5'-end, as even a fluorescein label was slowly removed on long incubations in the serum. In summary, these results demonstrate that ribozymes can be stabilized by chemical modification by factors of 1000 or more without sacrificing catalytic efficiency.

IV. *EX VIVO* INHIBITION OF GENE EXPRESSION BY CHEMICALLY MODIFIED OLIGORIBONUCLEOTIDES

Ribozymes are mainly introduced into cells by endogenous delivery and only a very limited number of reports have been so far published on their exogenous delivery. None of these have been carried out with chemically modified ribozymes. A summary of examples is presented in Heidenreich et al.[80] However, some examples exist for the exogenous application of antisense oligoribonucleotides. Thus, the antiviral activity of phosphorothioate oligoribonucleotides has recently been investigated.[48] An oligoribonucleotide targeted against the splice acceptor site of the HIV-1 tat gene was added to cell culture either simultaneously with the virus or 24 h postinfection. A sense and a mismatched oligoribonucleotide served as controls. When added simultaneously, all three oligoribonucleotides inhibited the viral replication independent of the sequence to the same extent. When added 24 h postinfection only the antisense oligoribonucleotide exhibited antiviral activity with a half-maximal inhibition (IC_{50}) at 5 μM, suggesting a sequence-specific effect. However, a phosphorothioate oligodeoxynucleotide of the same sequence inhibited viral replication with an IC_{50} of 1 μM. The authors explained this difference by the inability of the oligoribonucleotide to induce RNase H-mediated RNA cleavage.

2'-O-Methyloligonucleotides have been examined for the inhibition of HIV-1 replication in freshly infected MT4 cells.[81] Oligonucleotides containing unmodified phosphodiester linkages failed to inhibit HIV replication, whereas those containing either 5'-terminal or exclusively phosphorothioate linkages displayed significant antiviral activity. However, the observed effects did not seem to be sequence specific, as a phosphorothioate 2'-O-methyl-homooligomer containing only inosine was as effective as those oligonucleotides targeted against the primer binding site or the tat splicing acceptor site of HIV.

2'-O-Methyl phosphorothioate oligonucleotides were also employed to elucidate the inhibition mechanism of antisense oligonucleotides after exogenous delivery by lipofection.[82] Two phosphorothioate oligodeoxynucleotides targeted either against the translation initiation codon or the 3' untranslated region of the ICAM-1 mRNA inhibited efficiently the expression of ICAM-1. A phosphorothioate 2'-O-methyloligonucleotide targeted against the former site was three times less efficient. Another phosphorothioate 2'-O-methyloligonucleotide directed to the latter site failed to inhibit ICAM-1 expression. Because 2'-O-methyloligoribonucleotide-RNA duplexes are not substrates for RNase H,[64] this result suggests that in contrast to the 3' untranslated region the inhibition obtained with oligonucleotides targeted against the start codon was not due to RNase H-mediated cleavage.

Another study compared the influence of several 2'-modifications on the inhibition of Ha-ras expression.[19] As in the previous example, oligonucleotides were administered by lipofection. A 17-nucleotide-long phosphorothioate oligonucleotide containing only 2'-deoxynucleotides exhibited an IC_{50} of 150 nM, whereas one containing seven 2'-deoxynucleotides flanked by five 2'-O-methylnucleosides on each side had an IC_{50} of 20 nM. Replacement of the 2'-O-methylnucleosides by 2'-fluoronucleosides resulted in a further decrease of the IC_{50} to 10 nM. Incorporation of 2'-O-pentyl or 2'-O-propylnucleosides yielded IC_{50} values of 150 and 70 nM, respectively. As the affinity of these oligonucleotides to the target sequence decreases in the order 2'-fluoro > 2'-O-methyl > 2'-deoxy, these results suggest a direct correlation between the affinity of a particular oligonucleotide to its target sequence and the inhibition potential of this oligonucleotide. A phosphorothioate oligonucleotide containing exclusively 2'-O-methylnucleosides failed to inhibit Ha-ras expression, suggesting a strong dependence of the inhibition on RNase H-mediated cleavage.

V. CONCLUSIONS AND PERSPECTIVES

For the successful application of antisense oligoribonucleotides to the inhibition of gene expression, several problems have to be addressed: affinity to and specificity for the target sequence, extra- and intracellular stability, availability at the supposed place of action, and, moreover, an easy and efficient synthesis of such oligomers. Incorporation of modified building blocks into oligoribonucleotides might affect each of these topics in different ways. The examples discussed in this review demonstrate the possibility to synthesize oligoribonucleotides with an improved stability against degradation by nucleases without seriously affecting their affinity or catalytic activity. First *ex vivo* studies with modified antisense oligoribonucleotides illustrate their potential to interfere with gene expression. For ribozymes such studies have not been published yet. It still has to be demonstrated if the cleavage activity of the ribozymes makes them more efficient than antisense oligonucleotides in the inhibition of gene expression.

As *ex vivo* studies with chemically modified oligoribonucleotides are encouraging, the next step will be *in vivo* application to animal models. So far there is no information available about the *in vivo* activity or pharmacokinetics of such compounds. It has to be seen whether modifications to achieve high extra- and/or intracellular stability of an oligonucleotide are necessary or whether this advantage is outweighed by an increase in toxicity. The final judgment about the suitability of an oligonucleotide for *in vivo* applications will be determined by its successful and safe application.

ACKNOWLEDGMENT

Work in the authors´ laboratory was supported by the Deutsche Forschungsgemeinschaft and the Bundesgesundheitsamt/BMFT. We thank J. Thomson for critical reading of the manuscript.

REFERENCES

1. Stein, C. A. and Cheng, Y.-C., Antisense oligonucleotides as therapeutic agents — is the bullet really magical?, *Science,* 261, 1004, 1993.
2. Milligan, J. F., Matteucci, M. D., and Martin, J. C., Current concepts in antisense drug design, *J. Med. Chem.*, 36, 1923, 1993.
3. Rossi, J. J., Ribozymes, *Curr. Opion. Biotech.*, 3, 3, 1992.
4. Heidenreich, O., Pieken, W., and Eckstein, F., Chemically modified RNA: approaches and applications, *FASEB J.* 7, 90, 1993.
5. Usman, N. and Cedergren, R., Exploiting the chemical synthesis of RNA, *TIBS,* 17, 334, 1992.
6. Gait, M. J., Pritchard, C., and Slim, G., Oligoribonucleotide synthesis, in *Oligoribonucleotides and Analogues: a Practical Approach*, Eckstein, F., Ed., IRL Press, Oxford, 1991, 25.
7. Gait, M., Oligoribonucleotides, *Antisense Research and Applications,* CRC Press, Boca Raton, FL, 1993, 289.
8. Chou, J. Y. and Singer, M. F., Deoxyadenosine diphosphate as a substrate and inhibitor of polynucleotide phosphorylase of Micrococcus luteus, *J. Biol. Chem.*, 246, 7486, 1971.
9. Chamberlin, M. J. and Patterson, D. L., Physical and chemical characterization of the ordered complex formed between polyinosinic acid, polycytidylic acid and their deoxyribo-analogues, *J. Mol. Biol.*, 12, 410, 1965.
10. Saenger, W., *Principles of Nucleic Acid Structure,* Springer-Verlag, New York, 1984.
11. Chastain, M. and Tinoco, I., Jr., Structural elements in RNA, *Progr. Nucleic Acid Res. Mol. Biol.*, vol. 41, Cohn, W. E. and Moldave, K., Eds., Academic Press, London, 1991, 131.
12. Hall, K. B., NMR spectroscopy of DNA/RNA hybrids, *Curr. Opin. Biol.*, 3, 336, 1993.
13. Hall, K. B. and McLaughlin, L. W., Thermodynamic and structural properties of pentamer DNA:DNA, RNA:RNA, and DNA:RNA duplexes of identical sequence, *Biochemistry*, 30, 10606, 1991.
14. Roberts, R. W. and Crothers, D. M., Stability and properties of double and triple helices: dramatic effects of RNA or DNA backbone composition, *Science*, 258, 1463, 1992.
15. Freier, S. M., Kierzek, R., Jaeger, J. A., Sugimoto, N., Caruthers, M. H., Neilson, T., and Turner, D. H., Improved free-energy parameters for predictions of RNA duplex stability, *Proc. Natl. Acad. Sci. U.S.A.*, 83, 9373, 1986.

16. Turner, D. H., Sugimoto, N., and Freier, S. M., RNA structure and prediction, *Annu. Rev. Biochem. Biophys. Chem.*, 17, 167, 1988.
17. Breslauer, K. J., Frank, R., Blöcker, H., and Marky, L. A., Predicting DNA duplex stability from the base sequence, *Proc. Natl. Acad. Sci. U.S.A.*, 83, 3746, 1986.
18. Ecker, D. J., Vickers, T. A., Bruice, T. W., Freier, S. M., Jenison, R. D., Manoharan, M., and Zounes, M., Pseudo-half-knot formation with RNA, *Science*, 257, 958, 1992.
19. Monia, B. P., Lesnik, E. A., Gonzalez, C., Lima, W. F., McGee, D., Guinosso, C. J., Kawasaki, A. M., Cook, P. D., and Freier, S. M., Evaluation of 2'-modified oligonucleotides containing 2'-deoxy gap as antisense inhibitors of gene expression, *J. Biol. Chem.*, 268, 14514, 1993.
20. Guschlbauer, W. and Jankowski, K., Nucleoside conformation is determined by the electronegativity of the sugar substituent, *Nucleic Acids Res.*, 8, 1421, 1980.
21. Uesugi, S., Miki, H., Ikehara, M., Iwajashi, H., and Kyogoku, Y., A linear relationship between electronegativity of 2'-substituents and conformation of adenine nucleosides, *Tetrahedron Lett.*, 42, 4073, 1979.
22. Alderfer, J. M., Tazawa, I., Tazawa, S., and Ts'o, P. O. P., Comparative studies on homopolymers of adenylic acid possesing different C-2' substituents of the furanose. Poly(deoxyadenylic acid), poly(riboadenulic acid), poly(2'-O-methyladenylic acid), and poly(2'-O-ethyladenylic acid), *Biochemistry*, 13, 1615, 1974.
23. Inoue, H., Hayase, Y., Imura, A., Iwai, S., Miura, K., and Ohtsuka, E., Synthesis and hybridization studies on two complementary nona(2'-O-methyl)ribonucleotides, *Nucleic Acids Res.*, 15, 6131, 1987.
24. Kibler-Herzog, L., Zon, G., Uznanski, B., Whittler, G., and Wilson, W. D., Duplex stabilities of phosphorothioate, methylphosphonate, and RNA analogs of two DNA 14-mers, *Nucleic Acids Res.*, 19, 2979, 1991.
25. Lamond, A. I. and Sproat, B. S., Antisense oligonucleotides made of 2'-O-alkylRNA: their properties and application in RNA biochemistry, *FEBS Lett.*, 325, 123, 1993.
26. Lesnik, E. A., Guinosso, C. J., Kawasaki, A. M., Sasmor, H., Zounes, M., Cummins, L. L., Ecker, D. J., Cook, P. D., and Freier, S. M., Oligodeoxynuclotides containing 2'-O-modified adenosine: synthesis and effects on stability of DNA:RNA duplexes, *Biochemistry*, 32, 7832, 1993.
27. Iribarren, A. M., Sproat, B. S., Neuner, P., Sulston, I., Ryder, U., and Lamond, A. I., 2'-O-Alkyl oligoribonucleotides as antisense probes, *Proc. Natl. Acad. Sci. U.S.A.*, 87, 7747, 1990.
28. Janik, B., Kotick, M. P., Kreiser, T. H., Reverman, L. F., Sommer, R. G., and Wilson, D. P., Synthesis and properties of poly 2'-fluoro-2'-deoxyuridylic acid, *Biochem. Biophys. Res. Commun.*, 46, 1153, 1972.
29. Guschlbauer, W., Blandin, M., Drocourt, J. L., and Thang, M. N., Poly-2'-deoxy-2'-fluoro-cytidylic acid: enzymatic synthesis, spectroscopic characterization and interaction with poly-inosinic acid, *Nucleic Acids Res.*, 4, 1933, 1977.
30. Kakiuchi, N., Marek, C., Rousseau, N., Leng, M., De Clercq, E., and Guschlbauer, W., Polynucleotide helix geometry and stability, spectroscopin, antigenic, and interferon-inducing properties of deoxyribose-, ribose-, or 2'-deoxy-2'-fluororibose-containing duplexes of poly(inosinic acid):poly(cytidylic acid), *J. Biol. Chem.*, 257, 1924, 1982.
31. Ikehara, M., Fukui, T., and Kakiuchi, N., Synthesis and properties of poly(2'-deoxy-2'-fluoroadenylic acid), *Nucleic Acids Res.*, 5, 1877, 1977.
32. Aurup, H., Williams, D. M., and Eckstein, F., 2'-Fluoro- and 2'-amino-2'-deoxynucleoside 5'-triphosphates as substrates for T7 RNA polymerase, *Biochemistry*, 31, 9636, 1992.
33. Kawasaki, A. M., Casper, M. D., Freier, S. M., Lesnik, E. A., Zounes, M. C., Cummins, L. L., Gonzalez, C., and Cook, P. D., Uniformly modified 2'-deoxy-2'-fluoro phosphorothioate oligonucleotides as nuclease-resistant antisense compounds with high affinity and specificity for RNA targets, *J. Med. Chem.*, 36, 831, 1993.

34. Hobbs, J., Sternbach, H., Sprinzl, M., and Eckstein, F., Polynucleotides containing 2'-amino-2'-deoxyribose and 2'-azido-2'-deoxyribose, *Biochemistry*, 12, 5138, 1973.
35. Ikehara, M., Fukui, T., and Kakiuchi, N., Synthesis and properties of poly(2'-amino-2'-deoxyadenylic acid), *Nucleic Acids Res.*, 4, 989, 1977.
36. Aurup, H., Tuschl, T., and Eckstein, F., unpublished results.
37. Kuznetsova, L. G., Romanova, E. A., Volkov, E. M., Tashlitsky, V. N., Oretskaya, T. S., Krynetskaya, N. F., and Shabarova, Z. A., Oligonucleotides containing 2'-amino-2'-deoxypyrimidine nucleosides, *J. Bioorg. Chem.* (Moscow), 19, 455, 1993.
38. Debart, F., Rayner, B., Degols, G., and Imbach, J-L., Synthesis and base-pair properties of the nuclease-resistant α-anomeric dodecaribonucleotide α-[r(UCUUAACCCACA)], *Nucleic Acids Res.*, 20, 1193, 1992.
39. Morvan, F., Porumb, H., Degols, G., Lefebvre, I., Pompon, A., Sproat, B. S., Rayner, C., Malvy, C., Lebleu, B., and Imbach, J.-L., Comparative evaluation of seven oligonucleotide analogues as potential antisense agents, *J. Med. Chem.*, 36, 280, 1993.
40. Eckstein, F., Nucleoside phosphorothioates, *Annu. Rev. Biochem.*, 54, 367, 1985.
41. Cruse, W. B. T., Salisbury, S. A., Brown, T., Cosstick, R., Eckstein, F., and Kennard, O., Chiral phosphorothioate analogues of B-DNA; the crystal structure of Rp-d[Gp(S)CpGp(S)CpGp(S)C], *J. Mol. Biol.*, 192, 891, 1986.
42. LaPlanche, L. A., James, T. L., Powell, C., Wilson, W. D., Uznanski, B., Stec, W. J., Summers, M. F., and Zon, G., Phosphorothioate-modified oligodeoxynucleotides. III. NMR and UV spectroscopic studies of Rp-Rp, Sp-Sp, and Rp-Sp duplexes, [d(GG$_s$AATTCC)]$_2$, derived from disastereomeric O-ethyl phosphorothioates, *Nucleic Acids Res.*, 14, 9081, 1986.
43. Eckstein, F. and Jovin, T. M., Assignment of resonances in the phosphorus-31 nuclear magnetic resonance spectrum of poly[d(A-T)] from phosphorothioate substitution, *Biochemistry*, 22, 4546, 1983.
44. Cosstick, R. and Eckstein, F., Synthesis of d(GC) and d(CG) octamers containing alternating phophorothioate linkage: effects of the phosphorothioate group on the B-Z transition, *Biochemistry*, 24, 3630, 1985.
45. DeClercq, E., Eckstein, F., and Merigan, T. C., Interferon induction increased through chemical modification of a synthetic polyribonucleotide, *Science*, 165, 1137, 1969.
46. Eckstein, F. and Gindl, H., Polyribonucleotides containing a phosphorothioate backbone, *Eur. J. Biochem.*, 13, 558, 1970.
47. Morvan, F., Rayner, B., and Imbach, J.-L., Modified oligonucleotides. IV. Solid phase synthesis and preliminary evaluation of phosphorothioate RNA as potential antisense agents, *Tetrahedron Lett.*, 31, 7149, 1990.
48. Agrawal, S., Tang, J. Y., Sun, D., Sarin, P. S., and Zamecnik, P. C., Synthesis and anti-HIV activity of oligoribonucleotides and their phosphorothioate analogs, *Ann. N.Y. Acad. Sci.*, 600, 2, 1992.
49. Marshall, W. S. and Caruthers, M. H., Phosphorodithioate DNA as a potential therapeutic drug, *Science*, 259, 1564, 1993.
50. Slim, G. and Gait, M. J., Configurationally defined phosphorothioate-containing oligoribonucleotides in the study of the mechanism of cleavage of hammerhead ribozymes, *Nucleic Acids Res.*, 19, 1183, 1991.
51. Koizumi, M. and Ohtsuka, E., Effects of phosphorothioate and 2-amino groups in hammerhead ribozymes on cleavage rates and Mg^{++} binding, *Biochemistry*, 30, 5145, 1991.
52. Tsuji, H., Nomiyama, K., Murai, K., Akagi, K., and Fujishima, M., Comparison of the properties of ribonucleases in human liver and serum, *Eur. J. Clin. Chem. Biochem.*, 30, 339, 1992.
53. Eder, P. S., DeVine, R. J., Dagle, J. M., and Walder, J. A., Substrate specificity and kinetics of degradation of antisense oligonucleotides by a 3'-exonuclease in plasma, *Antisense Res. Dev.*, 1, 141, 1991.

54. Shaw, J.-P., Kent, K., Bird, J., Fishback, J., and Froehler, B., Modified deoxyoligonucleotides stable to exonuclease degradation in serum, *Nucleic Acids Res.*, 19, 747, 1991.

55. Yang, J.-H., Usman, N., Chartrand, P., and Cedergren, R., Minimum ribonucleotide requirement for catalysis by the RNA hammerhead domain, *Biochemistry*, 31, 5005, 1992.

56. Fisher, T. L., Terhorst, T., Cao, X., and Wagner, R. W., Intracellular disposition and metabolism of fluorescently-labeled unmodified and modified oligonucleotides microinjected into mammalian cells, *Nucleic Acids Res.*, 21, 3857, 1993.

57. Ott, J. and Eckstein, F., Protection of oligonucleotide primers against degradation by DNA polymerase I, *Biochemistry*, 26, 8237, 1987.

58. Spitzer, S. and Eckstein, F., Inhibition of deoxyribonucleases by phosphorothioate groups in oligodeoxyribonucleotides, *Nucleic Acids Res.*, 16, 11691, 1988.

59. DeClercq, E., Eckstein, F., Sternbach, H., and Merigan, T. C., The antiviral activity of thiophosphate-substituted polyribonucleotides *in vitro* and *in vivo*, *Virology*, 42, 421, 1970.

60. Black, D. R., Eckstein, F., DeClercq, E., and Merigan, T. C., Studies on the toxicity and antiviral activity of various polynucleotides, *Antimicrob. Agents Chemother.*, 3, 198, 1973.

61. Eckstein, F., Schulz, H. H., Rüterjans, H., Haar, W., and Maurer, W., Stereochemistry of the transesterification step of RNaseT1, *Biochemistry*, 11, 3507, 1972.

62. Moore, M. J. and Sharp, P. A., Evidence for two active sites in the splicesome provided by stereochemistry of pre-mRNA splicing, *Nature*, 365, 364, 1993.

63. Sproat, B. S., Lamond, A. I., Beijer, B., Neuner, P., and Ryder, U., Highly efficient chemical synthesis of 2'-O-methyloligoribonucleotides and tetrabiotinylated derivatives; novel probes that are resistant to degradation by RNA or DNA specific nucleases, *Nucleic Acids Res.*, 17, 3373, 1989.

64. Inoue, H., Hayase, Y., Iwai, S., and Ohtsuka, E., Sequence-dependent hydrolysis of RNA using modified oligonucleotide splints and RNase H, *FEBS Lett.*, 215, 327, 1987.

65. Seliger, H., Fröhlich, A., Gröger, G., Krist, B., Montenarh, M., Rösch, H., Rösch, R., and Ramalho Ortigao, F., Synthetic oligonucleotides for biomedical applications, *Nucleic Acids Res. Symp. Ser.*, 24, 193, 1991.

66. Debart, F., Rayner, B., and Imbach, J.-L., Sugar modified oligonucleotides. II. Solid phase synthesis of nuclease resistant a-anomeric uridylates as potential antisense agents, *Tetrahedron Lett.*, 31, 3537, 1990.

67. Hertel, K. J., Pardi, A., Uhlenbeck, E. C., Koizumi, M., Ohtsuka, E., Uesugi, S., Cedergren, R., Eckstein, F., Gerlach, W. L., Hodgson, R., and Symons, R., Numbering system for the hammerhead, *Nucleic Acids Res.*, 20, 3252, 1992.

68. Sarver, N., Cantin, E. M., Chang, P. S., Zaia, J. A., Ladne, P. A., Stephens, D. A., and Rossi, J. J., Ribozymes as potential anti-HIV-1 therapeutic agents, *Science*, 247, 1222, 1990.

69. Yu, M., Ojwang, J., Yamada, O., Hampel, A., Rapapport, J., Looney, D., and Wong-Staal, F., A hairpin ribozyme inhibits expression of diverse strains of human immunodeficiency virus type 1, *Proc. Natl. Acad. Sci. U.S.A.*, 90, 6340, 1993.

70. Buzayan, J. M., van Tol, H., Feldstein, P. A., and Bruening, G., Identification of a non-junction phosphodiester that influences an autolytic processing reaction of RNA, *Nucleic Acids Res.*, 18, 4447, 1990.

71. Ruffner, D. E. and Uhlenbeck, O. C., Thiophosphate interference experiments locate phosphates important for the hammerhead RNA self-cleavage reaction, *Nucleic Acids Res.*, 18, 6025, 1990.

72. Chowrira, B. M. and Burke, J. M., Extensive phosphorothioate substitution yields highly active and nuclease-resistant hairpin ribozymes, *Nucleic Acids Res.*, 20, 2835, 1992.

73. Taylor, N. R., Kaplan, B. E., Swiderski, P., Li, H., and Rossi, J. J., Chimeric DNA-RNA hammerhead ribozymes have enhanced *in vitro* catalytic efficiency and increased stability *in vivo*, *Nucleic Acids Res.*, 20, 4559, 1992.
74. Hendry, P., McCall, M. J., Santiago, F. S., and Jennings, P. A., A ribozyme with DNA in the hybridising arms displays enhanced cleavage ability, *Nucleic Acids Res.*, 20, 5737, 1992.
75. Goodchild, J., Enhancement of ribozyme catalytic activity by a contiguous oligodeoxynucleotide (facilitator) and by 2'-O-methylation, *Nucleic Acids Res.*, 20, 4607, 1992.
76. Shimayama, T., Nishikawa, F., Nishikawa, S., and Taira, K., Nuclease-resistant chimeric ribozymes containing deoxyribonucleotides and phosphorothioate linkages, *Nucleic Acids Res.*, 21, 2605, 1993.
77. Paolella, G., Sproat, B. S., and Lamond, A. I., Nuclease resistant ribozymes with high catalytic activity, *EMBO J.*, 11, 1913, 1992.
78. Pieken, W. A., Olsen, D. B., Benseler, F., Aurup, H., and Eckstein, F., Kinetic characterization of ribonuclease-resistant 2'-modified hammerhead ribozymes, *Science*, 253, 314, 1991.
79. Heidenreich, O. and Eckstein, F., Hammerhead ribozyme-mediated cleavage of the long terminal repeat RNA of human immunodeficiency virus type 1, *J. Biol. Chem.*, 267, 1904, 1992.
80. Heidenreich, O., Benseler, F., Fahrenholz, A., and Eckstein, F., High activity and stability of hammerhead ribozymes containing 2'-modified pyrimidine nucleosides and phosphorothioates, *J. Biol. Chem.*, 269, 2131, 1994.
81. Shibahara, S., Mukai, S., Morisawa, H., Nakashima, H., Kobayashi, S., and Yamamoto, N., Inhibition of human immunodeficiency virus (HIV-1) replication by synthetic oligo-RNA derivatives, *Nucleic Acids Res.*, 17, 239, 1986.
82. Chiang, M.-Y., Chan, H., Zounes, M. A., Freier, S. M., Lima, W. F., and Bennet, C. F., Antisense oligonucleotides inhibit intercellular adhesion molecule expression by two distinct mechanisms, *J. Biol. Chem.*, 266, 18162, 1991.

Chapter 11

Uptake and Localization of Phosphodiester and Chimeric Oligodeoxynucleotides in Normal and Leukemic Primary Cells

Arthur M. Krieg

CONTENTS

I. INTRODUCTION

For an oligodeoxynucleotide (ODN) to have an antisense or an antigene effect, it is self evident that it must first enter the target cell. Some aptamer ODN may work through binding soluble or cell-surface molecules, while others will require cell entry. ODN uptake in cell lines is saturable, sequence independent, and temperature and energy dependent (reviewed in References 1 to 3). There is some evidence to suggest that such uptake may occur through an 80-kDa membrane protein,[4,5] but this has not yet been cloned or further characterized. Unfortunately, the mechanism(s) of ODN uptake into primary cells has not been investigated as thoroughly as in cell lines. The focus of this chapter is the uptake and intracellular localization of various modified and unmodified ODN in primary lymphocytes and myeloid cells.

II. METHODS FOR STUDYING ODN UPTAKE AND INTRACELLULAR LOCALIZATION

One of the most popular methods for studying the cellular uptake of ODNs has been the use of radiolabeled ODNs. Since the results of experiments using 5' or 3' end-labeled ODNs can be markedly affected by cellular phosphatases, internally radiolabeled ODNs are preferable.[3,6] Another important caveat to the interpretation of studies using radiolabeled ODNs is that dead cells can have approximately 50-fold higher ODN uptake than live cells.[7,8] Even a few percent of dead cells can greatly increase apparent uptake of radiolabeled ODN in a cell population. Thus, differences in the uptake of radiolabeled ODNs between different cell populations could result from variations in the proportion of dead cells unless this has been carefully controlled. A further limitation of using radiolabeled ODNs for studies of heterogeneous cell populations such as primary cells is that this technique does not allow one to readily distinguish whether measured uptake results from many cells taking up a modest amount of ODN or from a cell subpopulation taking up a great deal.

An alternative to using radiolabeled ODNs in ODN uptake studies is flow cytometry with ODN conjugated to fluorochromes such as fluorescein isothiocyanate (FITC). After culture with FITC-conjugated ODNs, cells can be stained with monoclonal antibodies (mAb) that bind to subset-specific cell surface molecules and are conjugated to different dyes fluorescing at distinct wavelengths. Data are analyzed by gating on the desired combination of antibody staining characteristics to determine the relative levels of ODN uptake in each cell subpopulation of interest. Of course, conjugation of a fluorochrome to an ODN may alter ODN uptake. While this possibility can not be excluded, it is reassuring that uptake of an ODN conjugated to FITC parallels that of an ODN conjugated to a different fluorochrome, cyanine 3.18.[8]

An extremely important issue not always addressed in uptake studies is that the label used must be predominantly associated with ODN at the time points studied. ODNs can be largely degraded within hours in some cell types, and if the label is no longer present on them, the results are meaningless. For example, in spleen cells most intracellular phosphodiester (O-ODN) had been partially degraded within 4 h of cell culture.[8] In contrast, partial or complete FITC-phosphorothioate(S)- or FITC-methylphosphonate(MP)-O-ODNs had no detectable degradation at the same time point. As mentioned above, the ^{32}P on 5' and 3' radiolabeled O-ODN can be rapidly removed by serum or cellular phosphatases, rendering measurements of cell-associated cpm nearly useless unless it is verified that the cpm are predominantly on the ODN.

One limitation in studies of ODN uptake is the difficulty in distinguishing ODN bound to cell membranes from that which has been internalized. Cells can be washed in acidic buffers to remove cell membrane-bound ODN, but this harsh treatment reduces cell viability and is not useful for certain types of experiments. Any technique in which the cells are killed may encounter artifacts due to the preferential ODN uptake in dead cells referred to above.

Recently we observed that phosphorothioate ODNs have a much greater affinity for spleen cell membranes than phosphodiester ODNs.[8] This suggested the possibility that the phosphorothioates might be useful as a gentle way to displace cell membrane-bound phosphodiester ODN. Thus, by culturing cells with (F)-O-ODN, and then adding unconjugated S-ODN, it should be possible to quite simply quantitate the proportion of intracellular compared to cell membrane-bound (F)-ODN.

To test the possible utility of this technique, we incubated spleen cells on ice for 10 min in the presence or absence of FITC-O-ODN with or without competitor phosphorothioate. There is no detectable ODN internalization by confocal microscopy in cells cultured on ice (Figure 1).[9] Cells were then stained with T-cell- and B-cell-specific antibodies for flow cytometry and fluorescence quantitation in both of these cell subsets.

Under these conditions, the level of autofluorescence in B- and T-cells was 8.15 and 8.78, respectively (Table 1). Addition of 1 µg of (F)-ODN at 4°C increased the B- and T-cell fluorescence to 29.4 and 10.1, respectively. The difference between these sets of fluorescence levels represent the amount of membrane-bound ODN. Fluorescence was reduced essentially back to background if 10 µg of competitor S-ODN was added 5 min after the addition of the (F)-O-ODN (Table 1). This experiment confirmed that the amount of membrane-bound (F)-O-ODN can be quite simply determined as the difference of fluorescence levels between cells cultured with and without competitor S-ODN. This technique can also be used to calculate the rate of energy-dependent intracellular ODN uptake by incubating cells at 37°C for an arbitrary time period, and then comparing the level of fluorescence in cells with or without addition of competitor S-ODN. Thus, the difference in fluorescence intensity between the two samples represents membrane-bound ODN. The residual fluorescence in the sample containing S-ODN represents internalized (F)-O-ODN (after subtraction of background autofluorescence).

These calculations assume that cell membrane ODN binding at 4°C is the same as at 37°C. Our data indicate that this assumption is indeed correct for both B- and T-cells: the

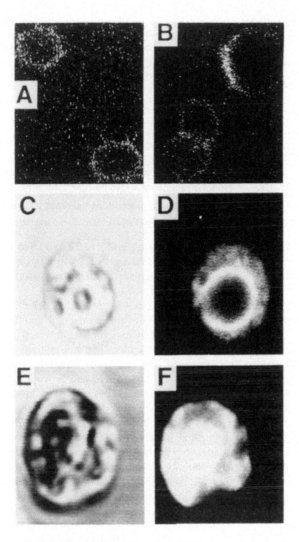

Figure 1 Oligonucleotide intracellular localization by confocal microscopy. Spleen cells were cultured with F-*O*-ODN (panels A, C, D) or with an F-*O*-ODN bearing a 5' cholesteryl moiety (panels B, E, F). Panels A and B show fluorescent images of cells cultured on ice for 15 min after ODN addition to demonstrate cell surface staining, and are a lower magnification than the other panels. C and E are phase contrast images; D and F are the corresponding fluorescence images.

reduction in fluorescence with addition of competitor *S*-ODN was essentially the same regardless of whether the cells had been cultured at 4°C or 37°C (Table 1).

Thus, addition of competitor *S*-ODN to cells cultured at 4°C reduced B- and T-cell fluorescence by 18.1 (29.4 to 11.3) and 1.8 (10.1 to 8.3), respectively; in cells cultured at 37°C, B- and T-cell fluorescence was reduced by 19.4 (52.0 to 32.6) and 1.4 (17.8 to 16.4), respectively. These data demonstrate that the level of B-cell membrane ODN binding is approximately tenfold higher than that by T-cells.

The relative levels of internalized FITC-*O*-ODN after 4 h at 37°C are determined by subtraction of autofluorescence from the fluorescence levels in cells after displacement of cell membrane bound FITC-*O*-ODN by competitor *S*-ODN. For Table 1, the data show

Table 1 **Calculation of membrane and intracellular oligonucleotide**

Oligonucleotide	Temp	Competitor[2]	Fluorescence Units[1] B-cells	T-cells
None (autofluorescence)	37°C	None	8.2	8.8
FITC-*O*-ODN	4°C	None	29.4 ± 2.8	10.1 ± 0.3
FITC-*O*-ODN	37°C	None	52.0 ± 2.0	17.8 ± 1.1
FITC-*O*-ODN	4°C	*S*-ODN	11.3 ± 0.1	8.3 ± 0.2
FITC-*O*-ODN	37°C	*S*-ODN	30.6 ± 1.2	16.4 ± 0.4

[1] Mean FITC fluorescence converted from channel number for the indicated cell population ± standard deviation of triplicate samples analyzed by flow cytometry.

[2] A tenfold molar excess of competitor 20 mer with an unrelated sequence was added (where indicated) 5 min after the FITC-*O*-ODN 20 mer.

average B- and T-cell intracellular fluorescence levels of 22.4 (30.6 to 8.2) and 7.8 (16.4 to 8.8), respectively. Thus, despite their tenfold higher level of ODN membrane binding, B-cells appear to have only a threefold increase in ODN internalization compared to T-cells.

These calculations may have to be corrected further for the quenching of intracellular FITC fluorescence caused by the acidic pH of the endosomal vesicles in which most intracellular ODN reside. This is supported by experiments in which cells cultured with chloroquine, which blocks endosomal acidification, had substantially increased cellular fluorescence (Figure 2). This increased fluorescence in chloroquine-treated cells may result from reduced quenching of FITC fluorescence and/or from reduced F-*O*-ODN degradation. It is also possible that different cell types may have different pathways for ODN uptake or that some membrane binding sites do not lead to internalization.

A separate but related problem to that of distinguishing membrane-associated from intracellular ODN is determining the intracellular compartment in which the ODN is located. Antisense ODN could in theory achieve their effects by hybridizing to their target RNA either in the cytoplasm or nucleus. Triplex ODN would need to enter the nucleus in sufficient concentration to work. Standard cell fractionation techniques are not well suited for analyses of ODN localization since the fractionation process is not immediate, and since ODN introduced into the cytoplasm rapidly moves to the nucleus.[10,11] In principle, the intracellular localization of ODNs can be investigated using autoradiography. However, because of technical problems with using autoradiography for this purpose, confocal microscopy has been more commonly used to determine intracellular localization of ODN, though of course artifacts conferred by the fluorochrome label are impossible to completely rule out. Dead cells can be avoided by adding propidium iodide (PI). Dead cells are unable to exclude PI and turn orange, allowing them to be distinguished from live cells using dual color confocal microscopy. ODN is frequently intranuclear in dead cells.

Such confocal studies generally indicate that if the ODN are taken up by cells through endocytosis, they remain in the cytoplasm.[8,9] Figure 1 shows by confocal microscopy that murine spleen cells cultured on ice have ODN only on the cell membrane (Figure 1, panels A,B). However, within 4 h of culture at 37°C, (F)-*O*-ODN is located in the cytoplasm (Figure 1, panels C,D). By adjusting the gain setting on the confocal microscope, the cytoplasmic localization can be seen to be typically punctate rather than diffuse, compatible with location within the endosomes as previously suggested in cell lines.[4] It has been suggested that the rate-limiting step in determining antisense efficacy may be escape of the ODN from the endosomes.[12] In our own studies, we have used confocal microscopy only on unfixed, living cells to avoid possible fixation artifacts. Unfortunately, this limits the image quality.

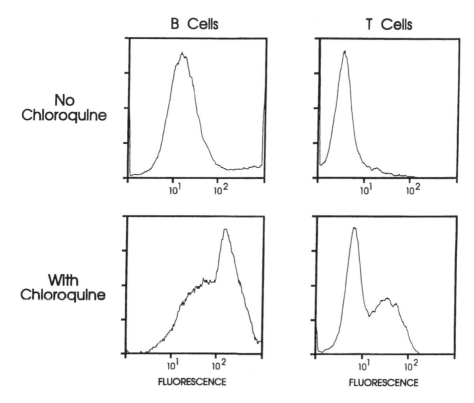

Figure 2 Effect of chloroquine on intracellular F-O-ODN. Fresh mouse spleen cells were cultured as previously described[14] for 4 h with 1 g of F-O-ODN in the presence or absence of chloroquine prior to flow cytometry. Data were gated on B-cells (6B2+) or T-cells (6B2⁻Ly1+) as indicated.

It seems likely that microinjected ODN may bypass the intracellular compartments through which endocytosed ODNs pass, and thus may have a different intracellular fate. Several ODN modifications can enhance endosomal exit and/or have been reported to increase nuclear localization. For example, conjugation of a 5' cholesteryl moiety onto ODNs not only substantially increases ODN uptake, but also causes increased intranuclear localization (Figure 2, panels E,F).[9] The importance of intracellular localization in antisense efficacy is suggested by the recent studies of Sullenger and Cech,[13] who found that ribozymes directed to a particular cell region cleaved target molecules in that region, but not in other cell regions.

We also studied the effect of the ODN backbone on intracellular localization using ODN bearing a 5' FITC in which the backbone was either O, S, S-O, or MP-O. Very little ODN could be detected in the nucleus with any of the backbones studied.[8] There was great cell-to-cell variability in fluorescence intensity, but no consistent differences in intracellular localization when comparing the different FITC-ODNs. Cytoplasmic staining was typically speckled, as shown in Figure 2, compatible with endocytic uptake.

Confocal microscopy in cells from the spleen, lymph nodes, and bone marrow (BM) have shown no tissue-specific difference in intracellular F-ODN localization.[13a]

III. ODN UPTAKE IN MURINE SPLEEN CELLS

In our initial studies, we used internally radiolabeled ODN to study ODN uptake in cultured murine spleen and lymph node cells.[14] By 48 h of culture, the number of cpm of

Table 2   **Radiolabeled oligonucleotide uptake by spleen cells**

	cpm in Spleen Cell Pellets[a]
Experiment 1[b]	
Fresh DBA/2	2,000
48 h DBA/2	32,000
Experiment 2[b]	
Fresh DBA/2	949
16 h DBA/2	1,443
24 h DBA/2	3,009
72 h DBA/2	12,775
Fresh BXSB	2,266
16 h BXSB	4,222

[a] Spleen single-cell suspensions were precultured for the indicated times (no preculture in the case of fresh cells) and radiolabeled oligonucleotide then added to all cells at the same time. Experiment 1 shows radioactivity taken up by 10^6 cells cultured with 2×10^6 cmp of internally radiolabeled oligonucleotide; experiment 2 shows radioactivity taken up by 7×10^5 cells cultured with 10^6 cpm. After a 2-h incubation, the cells were pelleted, washed three times in fresh medium, and dried onto glass fiber filters, and the incorporated radioactivity was determined by scintillation counting.

[b] DBA/2 mice used were 4-month-old females; BXSB mice were 2-month-old males.

Reproduced from Krieg, et al., 1991, with permission of the publisher.

radiolabeled ODN associated with the cell pellet had increased more than tenfold from that seen in fresh cells (Table 2). Autoradiography of these cells demonstrated that ODN uptake was very heterogenous: some cells had high levels and many had none detectable.

The spleen and lymph node cells used in those studies consisted primarily of B- and T-cells, which could be further divided into several subsets. To study possible variable ODN uptake by lymphoid subpopulations, we used FITC ODNs for uptake study and stained cells after culture for flow cytometry. As seen with the radiolabeled ODN, FITC-ODN uptake was markedly increased in cells preincubated for 48 h before incubation with ODN. However, this increased uptake was almost completely limited to the B-cell population. The percentage of B-cells positive for ODN uptake increased tenfold after 48 h preculture (Table 3). Flow cytometry with mAb to the T-cell subset markers CD4+ and CD8+ showed low uptake in both subsets. However, a high percentage of CD4−CD8− T-cells (which comprise 2 to 4% of all peripheral T-cells) were positive for ODN uptake.

To determine whether lymphocyte ODN uptake was affected by cell proliferation, spleen and lymph node cells were cultured for 48 h in either regular medium or medium supplemented with stimulatory doses of the B-cell mitogen lipopolysaccharide (LPS) or the T-cell mitogen concanavalin A (Con A). Spleen or lymph node cells cultured with LPS had dramatically enhanced ODN uptake which was limited to the B-cell population (Figure 3). In contrast, spleen or lymph node cell cultures treated with Con A for 48 h showed enhanced ODN uptake by T-, but not by B-cells (Figure 3).

In some experimental conditions it may be desirable to attempt to increase ODN uptake. Our studies indicated that such uptake is markedly and specifically induced by

Table 3   **Oligonucleotide uptake in lymphoid cell subpopulations**

Cell Source[a]	Percent Cells Positive for Oligonucleotide[b]	
	Ig[+]	Thy 1.2[+]
Fresh BXSB spleen	4	10
Fresh DBA/2 spleen	5	14
24 h BXSB spleen	25	14
24 h DBA/2 spleen	24	7
48 h BXSB spleen	59	19
48 h DBA/2 spleen	54	11
BXSB spleen Con A	56	68
DBA/2 spleen Con A	NS[c]	57
BXSB spleen LPS	93	23
DBA/2 spleen LPS	73	13
48 h BXSB LN	49	9
BXSB LN Con A	NS	85
BXSB LN LPS	89	10

[a] Data from the same experiments shown in Figures 1 and 2.

[b] Percent positive cells are indicated for Ig[+] cells or Thy-1.2[+] cells that were gated by staining with the appropriate antibody.

[c] NS, Insufficient cells for analysis (samples with fewer than 500 gated cells).

Reproduced from Krieg, et al., 1991, with permission of the publisher.

appropriate mitogens within 48 h. Our findings of heterogeneous cellular ODN uptake may also be relevant to attempts to apply antisense technology to *in vivo* therapy. For example, activated lymphocytes, which are the best hosts for HIV replication (and that of many other viruses), take up antisense more rapidly than other cells, thereby potentially improving the therapeutic index of antisense drugs.

IV.   ODN UPTAKE IN MURINE B-LINEAGE LYMPHOCYTES

As previously described by Hardy et al.,[15] B-lineage lymphocytes pass through several sequential stages with distinct expression of cell surface molecules as they develop from their hematopoietic progenitors. Among these markers are (i) B220, which is present on all cells in the B-lymphocyte lineage, but not in cells of other lineages such as myeloid cells; (ii) S7 (CD43), expressed on B-cells only during the early stages of differentiation; and (iii) BP-1, which is expressed on B-cells only during the intermediate stages of differentiation. Using four-color flow cytometry with FITC-ODNs and antibodies to these determinants we could readily distinguish FITC-ODN uptake among pre-Pro- and early Pro-B cells (B220[+]S7[+]BP-1[-]) from late Pro-B cells (B220[+]S7[+]BP-1[+]), Pre-B cells (B220[+]S7[-]BP-1[+]), and B-cells (B220[+]S7[-]BP-1[-]).[15] We found that uptake was quite low among the pre-Pro- and early Pro-B cells, increased in the late Pro-B and Pre-B cell populations, and was lower among mature B-cells.[16] Even in cells cultured on ice to block intracellular ODN uptake, there was a higher level of ODN present in the late Pro-B and pre-B cells than in the other B-cell stages, suggesting that the former cells have higher levels of membrane ODN binding sites.[16]

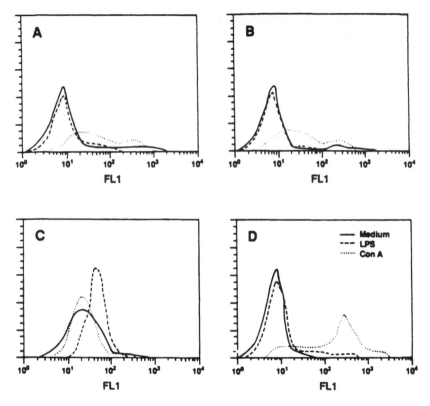

Figure 3 Induction of oligonucleotide uptake by T- or B-cell mitogens. Cells from the same mice shown in Figure 1 were cultured for 48 h in either regular medium or medium supplemented with 5µg/ml Con A or 10 mg/ml LPS, as indicated by the legend in D. Data were gated on either T-cells or B-cells by staining with the appropriate antibody, and are depicted by multiple histograms where increasing FL 1 florescence indicates increasing oligonucleotide uptake. Cells shown in the different panels are (A) BXSB spleen T-cells; (B) DBA/2 spleen T-cells; (C) BXSB spleen B-cells; and (D) BXSB LN T-cells. Reproduced from Krieg, et al., 1991, with permission of the publisher.

As we have reported previously in spleen cells,[8,9] intracellular (F)-ODN in living BM cells tends to show a somewhat stippled cytoplasmic localization, compatible with endocytotic uptake, and very little nuclear localization.[16]

We also examined the binding of a polyanionic dextran with similar molecular weight (MW) to the ODN because of our preliminary data indicating that it binds to the same cell surface sites as ODNs, and has similar intracellular localization by confocal microscopy. Cellular association of both ODNs and the polyanionic dextran was temperature dependent.[16] Both ODNs and the polyanionic dextran showed the same pattern of preferential association with the late Pro-B and Pre-B cells at both 0 and 37°C. These cells had slightly higher levels of autofluorescence than the pre-Pro-B and B cells, but this was still below the level seen with ODNs or dextran.

V. ODN UPTAKE IN HUMAN BONE MARROW CELLS

To determine whether or not ODN uptake may vary among different human bone marrow cell subpopulations, fresh bone marrow cells were prepared as above, cultured with FITC-labeled ODNs at 37°C for 4 h, stained with appropriate antibodies, and analyzed by flow cytometry (Table 4). The highest level of cellular ODN association was seen among

Table 4 ODN uptake among human cell subsets

	(F)-*O*-ODN Uptake[1]
Bone Marrow	
Unstained (background)	8.2
CD19+	24.2
CD7+	9.1
CD33+	38.6
CD19+CD34+sIgM-	15.3
CD19+CD34-sIgM-	23.7
CD19+Cd34-sIgM+	40.2
Peripheral blood mononuclear cells (PBMC)	
Unstained	6.5
CD8+	7.2
CD4+	7.0
CD22+	8.8
CD33+	82.6
Acute lymphocytic leukemia[2]	
Unstained	6.7
Nonleukemic cells	7.8
Leukemic cells	218
Acute myelogenous leukemia[3]	
Unstained	15.9
Nonleukemic cells	37.2
Leukemic cells	834

[1] Mean FITC fluorescence converted from channel number for the indicated cell subpopulations analyzed by flow cytometry. The different cell samples were analyzed on separate days, and so cannot be directly compared to one another.

[2] Results for one of three patients are shown. Leukemic cells were distinguished by their staining for CD7.

[3] Results for one of four patients shown. Leukemic cells were distinguished by their staining for CD13.

myeloid cells (CD33+), the lowest level among pre-T cells (CD7+), and an intermediate level was observed in B lineage cells (CD19+). Of note, ODN uptake varied among maturing B-cells with the lowest uptake among the most immature B-cells (CD19+, CD34+, surface immunoglobin M [sIgM]-), intermediate levels of uptake during the next step in differentiation (CD19+, CD34-, sIgM-), and higher levels among more mature B-cells (CD19+, CD34-, sIgM+).

Among peripheral blood cells, we observed low levels of ODN association in mature circulating CD22+ B-cells, CD8+ cytolytic T-cells, and CD4+ helper T-cells (Table 4). Peripheral CD33+ myeloid cells had higher levels of ODN association than T- or B-cells (Table 4).

Leukemic hematopoietic cells are far more sensitive to c-*myb* antisense ODNs than are normal cells.[17] To determine whether or not leukemic cells may differ in their level of ODN association from normal cells, we performed two-color flow cytometry using CD7 to identify leukemic cells in patients with acute lymphoblastic leukemia (ALL) and CD13 to identify leukemic cells in patients with acute myelogenous leukemia (AML) (Table 4). Both CD7+ ALL cells and CD13+ AML cells had markedly increased ODN association compared to the residual normal cells in the same patients. The level of ODN association in normal cells from leukemic patients was similar to that of the same cell type in normal controls.

Interleukin (IL)-3 has a broad range of growth and differentiation activities on he-
mopoietic progenitor cells[18] and like stem cell factor (SCF), stimulates proliferation of
CD34+ stem cells.[19] When bone marrow cells from normal human donors were cultured
with IL-3 (25 ng/ml) and recombinant SCF (50 ng/ml, kindly supplied by Amgen
Corporation) for 48 h, ODN uptake was increased approximately fourfold compared to
cells cultured in medium without these factors.

VI. THE EFFECTS OF ODN BACKBONE ON CELL BINDING
AND UPTAKE

In an effort to increase the nuclease resistance of O-ODNs, several backbone modifications such as the phosphorothioates (S-ODNs) and methylphosphonates (MP-ODNs)
have been developed. Both S- and MP-ODNs are extremely nuclease resistant in
cultured cells. To combine the good RNase H activation and hybridization properties
of O-ODNs with some of the nuclease resistance of S- and MP-ODNs, chimeric ODNs
modified with several phosphorothioate or MP linkages at just the 5' and 3' termini can
be used.

The effects of phosphorothioate (S-ODN) or terminal phosphorothioate-phosphodiester
(S-O-ODNs) or MP-phosphodiester (MP-O-ODNs) modifications on mouse spleen cell
surface binding, uptake, and degradation were studied using fluorescein (FITC)-conju-
gated ODNs synthesized at Amgen as described.[8] S-O- and MP-O-ODN had the indicated
modification (phosphorothioate or MP, respectively) at the first two 5' internucleotide
linkages and the last five 3' internucleotide linkages, since this was empirically found to
give the best balance between nuclease resistance and biological activity.

To determine whether FITC-O-, FITC-S-, FITC-S-O-, and FITC-MP-O-ODNs have
similar degrees of spleen cell binding, cells were briefly cultured on ice (to prevent
endocytosis) with FITC-conjugated ODNs, washed, and then studied by flow cytometry.[8]
Data were gated on B-cells which were prestimulated with LPS for 48 h to upregulate
ODN cell membrane binding sites. The highest level of cell binding was seen with the
S-ODN (Figure 4). FITC-S-O-ODNs were slightly lower than FITC-S-ODN, but still
gave brighter cell staining than did FITC-O-ODNs, which stained cells more brightly than
did FITC-MP-O-ODNs (Figure 4). Two different ODN sequences were studied for each
F-ODN backbone and gave essentially identical results. The level of S-ODN binding to
these activated B-cell membranes was more than ten times higher than that of FITC-O-
ODN. Our data cannot be extrapolated to predict the behavior of completely MP-
modified ODNs, the backbones of which would be uncharged.

Titration studies of FITC-ODN binding to B-cells performed over a range of ODN
concentrations from 0.3 to 10,000 nM showed an initial steep slope of increasing fluorescence with oligo concentration, compatible with a single type of high-affinity binding site
(Figure 5). This initial slope was steepest for the FITC-S-ODN, slightly lower for the
FITC-S-O-ODN, and lowest for the FITC-O-ODN (Figure 5).

If O-, S-O-, and S-ODNs bind to the same sites on the cell surface with different
affinities, it should be possible to differentially cross compete their binding. Therefore,
stimulated B-cells were preincubated with competitor unconjugated O-, S-O-, and
S-ODNs, before addition of FITC-ODNs. Flow cytometry showed that competitor S-
ODN completely blocked cell surface binding of FITC-O-ODN, even at a molar ratio of
<0.1, while competitor O-ODN gave essentially no decrease in FITC-S-ODN binding,
even at a molar ratio of 100 (Figures 6 and 7). Competitor S-O-ODN was a weak inhibitor
of S-ODN binding, and a moderate inhibitor of FITC-O-ODN binding (Figures 6 and 7).
Thus, the degree of phosphorothioate modification of an ODN is directly related to its
increased affinity for the cell surface sites bound by FITC-O-ODN. Under these experimental conditions, 50% of the B-cell surface binding of FITC-O-ODN was blocked by
just 1/100 the amount of S-ODN (Figure 6).

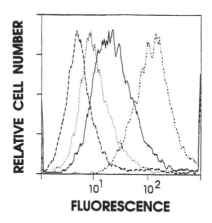

FLUORESCENCE

Figure 4 Binding of FITC-conjugated oligonucleotides to LPS-activated B-cell membranes. Spleen cells were prestimulated with LPS as in Figure 1B and then incubated on ice with the indicated oligonucleotides at 4°C for 30 min. D. were gated on B cells (6B2+ live cells). Oligonucleotides are dashed line, FITC-MP-*O*-oligonucleotide; dotted line FITC-*O*-oligonucleotide; solid line FITC-*S*-*O*-oligonucleotide; dot-dash line, FITC-*S*-oligonucleotide. Reproduced from **Zhao**, et al., 1993, with permission of the publisher.

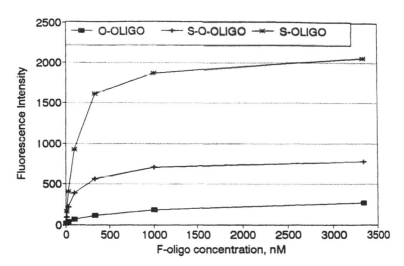

Figure 5 Titration of oligonucleotide surface binding. Flow cytometry was performed with mouse spleen cells essentially as described above except that cells were stained with oligonucleotide in a volume of 0.5 ml using oligonucleotide concentrations ranging from 0.3 to 10,000 nM. After 5 min cells were washed and stained with 6B2 for flow cytometry. The mean fluorescence intensity for the B-cell population at each oligonucleotide concentration is plotted on the y axis for the FITC-oligonucleotides shown. Data shown are from one of three representative experiments. Reproduced from Zhao, et al., 1993, with permission of the publisher.

Since *S*- and *S-O*-ODNs can completely prevent FITC-*O*-ODN binding to mouse B-cells, they must bind to the same cell membrane sites as FITC-*O*-ODN. The increased level of fluorescence in cells stained with FITC-*S*- or FITC-*S-O*-ODN may result from an increased affinity of these ODN for the same sites bound by FITC-*O*-ODN or from an

188

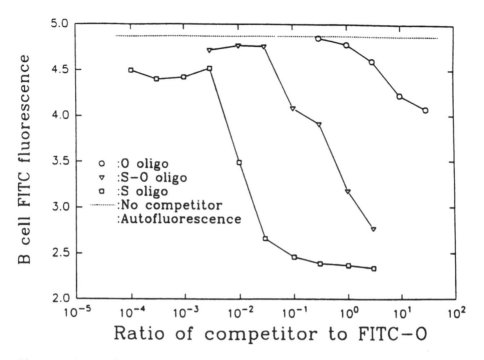

Figure 6 Cell-surface binding of FITC-*O*-oligonucleotide with or without competition with different oligonucleotide backbones (legend indicates type of competitor). Spleen cells were prepared and studied as in Figure 4 using 1 μg of FITC-*O*-oligonucleotide for each point. The *y* axis shows the mean level of B-cell fluorescence using an arbitrary scale where the baseline fluorescence with no added oligonucleotide is approximately 2.3. Reproduced from Zhao, et al., 1993, with permission of the publisher.

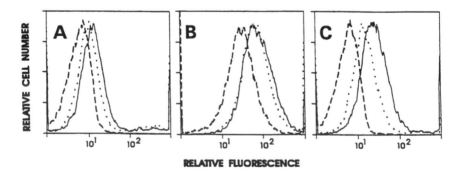

Figure 7 Competition studies with FITC oligonucleotides. Spleen cells were prestimulated with LPS, and then incubated on ice to assess cell-surface binding. Then 10 μg of competitor oligonucleotides were added 5 min before FITC oligonucleotides and data gated on B-cells. (A) Competition of FITC-*O*-oligonucleotide with: no competitor (solid line): *O*-oligonucleotide (dots); and *S*- or *S*-*O*-oligonucleotide (dashes). (B) Competition of FETC-*S*-oligonucleotide with: no competition or *O*-oligonucleotide (solid line); *S*-*O*-oligonucleotide (dots); or *S*-oligonucleotide (dashes). (C) Competition of FITC-*S*-*O*-oligonucleotide with: no competitor or *O*-oligonucleotide (solid line); *S*-*O*-oligonucleotide (dots); or *S*-oligonucleotide (dashes). Reproduced from Zhao, et al., 1993, with permission of the publisher.

ability to bind additional cell membrane sites not bound by FITC-*O*-ODN. Indeed, we have noticed decreased brightness by flow cytometry with fluorescent antibodies to several cell surface molecules in cells pretreated with 10 µg of *S*-ODN, suggesting that the *S*-ODN is competing with the antibodies for cell surface binding (data not shown). Of course, it is also possible that DNA normally present on cell membranes may partially block FITC-*O*-ODN binding, but not FITC-*S*-ODN binding. However, pretreatment of human or murine cells with DNase did not increase FITC-*O*-ODN binding (data not shown).

VII. SUMMARY

Antisense, antigene, and antiprotein ODN have great therapeutic potential and can be very useful research tools. Depending on the target, it may be desirable to try to direct the ODN to remain extracellular (such as aptamer ODN against cell surface proteins) or to various cytoplasmic or nuclear intracellular compartments. Much work will be required to develop this aspect of antisense technology, which has received far less attention to date than the areas of organic chemistry related to various DNA modifications and conjugates. Since there is good reason to believe that the limiting factor in *in vivo* efficacy will be intracellular uptake, many investigators are becoming increasingly interested in this area of research.

At present, it appears clear that for primary cells in the lymphoid and myeloid lineages, ODN uptake can vary greatly, depending on the cell type and stage of differentiation. Uptake is generally quite low in fresh cells, but increases during cell culture or upon treatment with mitogens or growth factors. In acute leukemia, ODN uptake by leukemic cells exceeds that by normal cells.

The ODN backbone and any conjugated molecules can greatly change ODN uptake and, for certain conjugated moieties such as cholesteryl, intracellular localization.

ACKNOWLEDGMENTS

We thank Vickie Weidner for outstanding assistance in preparing the manuscript. This work was supported by grants from the Arthritis Foundation, the Lupus Foundation of America, the Carver Charitable Trust, and the Veterans Administration.

REFERENCES

1. Jaroszewski, J. W. and Cohen, J. S., Cellular uptake of antisense oligodeoxynucleotides, *Adv. Drug Delivery Rev.,* 6, 235, 1991.
2. Akhtar, S., Shoji, Y., and Juliano, R. L., Pharmaceutical aspects of the biological stability and membrane transport characteristics of antisense oligonucleotides, in *Gene Regulation: Biology of Antisense RNA and DNA,* Erickson, R. P. and Izant, J. G., Eds., Raven Press, New York, 1992, 133.
3. Crooke, R. M., *In vitro* toxicology and pharmacokinetics of antisense oligonucleotides, *Anti-Cancer Drug Design,* 6, 609, 1991.
4. Loke, S. L., Stein, C. A., Zhang, X. H., Mori, K., Nakanishi, M., Subasinghe, C., Cohen, J. S., and Neckers, L. M., Characterization of oligonucleotide transport into living cells, *Proc. Natl. Acad. Sci. U.S.A.,* 86, 3474, 1989.
5. Yakubov, L. A., Deeva, E. A., Zarytova, V. F., Ivanova, E. M., Ryte, A. S., Yurchenko, L. V., and Vlassov, V. V., Mechanism of oligonucleotide uptake by cells: involvement of specific receptors?, *Proc. Natl. Acad. Sci. U.S.A.,* 86, 6454, 1989.
6. Zamecnik, P. C., Goodchild, J., Taguchi, Y., and Sarin, P. S., Inhibition of replication and expression of human T-cell lymphotropic virus type III in cultured cells by exogenous synthetic oligonucleotides complementary to viral RNA, *Proc. Natl. Acad. Sci. U.S.A.,* 83, 4143, 1986.

7. Thomas, L. L. and Price, C. M., Cellular uptake and degradation of phosphorothioate oligonucleotides by marine and freshwater ciliates, *Antisense Res. Dev.,* 2, 251, 1992.

8. Zhao, Q., Matson, S., Herrara, C. J., Fisher, E., Yu, H., Waggoner, A., and Krieg, A. M., Comparison of cellular binding and uptake of antisense phosphodiester, phosphorothioate, and mixed phosphorothioate and methylphosphonate oligonucleotides, *Antisense Res. Dev.,* 3, 56, 1993.

9. Krieg, A., Tonkinson, J., Matson, S., Zhao, Q., Saxon, M., Zhang, L.-M., Bhanja, U., Yakubov, L., and Stein, C. A., Modification of antisense phosphodiester oligodeoxynucleotides by a 5′ cholesteryl moiety increases cellular association and improves efficacy, *Proc. Natl. Acad. Sci. U.S.A.,* 90, 1048, 1993.

10. Chin, D. J., Green, G. A., Zon, G., Szoka, F. C., and Straubinger, R. M., Rapid nuclear accumulation of injected oligodeoxyribonucleotides, *New Biol.,* 2, 1091, 1990.

11. Leonetti, J. P., Mechti, N., Degols, G., Gagnor, C., and Lebleu, B., Intracellular distribution of microinjected antisense oligonucleotides, *Proc. Natl. Acad. Sci. U.S.A.,* 88, 2702, 1991.

12. Wagner, R. W., Matteucci, M. D., Lewis, J. G., Gutierrez, A. J., Moulds, C., and Froehler, B. C., Antisense gene inhibition by oligonucleotides containing C-5 propyne pyrimidines, *Science,* 260, 1510, 1993.

13. Sullenger, B. A. and Cech, T. R., Tethering ribozymes to a retroviral packaging signal for destruction of viral RNA, *Science,* 262, 1566, 1993.

13a. Krieg, A. M., Unpublished data.

14. Krieg, A. M., Gmelig-Meyling, F., Gourley, M. F., Kisch, W. J., Chrisey, L. A., and Steinberg, A. D., Uptake of oligodeoxyribonucleotides by lymphoid cells is heterogeneous and inducible, *Antisense Res. Dev.,* 1, 161, 1991.

15. Hardy, R. R., Carmack, C. E., Shinton, S. A., Kemp, J. D., and Hayakawa, K., Resolution and characterization of pro-B and pre-pro-B cell stages in normal mouse bone marrow, *J. Exp. Med.,* 173, 1213, 1991.

16. Zhao, Q., Waldschmidt, T., Fisher, E., Herrera, C. J., and Krieg, A. M., Stage specific oligonucleotide uptake in murine bone marrow B cell precursors, *Blood,* in press, Dec. 1, 1994.

17. Calabretta, B., Sims, R. B., Valtieri, M., Caracciolo, D., Szczylik, C., Venturelli, D., Ratajczak, M., Beran, M., and Gewirtz, A. M., Normal and leukemic hematopoietic cells manifest differential sensitivity to inhibitory effects of c-*myb* antisense oligodeoxynucleotides: an *in vitro* study relevant to bone marrow purging, *Proc. Natl. Acad. Sci. U.S.A.,* 88, 2351, 1991.

18. Burke, F., Naylor, M. S., Davies, B., and Balkwill, F., The cytokine wall chart, *Immunol. Today,* 14, 165, 1993.

19. Martin, F. H., Suggs, S. V., Langley, K. E., Lu, H. S., Ting, J., Okino, K. H., Morris, C. F., McNiece, I. K., Jacobsen, F. W., Mendiaz, E. A., Birkett, N. C., Smith, K. A., Johnson, M. J., Parker, V. P., Flores, J. C., Patel, A. C., Fisher, E. F., Erjavec, H. O., Herrera, C. J., Wypych, J., Sachdev, R. K., Pope, J. A., Leslie, I., Wen, D., Lin, C.-H., Cuppies, R. L., and Zsebo, K. M., Primary structure and functional expression of rat and human stem cell factor DNAs, *Cell,* 63, 203, 1993.

Chapter 12

Oligonucleotide Transport Across Membranes and Into Cells: Effects of Chemical Modifications

Jeffrey Hughes, Anna Avroutskaya, Henri M. Sasmor, Charles J. Guinosso, P. Dan Cook, and R. L. Juliano

CONTENTS

I. INTRODUCTION

Antisense and antigene oligonucleotides offer a promising new approach for therapy of cancer, inflammatory diseases, and viral diseases.[1,2] However, a number of issues remain before oligonucleotides can take their place as effective drugs. A major problem for oligonucleotide therapeutics concerns the limited ability of these compounds to gain access to their nucleic acid targets in the cytoplasm and nucleus. Due to the large molecular weight and polar nature of oligonucleotides, cellular uptake is potentially the limiting factor for therapeutic utilization.[3] Several mechanisms have been proposed to account for the cellular uptake of oligonucleotides. Phosphodiester oligonucleotides reportedly bind to an 80-kDa surface protein and enter cells via receptor-mediated endocytosis.[4,5] Phosphorothioate oligonucleotides can competitively inhibit phosphodiester uptake[5] and may be at least partially brought into cells through the same binding protein. However, other observations suggest that phosphorothioate uptake is nonsaturable and thus unlikely to involve a specific receptor.[5a] Methylphosphonates, which have an uncharged backbone, were originally thought to enter cells through passive diffusion.[6] However, more recent observations indicate that uptake of these compounds occurs by an endocytotic route.[7] Only a fraction of the cellular accumulated oligonucleotide actually reaches the cytoplasm, by an undefined process that may include diffusion, membrane destabilization, or leakage during vesicle fusion.[8] Once in the cytoplasm, oligonucleotides rapidly migrate to the nucleus and concentrate in that organelle.[9]

The physiochemical properties of oligonucleotides are likely to affect membrane transport. For example, modification of the internucleoside backbone influences both metabolic stability and cellular uptake.[10] One report demonstrated that purine-rich oligonucleotides have a greater cellular uptake as compared to pyrimidine-rich sequences,[11] while recent studies from our laboratory indicate that uptake of poly-G homo-oligomers is strongly favored over other homo-oligomers.[12] However, there are also reports in the literature demonstrating the lack of effect of sequence on cellular uptake.[4] Several approaches have been used to enhance cellular uptake of oligonucleotides. These include use of backbone modifications that reduce charge and contribute to stability, including methylphosphonates,[13,14] methylimino-linked backbones,[15] and polyamide nucleic acids.[16] Each of these backbone modifications increases stability, but none have proven to demonstrate clearly superior cellular transport characteristics. In attempts to improve

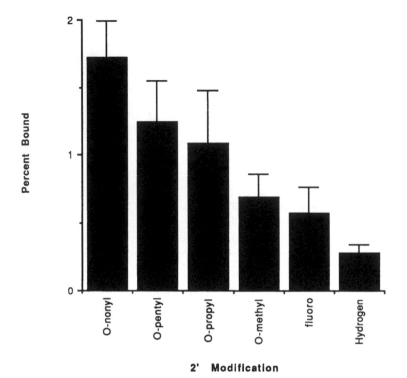

Figure 1 The binding of 2'-O-alkyl oligonucleotides to external lipid layer of blank liposomes at 37°C over 1 h. The graph represents three experimental runs. (From Hughes, J.A., et al., *J. Pharm. Sci.*, 83, 597, 1994. With permission.)

membrane transport, oligonucleotides have been coupled with lipophilic or membrane-destabilizing chemical moieties including cholic acid,[17] cholesterol,[18,19] and polylysine.[20] Moieties have been used to enhance oligonucleotide interaction with target nucleic acid sequences; for example, bleomycin,[21] acridine,[22] and psoralin[23] have been conjugated to oligonucleotides. Alternative approaches to improve intracellular delivery include liposomal entrapment of oligonucleotides[24] and the use of cationic lipid oligonucleotide particulates (Lipofectin®).[25] While these methods proved somewhat effective in tissue culture systems, there is a lack of basic information concerning the parameters that influence cellular uptake of oligonucleotides.

II. MEMBRANE ASSOCIATION AND PERMEATION OF 2'-O-ALKYL PHOSPHOROTHIOATE OLIGONUCLEOTIDES

One approach to influencing the membrane permeability of oligonucleotides is the introduction of lipophilic groups at the 2' position of the ribose sugar.[26,27] This modification may offer some advantages over pendant 5' or 3' modifications in that every residue can be modified, thus potentially resulting in a stronger impact on stability, on interaction with RNA or DNA targets, and on permeation properties as well. We have recently evaluated the effect of 2' alkyl modification on the ability of phosphorothioate oligonucleotides to (a) bind to lipid membranes (liposomes); (b) permeate across lipid membranes; and (c) accumulate in tissue culture cells. Details of procedures for liposome preparation and for the oligonucleotide binding and oligonucleotide efflux studies have been described previously.[28]

As indicated in Figure 1, 2'-O-alkyl oligonucleotides bound to lipid bilayer membranes to an increasing degree as the length of the 2' moiety increased, with the hydrogen

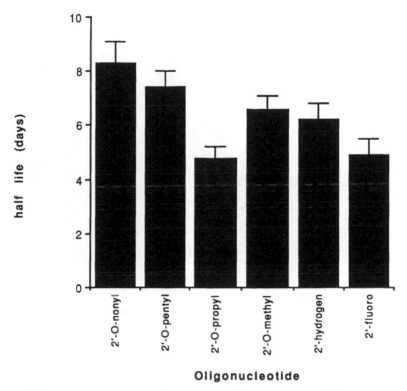

half life (days)

2'-O-nonyl 2'-O-pentyl 2'-O-propyl 2'-O-methyl 2'-hydrogen 2'-fluoro

Oligonucleotide

Figure 2 The percent of original amount of ^{14}C-sucrose, a fluid phase marker, retained within liposomes when co-incorporated with the 2'-*O*-alkyl oligonucleotides versus time. The nonyl and pentyl modifications increased the flux of ^{14}C-sucrose statistically over the other oligonucleotides ($p < 0.05$). (From Hughes, J.A., et al., *J. Pharm. Sci.*, 83, 597, 1994. With permission.)

oligonucleotide (i.e., no modification) giving the least binding and the nonyl oligonucleotide showing the greatest binding. Presumably, the alkyl side chains simply anchor the oligonucleotide to the surface of the liposome by intercalation into the hydrophobic region of the bilayer. This suggests that oligonucleotides with longer 2' alkyl modifications might destabilize lipid bilayers to some degree when present at high concentrations. This was indeed the case as shown in Figure 2, where the efflux of entrapped radiolabeled sucrose was measured in the presence or absence of 2'-*O*-alkyl oligonucleotides. Oligonucleotides having the nonyl and pentyl modifications had a definite effect on sucrose fluxes, whereas the other oligonucleotides were essentially without effect. This suggests a cautionary note for the design of oligonucleotides; an excess of amphiphilic character may impart toxic, detergent-like properties to oligonucleotides.

Evaluation of the ability of 2'-*O*-alkyl oligonucleotides to cross the lipid bilayer membrane provided an unexpected result. For small molecules there is a direct correlation between lipophilicity and rate of membrane permeation — this relationship is commonly known as Overton's Rule.[29] However, in the case of the modified oligonucleotides this was not the case (Figure 3). While the propyl and fluoro modifications showed efflux rates that were somewhat faster than that of the parent compound, the pentyl and nonyl compounds actually had slower flux rates than the parent compound. It should be noted that the efflux rates for all of these compounds were extremely slow, as has been the case for all other oligonucleotides we have tested.[30] The slow efflux from liposomes suggests that simple diffusion across the bilayer component of cell membranes is unlikely to be a major transport pathway for oligonucleotides. At this point we do not have a good

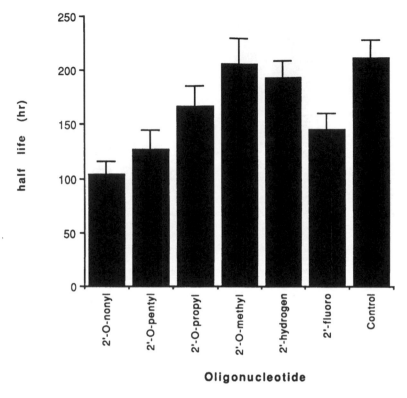

Figure 3 The efflux half-life of 2′-*O*-alkyl oligonucleotides from liposomes. The nonyl, propyl, and fluoro derivatives were statistically different ($p < 0.05$) from 2′-hydrogen control. (From Hughes, J.A., et al., *J. Pharm. Sci.*, 83, 597, 1994. With permission.)

explanation of the failure of the 2′-*O*-alkyl oligonucleotides to follow Overton's Rule; one possibility is the longer chain compounds simply bind too tightly to the lipid bilayer to allow rapid trans-bilayer movement.

We have also examined the uptake of 2′-*O*-alkyl oligonucleotides by cultured cells. In these previously unpublished studies, a fluorescein tag was used to follow the cell uptake of the oligonucleotides. Thus chimeric hamster ovary (CHO) cells were incubated with various concentrations of 2′-*O*-alkyl oligonucleotides; effects of temperature and duration of the incubation were also evaluated. The cellular accumulation of oligonucleotides was evaluated by flow cytometry as previously described.[12] In parallel studies, the intracellular distributions of oligonucleotides were evaluated by digital imaging fluorescence microscopy as previously described.[7] The presence of 2′-*O*-alkyl modifications had a strong effect on the total amount of oligonucleotides accumulated by the cells (Figure 4A). However, much of this was probably due to accumulation of oligonucleotides at the cell surface rather than internalization. Thus there was little evidence of a strong temperature dependence of cell association (Figure 4B), nor were metabolic inhibitors able to block cell association (data not shown); this would seem to rule out an active uptake process as the major determinant of cell association. The relatively small portion of 2′-*O*-alkyl oligonucleotides that did accumulate within the cell seemed to be present in cytoplasmic vesicles (Figure 5). Thus 2′-*O*-alkyl oligonucleotides basically seemed to be taken up by cells in the same manner as many other types of oligonucleotides — that is, by endocytosis.[8] However, the uptake process for the longer alkyl chain length was skewed by the fact that large amounts of these compounds bind to the cell surface.

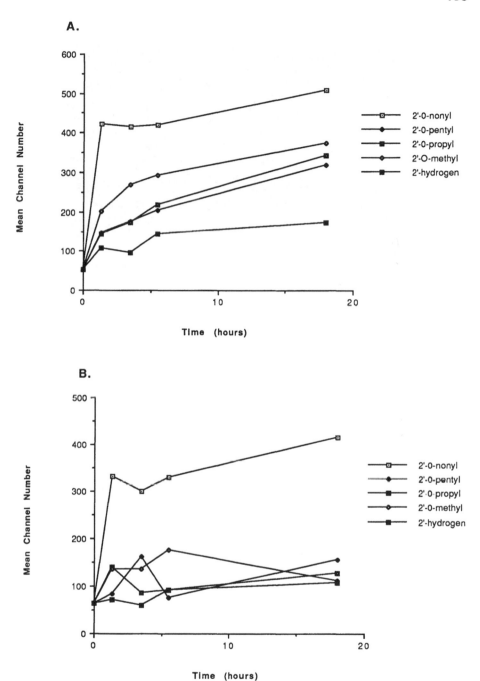

Figure 4 Cellular associated fluorescence of 2′-*O*-alkyl oligonucleotides as measured by flow cytometry. Graph A represents uptake and binding at 37°C, while graph B represent cellular binding at 4°C. In each case CHO cells (CH^{R}C5) were incubated with 2′-*O*-alkyl oligonucleotides (2.5 μ*M*).

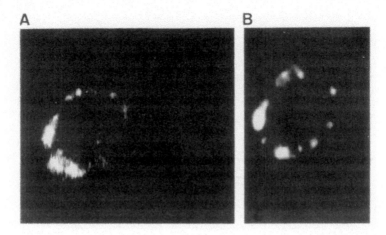

Figure 5 Subcellular distribution of (A) rhodamine-labeled dextran (40,000 MW) and 2'-*O*-propyl phosphorothioate fluroesceimated oligonucleotide. CHO cells were incubated with 0.5 mg/ml dextran and 1 μ*M* oligonucleotide for 4 h at 37°C in alpha-MEM (minimal essential medium) medium, then washed and visualized. A represents distribution of rhodamine-labeled dextran and B the distribution of fluorescein-labeled oligonucleotide.

III. CONCLUSIONS

Previous experience in our laboratory and in other laboratories suggests that most oligonucleotides initially accumulate in cells by binding to the cell surface, followed by endocytosis into cytoplasmic vesicles. Thus, internalized oligonucleotides still need to cross a membrane barrier, that of the endosome, in order to gain access to sites in the cytoplasm and nucleus. Our results have been consistent in showing that oligonucleotides, even uncharged compounds or compounds derivatized with lipophilic moieties, do not permeate across lipid bilayer membranes to any significant degree. Thus, other mechanisms, as yet undefined, must be responsible for entry of oligonucleotides into the cytoplasm and nucleus. The elucidation of such mechanisms is currently the main focus of our research.

REFERENCES

1. Crooke, S. T., Therapeutic applications of oligonucleotides, *Annu. Rev. Pharmacol. Toxicol.*, 32, 329, 1992.
2. Stein, C. A. and Cheng, Y.-C., Antisense oligonucleotides as therapeutic agents — is the bullet really magical, *Science*, 261, 1004, 1993.
3. Budker, V. G., Knorre, D. G., and Vlassov, V.V., Cell membranes as barriers for antisense constructions, *Antisense Res. Dev.*, 2, 177, 1992.
4. Yakubov, L. A., Deeva, E. A., Zarytova, V. F., Ivanova, E. M., Ryte, A. S., Yurchenko, L. V., and Vlassov, V. V., Mechanism of oligonucleotide uptake by cells: involvement of specific receptors?, *Proc. Natl. Acad. Sci. U.S.A.*, 86, 6454, 1989.
5. Loke, S. L., Stein, C. A., Zhang, X. H., Mori, K., Nakanishi, M., Subasinghe, C., Cohen, J. S., and Neckers, L. M., Characterization of oligonucleotide transport into living cells, *Proc. Natl. Acad. Sci. U.S.A.*, 86, 3474, 1989.
5a. Eads, C. and Juliano, R. L., unpublished observations.
6. Agris, C. H., Blake, K. R., Miller, P. S., Reddy, M. P., and Ts'o, P. O. P, Inhibition of vesicular stomatitis virus protein synthesis and infections sequence-specific oligodeoxyribonueoside methylphosphonates, *Biochemistry*, 25, 6228, 1986.

7. Shoji, Y., Akhtar, S., Periasamy, A., Herman, B., and Juliano, R. L., Mechanism of cellular uptake of modified oligodeoxynucleotides containing methylphosphonate linkages, *Nucleic Acids Res.*, 19, 5543, 1991.

8. Akhtar, S. and Juliano, R. L., Cellular uptake and intracellular fate of antisense oligonucleotides, *Trends Cell Biol.*, 2, 139, 1992.

9. Fisher, T. L., Terhorst, T., Cao, X., and Wagner, R. W., Intracellular disposition and metabolism of fluorescentyl-labeled unmodified oligonucleotides microinjected into mammalian cells, *Nucleic Acids Res.*, 21, 3857, 1993.

10. Zhao, Q., Matson, S., Herrera, C. J., Fisher, E., Yu, H., and Krieg, A. M., Comparison of cellular binding and uptake of antisense phosphorodiester, phosphorthioate, and mixed phosphorothioate and methylphosphonate oligonucleotides, *Antisense Res. Dev.*, 3, 53, 1993.

11. Iverson, P. L., Crouse, D., Zon, G., and Perry, G., Binding of antisense phosphorothioate oligonucleotides to murine lymphocytes is lineage specific and inducible, *Antisense Res. Dev.*, 2, 223, 1992.

12. Hughes, J. A., Avrutskaya, A. V., and Juliano, R. L., The influence of base composition on cellular transport and membrane binding of oligonucleotides, *Antisense Res. Dev.*, 4, 211, 1994.

13. Chem, T. L., M. P. S., Ts'o, P.O.P, and Colvin, O. M., Dispostion and metabolism of oligodeoxynucleoside methylphosphonate following a single iv injection in mice, *Drug Metab. Dispos.*, 18, 815, 1990.

14. Smith, C. C., Aurelian, L., Reddy, M. P., Miller, P. S., and Ts'o, P.O.P., Antiviral effect of an oligo(nucleoside methylphosphonate) complementary to the splice junction of herpes simplex virus type 1 immediate earyl pre-mRNAs 4 and 5, *Proc. Natl. Acad. Sci. U.S.A.*, 83, 2787, 1986.

15. Debart, F., Vasseur, J. J., Sanghvi, Y. S., and Cook, P. D., Synthesis and incorporation of methyleneoxy(methylimino) linked thymidine dimer into antisense oligonucleosides, *Bioorg. Med. Chem. Lett.*, 2, 1479, 1992.

16. Nielson, P. E., Egholm, M., Berg, R. H., and Buchardt, O., Peptide nucleic acids (PNAs): potential antisense and anti-gene agents, *Anti-Cancer Drug Design*, 8, 53, 1993.

17. Manoharan, M. et al., Chemical modifications to improve uptake and bioavailability of antisense oligonucleotides, *Ann. N.Y. Acad. Sci.*, 660, 306, 1992.

18. MacKellar, C., Graham, D., Will, D. W., Burgess, S., and Brown, T., Synthesis and physical properties of anti-HIV antisense oligonucleotide bearing terminal lipophilic groups, *Nucleic Acids Res.*, 20, 3411, 1992.

19. Saxon, M., Schieren, I., Zhang, L.-I., Tonkinson, J. L., and Stein, C. A., Stimulation of calcium influx in HL60 cells by cholesteryl-modified homopolymer oligodeoxynucleotides, *Antisense Res. Dev.*, 2, 243, 1992.

20. Degols, G., Leonetti, J.-P., and Lebleu, B., Antisense oligonucleotides as pharmacological modulators of gene expression, Springer-Verlag, New York, 1992, 100.

21. Zarytova, V. F., Sergeyev, D. S., and Godovikova, T. S., Synthesis of bleomycin A5 oligonucleotide derivatives and site-specific cleavage of the DNA target, *Bioconjugate Chem.*, 4, 189193, 1993.

22. Stein, C. A., Mori, K., Loke, S. L., Subasinghe, C., Shinozuka, K., Cohen, J. S., and Neckers, L. M., Phosphorothioate and normal oligodeoxyribonucleotides with 5'-linked acridine: characterization and preliminary kinectis of cellular uptake, *Gene*, 72, 333, 1988.

23. Kean, J. M. and Miller, P. S., Detection of psoralen cross-link sites in DNA modified by psoralen-conjugated oligodeoxyribonucleoside methylphosphonates, *Bioconjugate Chem.*, 4, 184, 1993.

24. Juliano, R. L. and Akhtar, S., Liposomes as a drug delivery system for antisense oligonucleotides, *Antisense Res. Dev.*, 2, 165, 1992.

25. Bennett, C. F., Chiang, M.-Y., Chan, H., Shoemaker, J. E. E., and Mirabelli, C. K., Cationic lipids enhance cellular uptake and activity of phosphorothioate antisense oligonucleotides, *Mol. Pharm.*, 41, 1023, 1992.
26. Keller, T. H. and Haner, R., Synthesis and hybridization properties of oligonucleotides containing 2'-*O*-modified ribonucleotides, *Nucleic Acids Res.*, 21, 4499, 1993.
27. Oberhauser, B. and Wagner, E., Effective incorporation of 2'-*O*-methyl-oligoribonucleotides into liposomes and enhanced cell association through modification with thiocholesterol, *Nucleic Acids Res.*, 20, 533, 1992.
28. Hughes, J. A., Bennett, C. F., Cook, P. D., Guinosso, C. J., Mirabelli, C. K., and Juliano, R. L., The lipid membrane permeability of 2' modified derivatives of phosphorothioate oligonucleotides, *J. Pharm. Sci.*, 83, 597, 1994.
29. Gennis, R. B., *Biomembranes: Molecular Structure and Function*, Springer-Verlag, New York, 1989.
30. Akhtar, S., Basu, S., Wickstrom, E., and Juliano, R. L., Interactions of antisense DNA oligonucleotide analogs with phospholipid membranes (liposomes), *Nucleic Acids Res.*, 19, 5551, 1991.

Chapter 13

Liposomes as a Delivery System for Antisense and Ribozyme Compounds

Alain R. Thierry and Garry B. Takle

CONTENTS

I. INTRODUCTION

The use of oligonucleotides as potential therapeutic agents has attracted widening interest over the last several years. Oligonucleotides can be designed to be complementary to regions of a gene or messenger RNA and act as specific blockers of transcription or translation. The regulation of expression of selected genes has been convincingly demonstrated in a variety of *in vitro* cell systems.[1-3] In general terms, three different classes of oligonucleotide-based therapeutics are currently under investigation: (1) the antisense oligodeoxynucleotides, which are DNA oligonucleotides designed to hybridize with a specific mRNA region, and consequently block translation; (2) the triple-helix-forming oligonucleotides which are formulated to be inserted into the DNA double helix and consequently block transcription; and (3) ribozymes which are RNA oligomers having the ability to select substrates by specific base pairing, cleave these substrates by a variety of mechanisms, and show turnover.

The therapeutic oligonucleotides that are the most thoroughly investigated are the antisense oligodeoxynucleotides, and this is possibly because they have the simplest structures. Their simple structure has allowed development of oligodeoxynucleotide analogs which, in turn, has made it possible to increase biological stability of antisense molecules, possibly also increasing their activity.[4,5] A small number of antisense molecules have shown therapeutic promise and are undergoing clinical trial.[6,7]

The *in vivo* application of this new class of therapeutics is hampered by several limitations. Factors such as oligonucleotide stability, cellular uptake, subcellular availability, and other pharmacokinetic parameters lead to a relatively poor delivery to target molecular sites at doses amenable to an affordable treatment. Thus, transport and intracellular delivery are important and fundamental considerations when developing an effective oligonucleotide-based therapy. Liposome delivery is one technique that addresses these concerns.

II. DELIVERY OF OLIGONUCLEOTIDES; DEVELOPMENT AND LIMITATIONS

The efficacy of oligonucleotide therapeutics appears to be affected by different pharmacological and biological properties mentioned above. Research efforts in antisense technology are often focused on the development of formulations conducive to the better delivery of functional oligonucleotides into living cells.

A. OLIGONUCLEOTIDE STABILITY

Oligonucleotide phosphodiester bonds are highly sensitive to enzymatic cleavage by nucleases in serum and in cells. Complete degradation of a nucleic acid can occur within 30 min of incubation in human serum at 37°C.[8] Thus investigation has focused on the design of more stable oligonucleotides having modified linkages or other decorations, such as phosphorothioates, methylphosphonates, phosphotriesters, or oligodeoxynucleotides.[8-10] These modifications can confer a dramatic increase in oligonucleotide stability in biological fluids or *in vivo*.[8-10] Akhtar et al.[8] demonstrated that degradation of methylphosphonate or phosphorothioate derivatives in human serum was detectable within 1 h, but corresponded to a very small portion (a few percent) of the total amount of oligomer present. These modified oligodeoxynucleotides showed no hydrolysis within 1 h in a nuclear extract or in cytoplasmic fractions.[10]

B. CELLULAR UPTAKE

Penetration of large molecules such as oligonucleotides into cells presents a major obstacle to their intended biological activity. The mechanisms of transport of oligonucleotide molecules into living cells remain unclear. However, recent indications suggest an active uptake process is occurring, whose rate is dependent on oligomer length, oligomer sequence, and any modifications.[2,7,11-13] Recent data suggest that uptake of phosphorothioated poly-C oligodeoxynucleotides may involve two distinct endocytic pathways: a fluid-phase pinocytosis and an adsorptive endocytosis.[10] The efficiency of oligonucleotide internalization is quite poor and, in efforts to improve it, artificial approaches bypassing the natural mechanism are being developed. An increase in oligomer accumulation in cells may be achieved by chemical conjugation to various ligands such as poly L-lysine, peptides, fatty acids, or cholesterol. Use of these conjugated oligomers has led to an increased inhibitory effect on gene expression by the delivered antisense molecule.[2,14,15]

C. INTRACELLULAR AVAILABILITY FOR ANTISENSE ACTIVITY

mRNAs are the main targets for antisense oligodeoxynucleotides or ribozymes, and consequently, for effective action, these should accumulate in the cytoplasm. However, the use of antisense oligonucleotides which are complementary to sites of transcription factor binding or splicing sites, for example, and the use of triplex-forming oligonucleotides, or the targeting of nascent transcripts, requires penetration of the nucleic acid therapeutic into the nucleus.

As recently demonstrated,[9,11,16,17] unmodified or phosphorothioate oligonucleotides are concentrated in endosomal vesicles following incubation in different *in vitro* cell systems. This localization pattern seems to indicate the involvement of an endocytic transport pathway in oligonucleotide cellular internalization. This vesicular transport may lead to a weak distribution in the cytoplasm or the nucleus, the intended sites of antisense or ribozyme activity. In addition, recent data have demonstrated the presence of an active efflux mechanism from the cells, resulting in a further decrease in intracellular concentration.[9,10] Thus, improved methods allowing increased intracellular availability for antisense activity by bypassing vesicular trafficking could significantly improve the efficacy of oligonucleotide therapeutics.

D. BIOAVAILABILITY

Therapeutic activity of antisense oligodeoxynucleotide compounds has been demonstrated in *in vivo* models.[18-20] Recent studies focusing on the biodistribution, pharmacokinetic properties, and toxicity of phosphorothioate oligodeoxynucleotides in animals have revealed that a therapeutic concentration may be obtained without major toxicity.[18,21] The liver, kidney, and lung are sites of significant accumulation, and free phosphorothioate oligonucleotides undergo rapid elimination from the blood. However, concentration in the micromolar range, which is sufficient for antisense activity, is achieved in some tissues and in plasma. Dosage levels in these experiments (3 to 50 mg/kg) is equivalent to those used for other therapies.[19] However, for use of this class of compound in a clinical setting, the dose administered should ideally be reduced due to the high cost of synthesis.

With this in mind, it is essential to develop an optimal delivery system that will enable low doses of drug to specifically accumulate to effective levels in the target cells or tissues.

III. LIPOSOME TECHNOLOGY

Drug delivery systems display considerable potential for improving therapeutic drug indices. Since their discovery in the early 1960s, liposome carriers appear to have many features of an optimal delivery system. Liposomes are closed structures comprised of phospholipidic membrane bilayers surrounding an internal aqueous space. They are versatile, nontoxic, and biodegradable, and have been studied extensively *in vitro* and *in vivo*. A wide variety of drugs have been encapsulated within liposomes. Liposomal drug delivery has been clinically used for the treatment of cancer, infectious disease, rheumatoid arthritis, for the detection of cancer, and for the preparation of vaccines.[22-25]

A. LIPOSOME STRUCTURE

The bipolar nature of phospholipid molecules used to form liposomes leads to the spontaneous formation, in excess water, of vesicles where hydrophilic heads face the external aqueous environment and hydrophobic tails face inward toward each other. In the resultant vesicular structures, or liposomes, water-soluble drugs can be encapsulated in the internal aqueous space and lipid-soluble drugs can be intercalated within the membrane itself. Peptides or proteins with hydrophobic domains may also be integrated into the liposome structure. Furthermore, antibodies may be attached via the lipid bilayer or covalently conjugated to the outer surface of the liposome.[26,27] Depending on the method of formation and the lipids used, liposomes can assume different structures: (i) multilamellar vesicles (MLVs) comprising several lipid bilayers separated by fluid, and (ii) unilamellar vesicles consisting of a single bilayer surrounding the internal fluid phase.

The unilamellar vesicles are typically characterized as being large or small (SUVs). Liposomes range in diameter from about 10 nm to 10 µm and have a minimum molecular weight of approximately 3×10^6 Da.[24] Considering the vast array of phospholipids and other polar lipids, and the various methodologies used for liposome preparation, there is a huge potential diversity in liposome structure. Consequently, each liposome-entrapped drug is likely to display specific pharmacological properties.

B. PHARMACOLOGICAL CONSIDERATIONS

Liposomes release their contents by interacting with cells using various different processes such as adsorption, lipid exchange, fusion, and endocytosis. Endocytosis is principally responsible for cellular uptake after systemic administration. Although lipid exchange or fusion have been demonstrated in *in vitro* experimental systems and may occur *in vivo*, these physicochemical events may not be important in an *in vivo* situation.

Most of the studies concerning the therapeutic and/or diagnostic effects of liposome-entrapped drugs have been conducted following intravenous administration. Intravenously administered liposomes are rapidly sequestered by mononuclear phagocytes of the reticuloendothelial system, the blood clearance rate depending on liposome charge, size, and lipid composition. In general terms, large neutral liposomes (>0.5 µm diameter) are confined primarily in the intravascular space after intravenous injection and are distributed in tissues or organs containing phagocytic cells, which in the liver, spleen, and bone marrow are mainly fixed-tissue macrophages. Conversely, small neutral liposomes (<0.1 µm diameter) have access to cells located in interstitial spaces, in addition to the phagocytic cells located in intravascular spaces. Despite this seemingly limited biodistribution, liposome-encapsulated drugs or imaging agents do circulate in the blood and reach their intended target tissues, as shown by successful cancer therapy and tumor radioimaging. As an example, the application of liposomal delivery systems in cancer therapy has clearly demonstrated better drug efficacy, reduced toxicity, and tissue targeting.[23] In general, liposome presentation gives an extended plasma half-life compared with free drug.[22-24] In some cases, the plasma half-life of a liposome-encapsulated drug may reach 8 to 15 h, indicating a possible sustained release.

Various routes of administration (subcutaneous, intraperitoneal, intramuscular, oral, dermal) for liposomal drugs are also being tested for a range of uses (anticancer, peritonitis, vaccines, dermatotherapies). In addition, the intra-arterial administration of liposomal drugs has shown potential advantages for treating hepatoma or liver metastasis and for gene therapy. Techniques of drug administration via catheter-balloon are developing, and since liposomes have a good extraction ratio from the blood circulation, they may be good candidates for this route of administration.

Increasing numbers of therapeutic agents presented in liposomes are under investigation. Some are undergoing clinical trials and are expected to be licensed for use in humans in the near future.

IV. LIPOSOMAL DELIVERY OF ANTISENSE OLIGODEOXYNUCLEOTIDES

The use of liposomes as carriers for oligonucleotides has recently received increasing attention. Successful cellular delivery and enhancement of specific antisense activity have been demonstrated by some laboratories.[28-33] Since different liposomal preparations vary in their mechanism of cellular delivery, it is worth first reviewing the data on delivery of the various antisense oligodeoxynucleotides by classifying the liposome types used: (A) conventional liposomes; (B) pH-sensitive liposomes; (C) antibody-targeted liposomes; and (D) cationic liposomes.

A. CONVENTIONAL LIPOSOMES

Conventional liposomes, as termed here, are liposomes whose phospholipidic membranes do not exhibit specific properties.

Burch and Mahan[28] explored the possibility of using antisense oligodeoxynucleotides in the design of anti-inflammatory therapeutics. They observed that free oligodeoxynucleotides complementary to sequences of the Interleukin-1 (IL-1) receptor gene partially inhibited IL-1-stimulated synthesis by murine fibroblasts in culture. However, when presented in liposomes, this antisense compound completely blocked the *de novo* synthesis of IL-1 receptor and IL-1-stimulated Prostaglandin E2 (PGE2).[28] Liposomes used for those experiments were prepared according to Papahadopoulos,[34] and consisted of cochleate lipid bodies derived from phosphatidylserine MLVs.

Akhtar et al.[35] investigated the interactions of DNA oligonucleotides with liposomes and the ability of liposomes to deliver nucleic acids. They showed that the rate of permeation across the liposome membrane of natural phosphodiester oligodeoxynucleotides, as well as analogs such as phosphorothioates or methylphosphonates, was relatively slow and sustained over a period of days. Liposomes used in this study were MLVs formulated from phosphatidylcholine, dimyristoyl phosphatidylglycerol (DMPG) and cholesterol and were prepared following lipid drying, hydration, and freeze-thaw cycles.

In our laboratory we have accumulated experience on the successful delivery of antisense oligodeoxynucleotides and ribozymes using conventional liposomes prepared by a newly developed method termed the minimal volume entrapment (MVE) method.[9,32,33,36] The characteristics and application of this delivery system will be detailed below.

B. pH-SENSITIVE LIPOSOMES

Following cellular uptake, liposomes usually enter the endosomal pathway and accumulate in endosomal vesicles.[30] Generally, when loaded with extracellular foreign material, endosomes fuse with lysosomes rich in degradative enzymes resulting in hydrolysis of the endosome contents. The ability of liposome to intracellularly deliver a drug is therefore greatly influenced by its capacity to escape the endosome before fusion with a lysosome. In an attempt to avoid this lysosomal degradation, and to use the natural acidity of endosomes, pH-sensitive liposomes have been developed.[30,37] Such liposomes are unstable when exposed to low pH. Acidic destabilization of the lipid membrane induces fusion with the endosome membrane and consequently facilitates the release of the liposome content into the surrounding cytoplasm. pH-Sensitive liposomes were employed by Ropert et al.[30] to transport antisense oligodeoxynucleotides directed to an env gene region of Friend retrovirus mRNA. Retrovirus replication was efficiently inhibited in an experimental cell system at nanomolar concentrations of oligo. A much lower modulation (two- to threefold less) was observed using pH-insensitive liposomes. No effect was noted using a free phosphodiester oligomer. Although pH-sensitive liposomes in some cases show better cellular delivery *in vitro*, this result has yet to be confirmed *in vivo*.[37]

C. IMMUNOLIPOSOMES

Curiously, one of the first successful liposomal oligonucleotide preparations employed one of the most intuitive liposome targeting technologies: the antibody-targeted liposomes, which are liposomes containing antibodies directed at specific cell surface markers.[25,26,31,38]

Bayard et al.[25] described in detail a method to anchor antibodies to liposomes encapsulating oligodeoxynucleotides. Oligonucleotides were first encapsulated in SUVs, then the phosphatidylethanolamine moiety in the liposome membrane was covalently coupled with protein A, which can serve as a basic adaptor for attachment at the liposome surface of any monoclonal antibody. The strength of this methodology lies in the fact that liposomes prepared by this method are ubiquitous and, depending on the choice of antibody, can be targeted to any cell surface determinant.

Antibody-targeted liposomes have been successfully used to deliver antisense oligomers[25,31,38] or antisense RNA.[39] Leonetti et al.[31] reported a 95% reduction of vesicular stomatitis virus replication in an experimental cell system using an antisense 15-mer DNA oligomer. An antisense-specific effect was dose dependent and occurred at a concentration 100 times lower than that required for unencapsulated oligodeoxynucleotides. Further, Zelphati et al.[38] demonstrated a nearly complete inhibition of HIV replication in acutely infected cultured cells using phosphodiester antisense oligodeoxynucleotides encapsulated in immunoliposomes targeted to MHC class I molecules. In this experimental cell system, the phosphodiester-linked version appeared significantly more sequence specific than the phosphorothioate analog. However, the converse was true in chronically infected cells and the reason for this apparent discrepancy is unclear.

The immunoliposome preparation previously described[25,31,38] does not allow a high oligonucleotide encapsulation efficiency. Methods are available to improve this and immunoliposomes may then provide a powerful tool to specifically deliver oligonucleotides. Despite the initial promise, successful use of antibody-targeted liposomes *in vivo* has not been demonstrated. The cost of preparation, especially for large-scale production, might hinder the use of immunoliposomes in the clinic. Furthermore, the presence of foreign proteins at the liposome surface may induce high levels of immunity and may thus lead to the inactivation of a significant portion of the administered dose.

D. CATIONIC LIPOSOMES

Recent papers[29,40-42] have reported the use of cationic liposomes as a delivery system for antisense oligodeoxynucleotides. In contrast to those positively charged liposomes used in the past (such as those containing stearylamine), in which drug is encapsulated in the liposome internal space, contemporary cationic liposomes are preformed and spontaneously interact with nucleic acids through the electrostatic interaction of the negatively charged sugar phosphate backbone and the positive charges at the surface of the liposome.

Cationic liposomes were first described by Felgner et al.[43] as a means of facilitating gene transfer into cells for transfection. Such liposomes have proved to be very effective at delivering nucleic acids, and commercial formulations such as Lipofectin® (Gibco BRL) and others, are available. Lipofectin® reagent contains liposomes constituted from the charged lipid, DOTMA (*N*-[1-(2,3-dioleyloxy)propyl]-*N,N,N*-trimethylammonium chloride) and dioleoyl phosphatidylethanolamine in a 1:1 (wt/wt) ratio.[43] Although this type of liposomal preparation was initially developed for gene transfer, it provides a very great potential for shorter nucleotidic sequences such as oligonucleotides.

The ability of Lipofectin® to deliver oligonucleotides was evaluated using phosphorothioate antisense sequences aimed at inhibiting ICAM adhesion molecules expression in human umbilical vascular endothelial cells.[29] Lipofectin® presentation was found to increase the activity of the antisense compound by at least 1000-fold. Oligonucleotide incorporation into cells was greatly enhanced (6- to 15-fold) when cationic liposomes were used. Increased activity seemed mostly due to better intracellular delivery, the oligomer appearing distributed in the cytoplasm as well as within the nucleus, whilst oligodeoxynucleotides presented in the free form located predominantly to the endosomes. There are, however, two reported drawbacks to the use of Lipofectin®: firstly, cytotoxicity and secondly, markedly decreased activity in the presence of serum. Nevertheless, new cationic liposomes may be formulated to be less toxic and to allow better protection for oligonucleotides.

Cationic liposomes have provided a new impetus to liposome technology, especially with regard to nucleic acid delivery. However, ultramicroscopic studies demonstrate that Lipofectin® reagent complexed with DNA is not constituted of membrane bilayer-containing vesicles, but is rather a lipid coating of particles that presumably contain nucleic acid. Thus Lipofectin® or other cationic reagents may not be liposomes in the true sense of the term; however, the fact that Lipofectin is a lipid complex need not necessarily mean that it has different pharmacological properties from liposomes.

E. MVE LIPOSOMES

We have developed a novel technique called the MVE technique for encapsulating high concentrations of oligonucleotides.[9,32,33,36] Preparation of MVE liposomes has proved to be simple. MVE liposomes are highly stable, they efficiently protect oligonucleotides from nuclease degradation, and they deliver functionally active antisense compounds.

1. Liposome Preparation

The MVE technique has been presented in detail.[9,33] Briefly, liposome-encapsulated oligonucleotides are prepared by hydrating a dried film of lipid with a highly concentrated aqueous oligodeoxynucleotides solution, followed by sonication. The resulting liposomes contain very high levels of nucleic acids. For the following discussion we will be concentrating on SUVs formed from a mixture of cardiolipin, phosphatidylcholine, and cholesterol at a 2:10:7 molar ratio. Two specific steps characterize MVE preparation: (i) in order to facilitate oligonucleotide adsorption to the lipid molecules, the oligonucleotide solution is added to the lipid film in a smaller volume than the volume of aqueous solution necessary to form liposomal vesicles and (ii) aqueous solution is then added to the mixture in a volume corresponding to the minimal volume necessary to obtain excess water. Unentrapped oligonucleotides are then removed by washing liposomes by ultra-centrifugation. The maximal entrapment rate was found to be 50 to 60% of the initial input dose for a final concentration of 60 to 70 mg oligonucleotides/mg lipid. The MVE liposomal oligonucleotide preparation is reliable and very stable, since it can be stored for at least 1 month. The high MVE encapsulation efficiency improves upon other liposome-entrapment procedures, appearing to be 6- and 11-fold greater than that observed in the preparation of antibody-targeted liposomes[25] and MLVs,[35] respectively.

In order to compare stability of MVE liposome-encapsulated unmodified and modified oligonucleotides, each preparation was incubated in culture medium containing fetal bovine serum (FBS) (10%), and the results are shown in Figure 1. Oligonucleotides were completely protected from degradation when encapsulated in MVE liposomes. Phosphorothioated analogs showed a dramatically improved stability for up to 48 h compared to phosphodiester oligonucleotides.

2. Cellular Delivery

Cellular uptake of antisense oligodeoxynucleotides is greatly increased by MVE liposomal encapsulation as demonstrated in Figure 2 and as shown previously.[32,33] Uptake varies greatly, depending on the cell line used, and may reach up to 27 pmol/10^6 cells in MOLT-3 lymphoma cells (Figure 2), which corresponds to an 18-fold enhancement compared to the uptake of the free phosphorothioated analog. The major route of cellular uptake for MVE liposomes is endocytosis,[32] and consequently cell lines with different endocytic activities may have varying capacities to internalize MVE liposomes. As demonstrated in Figure 3, undegraded phosphorothioated oligodeoxynucleotides are detectable in cells for more than 2 d following incubation with the liposomal preparation. When phosphodiesters oligodeoxynucleotides are encapsulated in liposomes and incubated with cells using the same conditions as the phosphorothioates, undegraded oligomers are hardly detectable, suggesting the higher intracellular stability of the thio-analogs.

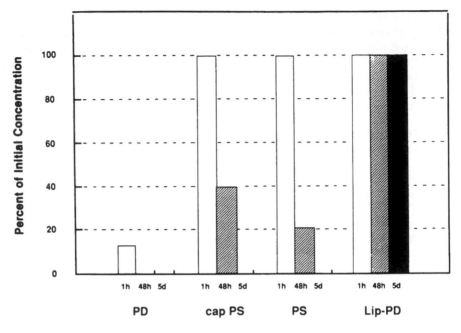

Figure 1 Stability of various forms of oligodeoxynucleotide in cell culture medium. 1 μM ^{32}P-end-labeled oligomers were incubated for 1 h, 2 and 5 d, in medium containing 10% heat-inactivated serum. Undegraded oligomers were detected following electrophoretic analysis and densitometry of the autoradiogram. Oligodeoxynucleotides: PD, phosphodiester; PS, phosphorothioated; cap PS, end-capped phosphorothioated; Lip-PD, liposome-encapsulated. Results are the mean of two or more experiments. SE ± 8%. (From Thierry, A. R. and Dritschilo, A., *Nucleic Acids Res.,* 20, 5691, 1992. With permission.)

In order to follow the intracellular pharmacokinetics of oligodeoxynucleotides in the lymphoma MOLT-3 cell line, two different parameters have been taken into consideration: (1) the level of intracellular undegraded oligomer and (2) cell efflux rate. MOLT-3 cells were incubated with 2 μM of a ^{32}P-end-labeled phosphorothioated 15 mer in liposomal or free form, and then rinsed and postincubated in fresh culture medium. As presented in Figure 4, the level of undegraded oligonucleotide decreased slowly up to 48 h following incubation. However, the level was constantly higher when oligomers were presented in liposomes (60% compared to 20% when delivered in the free form after 48 h). It is important to note that the total cell-associated radioactivity (corresponding to undegraded and degraded molecules) determined in cells treated with free phosphorothioated oligonucleotide diminished to exactly the same level as intracellular undegraded molecules.[9] Consequently, we suggested that the decrease in intracellular undegraded molecules corresponded to an efflux transport from the cells rather than an intracellular hydrolysis of the nucleotide chain.[9]

Akhtar et al.[8] have suggested that most oligodeoxynucleotide degradation occurs extracellularly when phosphorothioated analogs are added exogenously to cells. In contrast, when microinjected into oocytes, those analogs appeared to be rapidly degraded, with a half-life of approximately 3 h.[44] These observations suggest that exogenously added phosphorothioated oligonucleotides penetrate poorly into cytoplasm from the outside of the cell and that high levels of degradation may occur prior to penetration. As described for free oligomer uptake above, liposome uptake is accompanied by an efflux

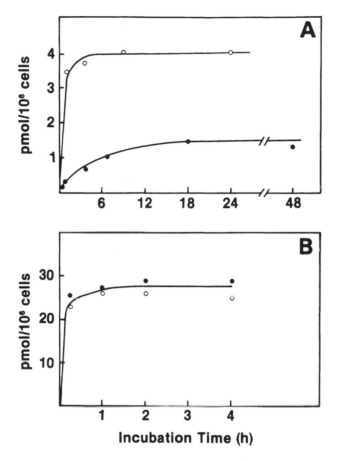

Figure 2 Cell-uptake kinetics of undegraded oligo(dn) ^{32}P-end labeled by leukemia MOLT-3 cells. Cells were incubated with 2 µM PS (●) or cap PS (○) in free (A) or liposomal (B) form. Intracellular undegraded oligomers were detected as described previously. Results are the mean of at least three experiments. SE ± 10%. (From Thierry, A. R. and Dritschilo, A., *Nucleic Acids Res.,* 20, 5691, 1992. With permission.)

mechanism (Figure 5), even though the efflux rate is much lower (40 and 79% efflux rate, following 48 h postincubation in cells treated with phosphorothioated oligomers presented in liposomal and in free form, respectively). In contrast to oligodeoxynucleotides presented in free form, oligodeoxynucleotides delivered by liposomes seem to be partially degraded in cells, suggesting a cytoplasmic rather than an endosomal distribution.

The biological effect of antisense oligodeoxynucleotides is influenced by their intracellular distribution. The latter is an important determinant, and we have examined it using laser-assisted confocal microscopy with FITC-labeled oligodeoxynucleotides (Figure 6). As previously described,[9,16,17] phosphorothioated analogs were distributed mainly in intracytoplasmic vesicles, being detectable at least 2 d postincubation (Figure 6E). When MOLT-3 cells were incubated with liposomally encapsulated analogs, oligomers first appeared in endocytic vesicles, but then relocated into the cytoplasm and subsequently concentrated in the nucleus (Figure 6D,F). A nuclear localization pattern in MOLT-3 cells (Figure 6D,F) and in human ovarian carcinoma SKVLB cells[33] following liposome treatment provided evidence that a significant release of oligonucleotides from endocytic vesicles into the cytoplasm was taking place. When oligonucleotides were

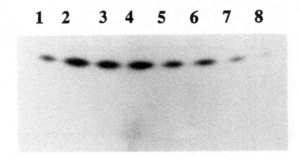

Figure 3 Intracellular content of PS (15-mer fully modified phosphorothioate oligo(dn)) in MOLT-3 cells during and after incubation of cells with 2 μM liposomally encapsulated ^{32}P-end-labeled PS. After treatment, cells were lysed and extracted with phenol, and extracts were analyzed by denaturing gel electrophoresis followed by autoradiography. Lane 1, control free PS; lanes 2, 3, and 4, cells treated for 15 min, 1 h, and 4 h with Liposomal PS, respectively; lanes 5, 6, 7, and 8, cells postincubated in drug-free medium after 4-h exposure to Lip-PS, for 1 h, 4 h, 24 h, and 48 h, respectively. (From Thierry, A. R. and Dritschilo, A., *Nucleic Acids Res.*, 20, 5691, 1992. With permission.)

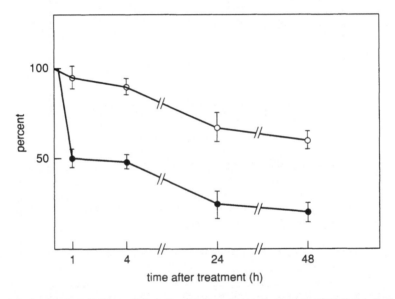

Figure 4 Reduction of intracellular oligo(dn) content in leukemia MOLT-3 cells following cell treatment. Cells were exposed to 2 μM ^{32}P-end-labeled PS (A) oligo(dn) in free (●) or liposomal (○) form. Intracellular undegraded oligomer content was estimated as previously described.[9] Results are expressed as percent of intact oligo(dn) initially incorporated after cell treatment. Results are the mean of two or more experiments. SE ± 10%.

delivered to cells in MVE liposomes containing only phosphatidylcholine and cholesterol without cardiolipin, release from endosomes was barely observed. This observation led us to assume that cardiolipin is mostly responsible for the fusogenic properties described above.

When microinjected into cells, phosphodiester or phosphorothioated oligodeoxynucleotides were rapidly transported to the nucleus.[16,17,45] Hence, liposomal presentation facilitates the release of antisense molecules from endosomes into the cytosol, while

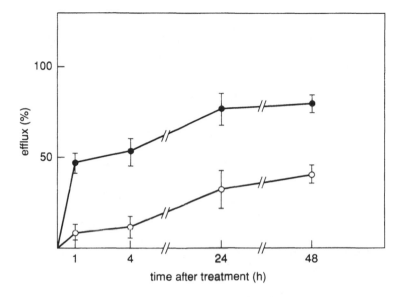

Figure 5 Relative efflux of oligo(dn) from MOLT-3 cells treated with 2 μM ^{32}P-end-labeled PS. Results are expressed as percent of oligomers initially incorporated in cells and are derived from measurement of cell-associated radioactivity. Data are the mean of two or more experiments. SE ± 10%.

direct injection seems to promote nuclear accumulation. Furthermore, the use of MVE liposomes appears to promote intracellular transport of phosphodiester oligomers, as demonstrated by the detection of unmodified oligonucleotides in both the cytoplasmic and nuclear compartments.[9]

3. Application to the Inhibition of Multidrug Resistance in Cancer Cells

We have used MVE liposomes for the delivery of antisense oligodeoxynucleotides in an attempt to inhibit multidrug resistance (MDR) in cancer cells.[33] The MDR phenotype is associated with the overexpression of the mdr-1 gene, which encodes a plasma membrane glycoprotein (P-gp) thought to act as a drug-efflux pump that is involved in the control of cellular drug accumulation.[46] End-capped phosphorothioate 15-mer oligodeoxynucleotides (with two modifications at the 5′ and 3′ ends) complementary to a loop-forming site in the mdr-1 mRNA (5′-AAAGAATACAGTGAG-3′) were encapsulated in MVE liposomes. Multidrug-resistant human ovarian carcinoma SKVLB cells treated with 5 μM of the liposomally encapsulated oligonucleotides exhibited a large decrease in P-glycoprotein synthesis and in doxorubicin resistance (Figure 7). A reduced effect was observed when free oligodeoxynucleotides were used. Compared to other MDR inhibitors, such as calcium channel blockers,[46] antisense compounds may circumvent MDR comparably and are likely to be more specific.

V. LIPOSOMAL DELIVERY OF RIBOZYMES

A. RIBOZYME TECHNOLOGY

Work from the laboratories of Cech et al.[47] and Altman et al.[48] in the early 1980s[47,48] on two unrelated systems established the ability of RNAs to catalyze the covalent cleavage of substrate RNA either in an intramolecular manner or in *trans*. Cech et al. coined the term "ribozyme" when referring to the self-splicing activities of the *Tetrahymena* ribosomal intervening sequence (IVS). However, it was recognized that the IVS RNA was not a true enzyme in that it did not undergo turnover and was itself modified during the

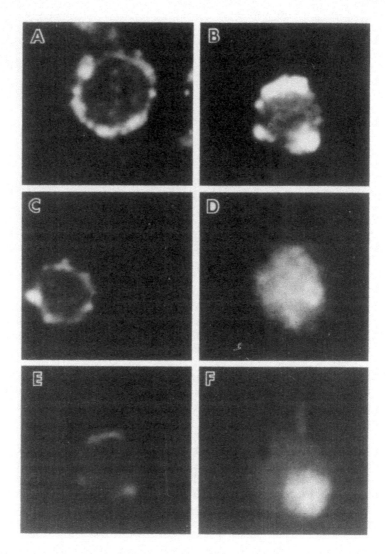

Figure 6 Intracellular distribution of FITC-end labeled cap PS in leukemia MOLT-3 cells following cap PS (A,C,E) and Lip-cap PS (B,D,F). Cells were exposed to 2 μ*M* cap PS for 24 h (A) and then incubated in drug-free containing medium for 24 h (C) or 48 h (E). Cells were treated with 2 μ*M* Lip-cap PS for 4 h (B) and then postincubated in drug-free containing medium for 24 h (D) or for 48 h (F). Each photograph represents images from laser-assisted confocal microscopy. (Magnification × 320.) (From Thierry, A. R. and Dritschilo, A., *Nucleic Acids Res.*, 20, 5691, 1992. With permission.)

reaction, but nonetheless it exhibited the capability of covalent bond breakage and formation.[49]

The first ribozyme to demonstrate turnover was the RNA component of RNase P. RNase P is an endoribonuclease that cleaves precursor tRNAs to give their mature 5′ termini. In Eubacteria such as *Escherichia coli* or *Bacillus subtilis* RNase P is composed of a protein subunit of 14,000 Da and one single-stranded RNA molecule of 377 or 401 nucleotides, respectively. The RNA component forming approximately 90% by weight of

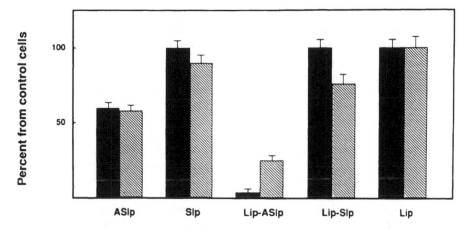

Figure 7 Inhibition of multidrug resistance phenotype in SKVLB cells by AS*lp* and S*lp* sequences presented in free or liposomal form. Cells were treated with 5 μ*M* end-capped phosphorothioate antisense (AS) or sense (S) oligodeoxynucleotides in free or liposomal (Lip-) form for 4 d. Lip, empty liposomes. Cellular P-glycoprotein content (black bar) was determined by immunoflow cytometry using the murine monoclonal antibody to P-glycoprotein, C219 as primary antibody, and an antimouse IgG fluorescein conjugate as second antibody.[33] Doxorubicin resistance (hatched bar) was ascertained by a clonogenic assay, following cell treatment for 1 h at a concentration corresponding to the drug IC$_{50}$ of SKVLB cells (20 μg/ml). Results are the mean of at least two experiments carried out in duplicate. Bars represent SE. (From Thierry, A. R., Rahman, A., and Dritschilo, A., *Biochem. Biophys. Res. Commun.*, 190, 952, 1993. With permission.)

the holoenzyme.[50] The protein and nucleic acid components of bacterial RNase P can be separated by incubation in 7 *M* urea and are thus not covalently linked.[51] Subsequent experiments demonstrated that the enzymic activity of bacterial RNase P was due exclusively to the RNA moiety and that the RNase P RNA component cleaved optimally in the presence of high-ionic strength buffers, presumably due to the stabilization of the three-dimensional structure of both the substrate and RNase P RNA.[52] At high ionic strength the Km of the catalytic RNA against a pre-tRNA is approximately the same as for the holoenzyme.[53]

Following these initial observations, numerous pathogenic self-replicating RNAs were shown to undergo self-cleavage,[54,55] as have the transcripts from a satellite DNA of the newt[56] and a plasmid for *Neurospora*.[57] Unlike RNase P, which cleaves hydrolytically, and the *Tetrahymena* Group I intron, which requires guanosine or a guanosine nucleotide cofactor for cleavage, the self-replicating RNAs undergo phosphoryl transfer in the presence of divalent cations to leave a 5′ hydroxyl and a 2′3′ cyclic phosphate.

The potential for these catalytic RNAs as sequence-specific therapies for disease associated with the presence of aberrant RNAs was quickly recognized, as was the extent of the spectrum of such relevant diseases, extending from viral and bacterial diseases through to other parasitic diseases and cancers. In fact, ribozymes are potentially useful in every case where a reduction or ablation of an mRNA is the intention. Conceptually, ribozymes also offer distinct advantages over antisense molecules because they are inherently catalytic. The recognition of this increased potential for gene inactivation has heralded the pursuit of methods to modify the catalytic RNAs into molecules that recognize target sequences, that cleave in *trans*, that can be delivered in *trans* to cells, and that exhibit turnover.

The first ribozyme to be developed as a *trans*-acting mRNA inactivator or RNA restriction enzyme was a reduced version of the *Tetrahymena* IVS. This ribozyme, composed of the IVS truncated by 19 nucleotides at the 5' end, contains an internal guide sequence directed to base pair with a specific sequence present on the target substrate RNA.[57] Cleavage of globin mRNA and a transcript from the plasmid pBR322 was observed in a ratio that demonstrated that the ribozyme was unchanged at the end of the reaction and had a turnover of greater than one.[58] Further modifications of the *Tetrahymena* IVS have shown that it is possible to reduce the length of the ribozyme still further,[59] but that a sequence of approximately 400 nucleotides was still required for activity. The complete structure of the hammerhead ribozymes of the self-replicating RNAs are only approximately 50 nucleotides in length and thus immediately lend themselves to easy analysis of structure and easy chemical synthesis, and have therefore become one of the many foci of attempts to generate a ribozyme therapeutic. This endeavor has been greatly facilitated by the work of Sheldon and Symons,[61] who mutated the hammerhead sequence to determine a consensus requirement for self-cleavage, by Uhlenbeck,[62] who succeeded in defining catalytic and substrate domains in the hammerhead structure required for self-cleavage, and by Haseloff and Gehrlach,[63] who extended previous work by Uhlenbeck,[62] showing that the basic requirements of the hammerhead were three base-paired stems, to demonstrate that the catalytic core of the hammerhead ribozyme could be separated and used as an RNA restriction enzyme (Figure 8) designed to cleave any substrate RNA containing an appropriate NUH cleavage site selected by two adjacent sequence-specific hybridizing arms. Further developments have revealed that the basic hammerhead structure can be reduced to a stretch of approximately 35 nt without compromising *trans*-cleavage activity, giving a group of molecules that have become known as minizymes.[64]

Two variants of the self-cleaving RNAs also found in the genomes of RNA viruses are the "hairpin" ribozyme of tobacco ringspot virus[65] and the "axehead" ribozyme of hepatitis delta.[65-67] Both have been mutated and truncated to give molecules that are capable of cleavage of substrate RNAs in *trans,* and recently hairpin ribozymes have been used to inactivate HIV-1 in cell culture.[68]

Continued work on cleavage of target tRNAs using the human RNase P has demonstrated that it is possible to reduce the length of the tRNA target to one half turn of an RNA helix and a 3' proximal CCA sequence and that it is possible to separate the 5' end from the 3' end of the RNA helix without inhibiting cleavage.[69] Thus the 3' end portion of the tRNA molecule can be considered as an external guide sequence (EGS) that can be designed to direct the cleavage by RNase P of any target mRNA that contains a complementary 5' sequence.[70] *In vitro* and cell culture data have demonstrated that sequence-specific cleavage by RNase P can be mediated by custom-designed EGSs.[71]

In brief, there are now a number of catalytic RNA systems that are amenable to use as *trans*-cleavers of specific target mRNAs that are in the process of being evaluated in cell culture and *in vivo* for ultimate use in humans. Therapeutics based on the action of RNA have an advantage in that there are two possible forms for delivery, either of which may lend itself to a particular application. RNAs can either be chemically synthesized, with or without modification that may improve stability and catalytic action, and then delivered as such to the cells, tissues, or organs of interest. Alternatively, the RNA therapeutic can be transcribed from a replicon delivered to cells in the same manner as for the chemically synthesized versions.

Delivery of synthetic RNAs or RNA analogs is pharmacologically well characterized, and the delivery of such molecules can be very well controlled to achieve high concentrations in the target cell or tissue. Chemical modifications that could improve the catalytic ability and/or stability of ribozymes are still being defined. However, the modifications of the sugar-phosphate backbone commonly used for antisense molecules such as phosphorothioate linkages or analogs of the ribose moiety are just as useful for

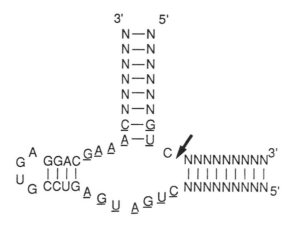

Figure 8 Schematic diagram of the generalized hammerhead structure described by Haseloff and Gehrlach[63] demonstrating the *trans*-acting ribozyme functioning as an endoribonuclease to cleave (at arrow) a selected base-paired substrate 3' to a GUC triplet. Conserved nucleotides are underlined.

ribozymes. The only potential drawback to the use of chemically altered ribozymes is the need to maintain a catalytically active three-dimensional structure. The combination of increased nuclease resistance and high catalytic activity in a modified ribozyme can quite readily be achieved. The catalytic RNA system that seems most appropriate to the use of modified RNA analogs is the EGS-directed cleavage by RNase P. The modified EGS molecule delivered *in trans* to the cells of interest is not itself catalytic, it merely directs the cleavage by endogenous RNase P by selection of a specific target by base pairing. Thus there may be a less stringent requirement for maintenance of the structure of an EGS, although recognition by RNase P of the EGS-substrate complex may be reduced by chemical modification of the EGS.

Many groups have reported the use of ribozymes and EGSs carried on replicons for the inactivation of various cellular and viral genes in tissue culture[66,71,73-77] with considerable apparent success, and some antiviral ribozymes are presumably currently undergoing evaluation in animal trials. Vector-encoded ribozymes have been delivered to cells using a variety of techniques including liposomes,[76] electroporation,[75,78] calcium phosphate,[61,64,66] or microinjection.[74] The choice of a delivery system for a replicon in tissue culture is clearly less stringent than for whole-animal experiments; however, the selection of a delivery vehicle for cell culture experiments that will be ultimately usable in animals would reduce the time lag between cell and animal testing. One possible advantage of using a virus-based ribozyme system is its potential for directing the virally encoded ribozyme to the same cellular compartment as the target viral RNA, and colocalization of this type has been shown to be essential for effective ribozyme action.[78] Certain liposome formulations have the capacity to direct the accumulation of polynucleotides in the nucleus,[78a] and it is likely with continued investigation of liposome components will come an understanding of formulations that can be tailored to deliver their contents to specific intracellular sites of action.

B. PRELIMINARY RESULTS ON MVE LIPOSOME-ENCAPSULATED RIBOZYMES

Results obtained and described earlier for liposomal delivery of antisense oligonucleotides using MVE liposomes can be extrapolated readily when considering ribozymes. All-RNA ribozymes are more susceptible to degradation than their DNA antisense

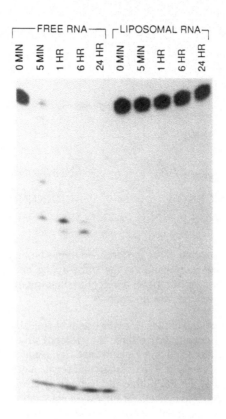

┌──── FREE RNA ────┐ ┌─ LIPOSOMAL RNA ─┐

0 MIN   5 MIN   1 HR   6 HR   24 HR   0 MIN   5 MIN   1 HR   6 HR   24 HR

Figure 9 Demonstration that encapsulating ribozymes with an MVE liposome protects them from the action of serum nucleases. A ^{32}P-labeled all-RNA hammerhead ribozyme was either encapsulated within a liposome composed of DMPG, phosphatidylethanolamine, and cholesterol (5:5:7) and then incubated for varying times in 7% heat-inactivated calf serum or incubated without encapsulation and run on a 15% Page gel. After approximately 5 min, all of the ribozyme has been degraded when not protected by the liposome, while the ribozyme remains intact to the end of the experiment (24 h) when complexed with the liposome. (P. Narayan et al., unpublished data.)

counterparts, and this increased lability has been counteracted by a number of means, including the substitution of certain nucleosides with resistant analogs as mentioned previously, or by encapsulation within liposomes.

We have investigated features of liposomal encapsulation of ribozymes using the MVE technique, particularly the protective effect provided by encapsulation with a lipid layer. Figure 9 clearly demonstrates the increased resistance to serum ribonucleases of a ribozyme encapsulated within liposomes containing DMPG, phosphatidylethanolamine, and cholesterol for at least 24 h, whilst free ribozyme is rapidly degraded in the presence of serum. This type of result has been obtained by others, e.g.,[79] and it seems that many liposome types are capable of preventing exposure of their contained oligoribonucleotides to external nucleases.

Protection from the action of serum nucleases by liposomes may allow a ribozyme, or any other encapsulated nucleic acid therapeutic, time to accumulate at its intended site of action. However, for the therapeutic to be effective in most cases it must transverse the target cell membrane and accumulate intracellularly. In cell culture numerous studies

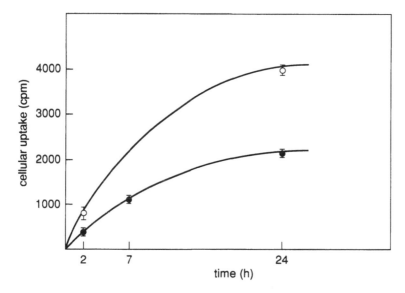

Figure 10 Time course of binding of MVE liposomes to hepatoma cells. A ^{32}P-labeled ribozyme was encapsulated by DMPG, phosphatidylethanolamine, and cholesterol (5:5:7) (open circles) or cardiolipin, phosphatidylcholine, and cholesterol MVE liposomes (2:10:7) (filled circles) and added to HepG2 cells at 70% confluency in six-well dishes. Cell-associated radioactivity was determined as described previously[9] over 24 h. Bound radioactivity after 24 h for the DMPG:PE:Ch liposomes corresponds to approximately 1.5% of the added dose.

have shown that material encapsulated within liposomes is delivered to cells to a greater extent than unencapsulated material,[9,23,32] and our data shown in Figure 10 demonstrate that an MVE-prepared liposome formulation can increase the accumulation of cell-associated ribozyme to up to 1.5% of the added dose after 24 h. Taken together, these data demonstrate that liposomes possess two of the most important features of any ribozyme delivery system, they protect their contents and they accumulate at a cellular site of action.

Reports of successful liposomal delivery of ribozymes have been published. One paper has shown that a liposomally delivered anti-HIV-1 *gag* hammerhead ribozyme accumulated undegraded within cells, and that specific intracellular degradation of the target RNA occurs to reveal expected cleavage products.[79] This report is encouraging, since intracellular ribozyme cleavage products would be expected to be degraded very rapidly, and demonstrates that the specific catalytic action of ribozymes on RNAs can indeed be detected intracellularly.

One further report of the successful use of a liposome-delivered ribozyme, again a hammerhead, described the inhibition of the *bcr-abl* fusion mRNA in a Philadelphia chromosome-positive cell line derived from a patient with chronic myelogenous leukemia.[80] In this work the cationic liposome Transfectam® (lipopolyamine, Promega, Madison, WI) was used to deliver the first dose of ribozyme, and thereafter unencapsulated ribozyme was delivered to the cells. Stability of the ribozyme was increased by substituting part of the structure with deoxyribonucleotides, making it a DNA-RNA hybrid.

Further work needs to be carried out before it becomes possible to select a liposome of choice to fulfill all specific delivery requirements for ribozymes. The preliminary work described here hints that such a specific delivery system will be achievable in the near future.

Table 1 **Capability on various aspects of the use of different liposome delivery systems for oligonucleotide-based therapeutics**

	Protection from Degradation	Intracellular Availability for Antisense Activity	Cellular Specificity	Toxicity	Clinical Feasibility
Conventional liposomes	+++	+	+	+	+++
Immunoliposomes	++	++	+++	+	+
Cationic liposomes	+	+++	+	+++	++
pH-Sensitive liposomes	++	++	+	++	++
MVE liposomes	+++	++	+	++	+++

Note: Estimated relative capability has been expressed as low (+), medium (++), and high (+++).

VI. CONCLUSION

The different liposomes employed as oligonucleotide carriers have been shown to stabilize DNA or RNA molecules in a biological setting and to potentially improve intracellular availability for antisense activity. Liposomal delivery has clearly improved the efficacy of some antisense molecules, as observed in various experimental cell systems. Table 1 summarizes the relative capabilities of the different liposomal preparations discussed herein. No *in vivo* application of liposomal delivery for antisense compounds has been yet reported. However, it seems likely that this type of delivery system might improve the drug bioavailability and allow sustained drug delivery to target tissues. With this in mind, we are investigating the effectiveness of liposome-entrapped unmodified antisense oligonucleotides in a Kaposi's Sarcoma mouse experimental model.[80a]

The use of oligodeoxynucleotide analogs, in particular phosphorothioates, results in improvement of the *in vitro* antisense activity and positive results in this area have allowed the progression to *in vivo* studies in animal models and clinical trials.[14] However, modifications of the oligonucleotide backbone alter the kinetics and the specificity of association with target sequences. Phosphorothioate oligomers are chiral and their hybridization to duplex DNA-RNA is weaker than natural phosphodiester compounds.[14] In addition, phosphorothioate derivatives exhibit nonsequence-specific inhibition of RNase H, thus decreasing antisense activity and bringing about a nonspecific effect that is not seen with phosphodiester oligomers.[81]

The presence of phosphorothioate modifications to nucleic acids leads to increased nonspecific interaction with serum and cellular proteins.[8,10] Gao et al.[10] suggested that phosphorothioated oligonucleotides may bind extensively to intracellular proteins, and this binding leads to the characteristic long retention time and the low biological activity. In addition, recent data have indicated that all oligonucleotides, especially phosphorothioate derivatives, may bind to serum proteins to a significant extent. It has been suggested that phosphorothioated oligonucleotides bind to albumin and that this contributes to the long β-half-life of elimination in animal models.[82,82a] It is not clear whether serum protein binding improves the pharmacologic effect or causes inactivation. Furthermore, it has been recently reported that phosphorothioated oligomers may be immunogenic.[83,84] Thus, taken together, these observations show that except for increased stability, phosphorothioated oligonucleotides have numerous limitations.

If one is therefore required to use phosphodiester, one way of getting around the inherent instability of phosphodiester oligodeoxynucleotides would be to package them into liposomes.

Liposomes might be successfully used to deliver a recently developed oligomer analog, peptide nucleic acids (PNA). PNAs are oligonucleotide-like compounds in which the entire sugar-phosphate unit is replaced by an appropriately sized polyamide structure.[85] These derivatives are nuclease resistant and show very high RNA or DNA hybridization potency.[85] However, one problem is that PNAs are very poorly taken up by cells. The use of a liposomal carrier might improve their transmembrane transport and allow evaluation in a cell system.

In the case of ribozymes, modified analogs are still being developed with a mind to the preservation of secondary structure. The use of a ribozyme compound as a therapy relies on the development of an effective delivery system, and in this regard liposomal carriers are promising candidates. As for antisense therapeutics, ribozymes can be chemically synthesized and delivered as such, or can be transcribed from a gene in a transient or stable expression system. Both approaches can be designed to bring about greater concentrations of therapeutic in specific intracellular compartments where activity occurs.

Hence, both "drug-like" therapy approaches using synthetic nucleic acids, or gene therapy approaches may be considered for oligonucleotide-based therapeutics. Each of these is accompanied by attendant shortcomings, and the paucity of comparative data on the use of both types of oligonucleotide in cell culture against target mRNAs does not yet permit a valid judgment of the best choice. Use of a vector system would alleviate the need for stringent protection of an RNA or DNA from the action of serum nucleases, it would allow expression of RNA, might permit accumulation in the correct cellular compartment, and may facilitate targeting. However, the introduction of foreign, relatively uncharacterized genetic elements into humans has an unpredictable outcome.

The reported liposome delivery systems so far developed for oligonucleotide-based therapeutics clearly improve biological activity. Liposomal transport presents an alternative to the use of chemically modified derivatives for improving the stability of delivered oligonucleotides. Despite a great early potential, liposomes have not provided the expected spectrum of highly effective delivery vehicles for a variety of therapeutics. However, recent research in the delivery of nucleic acids by liposomes and lipid complexes has provided a new impetus for the field, and within this area improved methods will undoubtedly be developed in the future. However, it will only be from an *in vivo* evaluation of liposomes as a delivery system for oligonucleotide-based therapeutics that a real conclusion of their benefits can be drawn.

ACKNOWLEDGMENTS

We thank Drs. Umberto Pace and Peter Rabinovich for their critical reading of the manuscript, and Dr. Prema Narayan for her permission to use unpublished data.

REFERENCES

1. Rothenberg, M., Johnson, G., Laughlin, C., et al., Oligodeoxynucleotides as antisense inhibitor of gene expression: therapeutic implications. *J. Natl. Cancer Inst.*, 81, 1539, 1989.
2. Helene, C. and Toulme, J. J., Specific regulation of gene expression by antisense, sense and antigene nucleic acids. *Biochim. Biophys. Acta*, 1049, 99, 1990.
3. Erickson, R. P. and Yzant, J., *Gene Regulation, Biology of Antisense RNA and DNA I*. Raven Press, New York, 1992.
4. Miller, P. S., Me Parland, K. B., Jayaraman, N., and Ts'o, P., Biochemical and biological effects of nonionic nucleic acid methylphosphonates. *Biochemistry*, 20, 1874, 1981.

5. Zon, G. and Stec, W. J., *Oligonucleotides and Analogues — a Practical Approach,* Eckstein, F., IRL Press, Oxford, 87, 1991.

6. Bayever, E., Iversen, P., Smith, L., and Zon, G., Systemic human antisense therapy begins. *Antisense Res. Dev.,* 2, 109, 1992.

7. Thierry, A. R., Bradley, M. O., and Chrisey, L. A., Progress in antisense research and development. *Antisense Res. Dev.,* 3, 411, 1993.

8. Akhtar, S., Kole, R., and Juliano, R. L., Stability of antisense DNA oligodeoxynucleotide analogs in cellular extracts and sera. *Life Sci.,* 49, 1793, 1991.

9. Thierry, A. R. and Dritschilo, A., Intracellular availability of unmodified, phosphorothioated and liposomally encapsulated oligodeoxynucleotides for antisense activity. *Nucleic Acids Res.,* 20, 5691, 1992.

10. Gao, W. Y., Storm, C., Egan, W., and Cheng, Y. C., Cellular pharmacology of phosphorothioate homooligodeoxynucleotides in human cells. *Mol. Pharmacol.,* 45, 45, 1993.

11. Shoji, Y., Akhtar, S., Periasamy, A., Herman, B., and Juliano, R. L., Mechanism of cellular uptake of modified oligodeoxynucleotides containing methylphosphonate linkage. *Nucleic Acids Res.,* 19, 5543, 1991.

12. Spiller, D. G. and Tidd, D. M., The uptake kinetics of chimeric oligodeoxynucleotide analogs in human leukaemia MOLT-4 cells. *Anticancer Drug Design,* 7, 115, 1992.

13. Iversen, P. L., Zhu, S., Meyer, A., and Zon, G., Cellular uptake and subcellular distribution of phosphorothioate oligodeoxynucleotides into cultured cells. *Antisense Res. Dev.,* 2, 211, 1992.

14. Stein, C. A. and Cheng, Y. C., Antisense oligonucleotides as therapeutic agents — is the bullet really magical? *Science,* 261, 1004, 1993.

15. Degols, G., Machy, P., Leonetti, J., Leserman, L., and Lebleu, B., Transmembrane passage and cell targeting of antiviral synthetic oligonucleotides, in *Antisense RNA and DNA.* Wiley-Liss, London, 1992, 255.

16. Leonetti, J., Mechti, N., Degols, G., and Lebleu, B., Intracellular distribution of microinjected antisense oligonucleotides. *Proc. Natl. Acad. Sci. U.S.A.,* 88, 2702, 1991.

17. Chin, D. J., Zon, G., Szoka, F. C., and Straubinger, F. M., Rapid nuclear accumulation of injected oligodeoxynucleotides. *New Biol.,* 2, 1091, 1990.

18. Mirabelli, C. K., Benett, C. F., and Crooke, S. T., *In vitro* and *in vivo* pharmacologic activities of antisense oligonucleotides. *Anticancer Drug Design,* 6, 647, 1991.

19. Iversen, P., In vivo studies with phosphorothioate oligonucleotides: pharmacokinetics prologue. *Anticancer Drug Design,* 6, 531, 1991.

20. Offensperger, W. B., Igloi, G., Blum, H., and Gerok, W., *In vivo* inhibition of duck hepatitis B virus replication and gene expression by phosphorothioate modified antisense oligodeoxynucleotides. *EMBO J.,* 12, 1257, 1993.

21. Agrawal, S., Temsamani, J., and Tang, J. U., Pharmokinetics biodistribution and stability of oligodeoxynucleotide phosphorothioates in mice. *Proc. Natl. Acad. Sci. U.S.A.,* 88, 7595, 1991.

22. Ostro, M. J. and Cullis, P. R., Use of liposomes as injectable-drug delivery systems. *Am. J. Hosp. Pharm.,* 46, 1576, 1989.

23. Sugarman, S. M. and Perez-Soler, R., Liposomes in the treatment of malignancy: a clinical perspective. *Crit. Rev. Oncol. Hematol.,* 12, 231, 1992.

24. Mayhew, F. G., Liposomes and delivery of chemotherapeutic agents. *Adv. Oncol.,* 9, 3, 1993.

25. Bayard, B., Leserman, L. D., Brisbal, E., and Lebleu, B., Antiviral activity in L1210 cells of liposome-encapsulated (2'5') oligo(adenylate) analogues. *Eur. J. Biochem.,* 151, 319, 1985.

26. Betageri, G. V., Black, V., Wahl, L. M., and Weinstein, J. N., Fc-Receptor-mediated targeting of antibody-bearing liposomes containing dideoxycytidine triphosphate to human monocytes/macrophages. *J. Pharm. Pharmacol.,* 45, 48, 1993.

27. Hwang, K. L. and Beaumier, P. L., Disposition of liposomes *in vivo* in Liposomes as Drug Carriers, Gregoriadis, G., Ed., John Wiley & Sons, New York, 1988, chap. 2.

28. Burch, R. M. and Mahan, L., Oligonucleotides antisense to the interleukin-1 receptor mRNA block the effects of interleukin in cultured murine and human fibroblasts and in mice. *J. Clin. Invest.,* 88, 1190, 1991.

29. Bennett, C. F., Chiang, M. Y., Chan, H., and Mirabelli, C. K., Cationic lipids enhance cellular uptake and activity of phosphorothioate antisense oligonucleotides. *Mol. Pharmacol.,* 41, 1023, 1992.

30. Ropert, C., Lavignon, M., Couvreur, P., and Malvy, C., Oligonucleotides encapsulated in pH sensitive liposomes are efficient towards Friend Retrovirus. *Biochem. Biophys. Res. Commun.,* 183, 879, 1992.

31. Leonetti, J. P., Machy, P., Degols, G., Lebleu, B., and Leserman, L., Antibody-targeted liposomes containing oligodeoxyribonucleotides complementary to viral RNA selectively inhibit viral replication. *Proc. Natl. Acad. Sci. U.S.A.,* 87, 2448, 1990.

32. Thierry, A. R., Rahman, A., and Dritschilo, A., Liposomal delivery as a new approach to transport antisense oligonucleotides, in *Gene Regulation: Antisense RNA and DNA,* Erickson, R. P. and Yzant, J., Eds., Raven Press, New York, 1992, 147.

33. Thierry, A. R., Rahman, A., and Dritschilo, A., Overcoming multidrug resistance in human tumor cells using free and liposomally encapsulated antisense oligodeoxynucleotides. *Biochem. Biophys. Res. Commun.,* 190, 952, 1993.

34. Papahadjopoulos, D., Wail, W. J., Jacobson, K., and Poste, G., Cochleate lipid cylinders: formation by fusion of unilamellar lipid vesicles, *Biochem. Biophys. Acta,* 394, 483, 1975.

35. Akhtar, S., Basu, S., Wikstrom, E., and Juliano, R. L., Interactions of antisense DNA oligonucleotide analogs with phospholipid membranes (liposomes). *Nucleic Acids Res.,* 19, 5551, 1991.

36. Thierry, A. R. and Dritschilo, A., Liposomal delivery of antisense oligodeoxynucleotides — application to the inhibition of the multidrug resistance in cancer cells. *Ann. N.Y. Acad. Sci.,* 660, 300, 1992.

37. Reddy, R., Zhou, F., Huang, L., and Rouse, B. T., *In vivo* cytotoxic T lymphocyte induction with soluble proteins administered in liposomes. *J. Immunol.,* 148, 1585, 1992.

38. Zelphati, O., Zon, G., and Leserman, L., Inhibition of HIV-1 replication in cultured cells with antisense oligonucleotides encapsulated in immunoliposomes. *Antisense Res. Dev.,* 3, 323, 1993.

39. Reinessen, K., Leserman, L., Shroder, H. C., and Muller, W., Inhibition of expression of human immunodeficiency virus-1 *in vitro* by antibody-targeted liposomes containing antisense RNA to the env region. *J. Biol. Chem.,* 265, 16337, 1990.

40. Itoh, H., Mukoyama, M., Pratt, R. E., and Dzau, V. J., Specific blockade of basic fibroblast growth factor gene expression endothelial cells by antisense oligonucleotide. *Biochem. Biophys. Res. Commun.,* 188, 1205, 1992.

41. Cumin, F., Asselbergs, R., Lartigot, M., and Felder, E., Modulation of human progenin gene expression by antisense oligonucleotides in transfected CHO cells. *Eur. J. Biochem.,* 212, 347, 1993.

42. Yu, A. C., Lee, Y. L., and Eug, L. F., The model and the effect of antisense oligonucleotides on glial fibrillary acidic protein synthesis. *J. Neurosci. Res.,* 34, 295, 1993.

43. Felgner, P. L., Gadek, T. R., Holm, M., et al., Lipofection: a highly efficient, lipid-mediated DNA-transfection procedure. *Proc. Natl. Acad. Sci. U.S.A.,* 84, 7413, 1987.

44. Cazenave, C., Stein, E., Loreau, N., et al., Comparative inhibition of rabbit globin mRNA translation by modified antisense oligodeoxynucleotides. *Nucleic Acids Res.,* 17, 4255, 1989.
45. Fisher, T. L., Terhost, T., Cao, X., and Wagner, R., Intracellular disposition and metabolism of fluorescently labeled unmodified and modified oligonucleotides into mammalian cells. *Nucleic Acids Res.,* 21, 3857, 1993.
46. Endicott, J. A. and Ling, V., The biochemistry of P-glycoprotein-mediated multidrug resistance. *Annu. Rev. Biochem.,* 58, 137, 1989.
47. Kruger, K., Grabowski, P. J., Zaug, A. J., Sands, J., Gottshling, D. E., and Cech, T. R. Self-splicing RNA: autoexcision and autocyclisation of the ribosomal RNA intervening sequences of *Tetrahymena. Cell,* 31, 147, 1982.
48. Guerrier-Takada, C., Gardiner, K., Marsh, T., Pace, N., and Altman, S., The RNA moiety of ribonuclease P is the catalytic subunit of the enzyme. *Cell,* 35, 849, 1983.
49. Cech, T. R. and Bass, B. L., Biological catalysis by RNA. *Annu. Rev. Biochem.,* 55, 599, 1986.
50. Baer, M. F., Arnez, J. G., Guerrier-Takada, C., Vioque, A., and Altman, S., Preparation and characterization of RNase P from *Escherichia coli. Methods Enzymol.,* 181, 569, 1990.
51. Altman, S., Ribonuclease P: an enzyme with a catalytic RNA subunit. *Adv. Enzymol.,* 62, 1, 1987.
52. Waugh, D. S., Green, C. J., and Pace, N. R., The design and catalytic properties of a simplified ribonuclease P RNA. *Science,* 244, 1569, 1989.
53. Pace, U. and Smith, D., Ribonuclease P: function and variation. *J. Biol. Chem.,* 265, 3587, 1990.
54. Symons, R. H., Self cleavage of RNA in the replication of small pathogens of plants and animals. *Trends Biochem. Sci.,* 14, 445, 1989.
55. Breuning, G., Compilation of self-cleaving sequences from plant virus satellite RNAs and other sources. *Methods Enzymol.,* 180, 546, 1989.
56. Epstein, L. M. and Gall, J. G., Self-cleaving transcripts of satellite DNA from the newt. *Cell,* 48, 535, 1987.
57. Saville, B. J. and Collins, R. A., A site-specific self-cleavage reaction performed by a novel RNA in *Neurospora* mitochondria. *Cell,* 61, 685, 1990.
58. Cech, T. R., Self-splicing of group 1 introns. *Annu. Rev. Biochem.,* 59, 543, 1990.
59. Zaug, A. J., Been, M. D., and Cech, T. R., The *Tetrahymena* ribozyme acts as an RNA restriction enzyme. *Nature,* 324, 429, 1986.
60. Zaug, A. J., Grosshams, C. A., and Cech, T. R., Sequence-specific endoribonuclease activity of the Tetrahymena ribozyme: enhanced cleavage of certain oligonucleotide substrates that form mismatched ribozyme-substrate complexes. *Biochemistry,* 27, 8924, 1988.
61. Sheldon, C. C. and Symons, R. H., Mutagenesis analysis of self-cleaving RNA. *Nucleic Acids Res.,* 17, 5679, 1989.
62. Uhlenbeck, O. C., A small catalytic oligonucleotide. *Nature,* 328, 596, 1987.
63. Haseloff, J. and Gehrlach, W. L., Simple RNA enzymes with new and highly specific endoribonuclease activities. *Nature,* 334, 585, 1988.
64. McCall, M. J., Hendry, P., and Jennings, P. A., Minimal sequence requirements for ribozyme activity. *Proc. Natl. Acad. Sci. U.S.A.,* 89, 5710–5714, 1992.
65. Hampl, A., Tritz, R., Hicks, M., and Cruz, P., "Hairpin" catalytic RNA model: evidence for helics and sequence requirements for substrate RNA. *Nucleic Acids Res.,* 18, 299, 1990.
66. Sharmeen, L., Kuo, M. Y. P., Dinter-Gottleib, G., and Taylor, J., Antigenomic RNA of human hepatitis delta virus can undergo self-cleavage. *J. Virol.,* 62, 2674, 1988.

67. Wu, H. N., Lin, Y. J., Lin, F. P., Makino, S., Chang, M. F., and Lai, M. M., Human hepatitis delta virus RNA subfragments contain an autocleaving activity. *Proc. Natl. Acad. Sci. U.S.A.,* 86, 6821, 1989.

68. Branch, A. D., Levine, B. J., and Robertson, H. D., The brotherhood of circular RNA pathogens: viroids, circular satellites, and the delta agent. *Semin. Virol.,* 1, 143, 1990.

69. Yu, M., Ojwang, J., Yamada, O., Hampel, A., Rapaport, J., Looney, D., and Wong-Staal, F., A hairpin ribozyme inhibits expression of diverse strains of human immunodeficiency virus type 1. *Proc. Natl. Acad. Sci. U.S.A.,* 90, 6340, 1993.

70. McClain, W. H., Guerrier-Takada, C., and Altman, S., Model substrates for an RNA enzyme. *Science,* 238, 527, 1987.

71. Forster, A. C. and Altman, S., External guide sequences for an RNA enzyme. *Science,* 249, 783, 1990.

72. Yuan, Y., Hwang, E., and Altman, S., Targeted cleavage of mRNA by human RNaseP. *Proc. Natl. Acad. Sci. U.S.A.,* 89, 8006–8010, 1992.

73. Xing, Z. and Whitton, J. L., An anti-lymphocytic Choriomeningitis virus ribozyme expressed in tissue culture cells diminishes viral RNA levels and leads to a reduction in infectious virus yield. *J. Virol.,* 67, 1840–1847, 1993.

74. Zhao, J. J. and Pick, L., Generating loss-of-function phenotypes of the *fushi tarazu* gene with a targeted ribozyme in *Drosophila. Nature,* 365, 448–451, 1993.

75. Cameron, F. H. and Jennings, P. A., Specific gene suppression by engineered ribozymes in monkey cells. *Proc. Natl. Acad. Sci. U.S.A.,* 86, 9139–9143, 1989.

76. L'Huillier, P. J., Davis, S. R., and Bellamy, A. R., Cytoplasmic delivery of ribozymes leads to efficient reduction in α-lactalbumin mRNA levels in C1271 mouse cells. *EMBO J.,* 11, 4411–4418, 1992.

77. Cotten, M. and Birnstiel, M. L., Ribozyme mediated destruction of RNA *in vivo. EMBO J.,* 8, 3861–3866, 1989.

78. Sullenger, B. A. and Cech, T. R., Tethering ribozymes to a retroviral packaging signal for destruction of viral RNA. *Science,* 262, 1566–1569, 1993.

78a. Thierry, A. R. and Takle, G. B., unpublished observations.

79. Rossi, J. J., Elkins, D., Zaia, J. A., and Sullivan, S., Ribozymes as anti-HIV therapeutic agents: principles, applications and problems. *Aids Res. Hum. Retroviruses,* 8, 183–189, 1992.

80. Snyder, D. S., Wu, Y., Wang, J. L., Rossi, J. J., Swiderski, P., Kaplan, B. E., and Forman, S. J., Ribozyme-mediated inhibition of *ber-abl* gene expression in a Philadelphia chromosome-positive cell line. *Blood,* 82, 600–605, 1993.

80a. Lunardi-Iskandar, et al., submitted.

81. Ghosh, M. L., Ghosh, K., and Cohen, J. S., Translation inhibition by phosphorothioate oligodeoxynucleotides in cell-free systems. *Antisense Res. Dev.,* 2, 111, 1992.

82. Vlasov, V. V., Pautova, L. V., Ryhova, E. Y., and Yakubov, L. A., Oligonucleotide interaction with serum proteins, *Biokhimia,* 58, 1247, 1993.

82a. Yakubov, L., personal communication.

83. Branda, R. F., Moore, A. L., Mathews, L., McCormack, J. J., and Zon, G., Immune stimulation by an antisense oligomer complementary to the "rev" gene of HIV-1. *Biochem. Pharmacol.,* 45, 2037, 1993.

84. Pistesky, D. S. and Reich, C., Stimulation of *in vitro* proliferation of murine lymphocytes by synthetic oligodeoxynucleotides. *Mol. Biol. Rep.,* 18, 217, 1993.

85. Nielsen, P. E., Egholm, M., Berg, R. H., and Buchardt, O., Sequence-selective recognition of DNA by strand displacement with a thymine-substituted polyamide. *Science,* 254, 1997, 1991.

86. Morishita, R., Gibbons, G. H., Ellison, K. E. et al., Single intramural delivery of antisense cdc kinase and proliferating-cell nuclear antigen oligonucleotides results in chronic inhibition of neointimal hyperplasia. *Proc. Natl. Acad. Sci. U.S.A.,* 90, 8474, 1993.

Chapter 14

Intracellular Delivery of Oligonucleotides with Cationic Liposomes

C. Frank Bennett

CONTENTS

I. INTRODUCTION

Oligonucleotides represent a chemical class of agents currently being explored for their potential therapeutic applications. Although the interaction of oligonucleotides with nonnucleic acids molecules could prove to have therapeutic utility, it is the predictable hybridization of oligonucleotides with nucleic acids targets which has received the greatest amount of attention. As is evident from several recent review articles, numerous investigators have used oligonucleotides to inhibit the expression of a variety of cellular and viral gene products, primarily as antisense agents.[1-3] To be effective as an antisense or antigene agent, the oligonucleotide must reach its nucleic acid receptor, i.e., RNA or DNA, inside the cell. The process by which oligonucleotides gain access to their receptors is poorly understood and may differ between cells in culture and cells within tissues. We have found that cationic liposome preparations markedly increase the potency of phosphorothioate antisense oligonucleotides in cell culture-based assays.[4,5] In this chapter the use of cationic liposomes to facilitate delivery of antisense oligonucleotides to cells shall be discussed.

II. EFFECT OF CATIONIC LIPOSOMES ON THE CELLULAR PHARMACOKINETICS OF OLIGONUCLEOTIDES

Cationic lipids were originally synthesized as general drug delivery vehicles which would interact with the negative charge on the cell surface. Felgner et al.[6] demonstrated that liposomes prepared from the synthetic cationic lipid, N-[1-(2,3-dioleyloxy)propyl]-N,N,N-trimethylammonium chloride (DOTMA), and dioleoyl phosphatidylethanolamine (DOPE) facilitated the intracellular delivery of DNA and RNA to cultured cells.[7] Additional cationic lipid species have been described which enhance nucleic acid transfection.[8-13] In addition, cationic liposomes have also been used to deliver proteins into cells.[14,15]

We have previously reported that liposomes composed of DOTMA/DOPE enhanced the biological activity of a phosphorothioate oligonucleotide by at least a factor of 1000 in human umbilical vein endothelial cells.[4,5] Subsequent studies have demonstrated that

0-8493-4778-5/95/$0.00+$.50

DOTMA/DOPE liposomes enhanced the activity of oligonucleotides targeting several gene products in numerous cell lines.[16-27] Cationic liposomes appear to enhance the biological activity of phosphorothioate oligonucleotides in part by increasing the amount of oligonucleotide which associates with the cell. Studies using human umbilical vein endothelial cells demonstrated that liposomes composed of DOTMA/DOPE enhanced the rate of association of an 18-mer phosphorothioate oligonucleotide targeting human intracellular adhesion molecule 1 (ICAM-1) mRNA with cells by 20-fold and the extent of cellular association by 10- to 15-fold.[5] In addition, incubation of cells with oligonucleotides in the presence of cationic liposome vesicles markedly changed the subcellular distribution of a fluorescently labeled oligonucleotide, as determined by fluorescence microscopy. In the absence of cationic liposomes the oligonucleotide appeared to concentrate within punctate structures in the cytoplasm of the cell. In the presence of cationic liposomes, the oligonucleotide localized predominantly within the nucleus, but not the nucleolus, with some oligonucleotide also found in cytoplasmic vesicles. The oligonucleotide appeared in the nucleus as early as 15 min following application to the cells and remained within the nucleus for at least 48 h. In the presence of cationic liposome, the appearance of oligonucleotide in the nucleus of the cell correlated with pharmacological activity. These results suggest that cationic liposomes changed the cellular pharmacokinetics of phosphorothioate oligonucleotides.

Optimal activity of cationic liposomes was observed at an equal molar ratio of positive charge from the cationic lipid and negative charge from the oligonucleotide (charge neutral). Pharmacological activity decreased if negative charge contributed by the oligonucleotide was in molar excess over the positive charge contributed by the cationic lipid, although it did not completely reverse the activity.

There are marked differences in the sensitivities of different cell lines to cationic liposome-enhanced oligonucleotide delivery. Some cells such as primary human endothelial cells or the bladder carcinoma cell line T24 are relatively sensitive to cationic liposome-mediated oligonucleotide delivery, with inhibition of gene expression occurring between 10 and 100 nM oligonucleotide concentration. Primary fibroblasts, smooth muscle cells, keratinocytes and various carcinoma cells required between 100 and 500 nM phosphorothioate oligonucleotide for inhibition of target gene expression in the presence of DOTMA/DOPE liposomes.[4,5,19,22,28] Cell lines derived from myeloid lineages, such as the human promyelocytic leukemia line, HL-60, or the human T-cell leukemia line, Jurkat, do not appear to be susceptible to cationic liposome-mediated oligonucleotide delivery. The reason for the marked differences in the sensitivities of various cell lines to cationic liposomes is unknown. It is possible that differences in membrane lipid compositions could affect the ability of the cationic lipid vesicles to fuse with cellular membranes. Alternatively, differences in the cellular biology of the different cell types, such as processing of endosome vesicles or the presence of specific cell surface molecules which interact with cationic liposomes, could account for the differential sensitivity. Further investigations into the interactions of cationic liposomes with cellular membranes should provide insights into the reason for these differences.

To date, only anionic oligonucleotides, such as phosphorothioate oligodeoxynucleotides, have been reported to be enhanced by cationic liposomes. We have found cationic liposomes to be useful for the intracellular delivery of other anionic oligonucleotide analogs. In particular, phosphorothioate oligonucleotides modified at the 2′ position of the ribose sugar have been shown to exhibit activity in several systems.[4,28] Itoh et al.[18] demonstrated inhibition of basic fibroblast growth factor expression using unmodified phosphodiester oligodeoxynucleotides in the presence of DOTMA/DOPE liposomes. In contrast, we have been unable to demonstrate inhibition of gene expression by uniform phosphodiester oligodeoxynucleotides directed towards a number of cellular targets in the presence of cationic lipids. The lack of activity appears to be due to nuclease degradation of the phosphodiester backbones rather than a unique interaction between the phosphorothioate backbone and the cationic liposome, as more nuclease-stable 2′-modified oligonucleotides with a phosphodiester backbone exhibit activity in cellular assays.[28a] Thus, cationic liposome formulations are useful for the delivery of a large number of chemical classes of oligonucleotides.

III. CATIONIC LIPID CHEMISTRY

All cationic lipids used for DNA transfections are synthetic, as there are no known naturally occurring cationic lipids which form stable bilayer structures. A number of different cationic lipid species have been described which facilitate DNA delivery to cells.[6,9-13,29,30] Based upon the published literature it is difficult to assess the relative value of one cationic lipid preparation to another as there have been relatively few comparative studies. Leventis and Silvius[29] compared several cationic lipid dispersions for DNA transfection efficiency. The cationic lipid used in the study were synthesized with ester bonds linking the hydrophobic moiety to the cationic moiety, which was postulated to be more readily metabolized in biological assay systems than ether linkages. Several lipids were found to more efficiently transfect cells than DEAE dextran transfection. Lipids based on a dioleoylglycerol hydrophobic group were more effective than lipids containing the same quaternary ammonium group derived from cholesterol.[29] The rank order potency of the lipid preparations varied, depending on the cell line examined. In CV-1 cells 1,2-dioleoyl-3-(4′-trimethylammonium)butanoyl-sn-glycerol (DOTB) was the most effective, while in 3T3 cells 1,2-dioleoyloxy-3-(trimethylammonium)propane (DOTAP) was five times more efficient than DOTB.[29] These studies demonstrate that it is risky to draw conclusions regarding the relative efficiency of cationic liposome preparations from studies in one cell line, a conclusion which is also borne out by our studies with phosphorothioate oligodeoxynucleotides. Gao and Huang[11] synthesized a cationic derivative of cholesterol, 3β[N-(N′,N′-dimethylaminoethane)-carbamoyl] cholesterol, which was threefold more active than liposomes containing DOTMA and approximately fourfold less toxic in human epidermal cells, A431.

More recently, Felgner et al.[12] synthesized a series of novel 2,3-dialkyloxypropyl quaternary ammonium compounds containing different hydroxyalkyl moieties on the quaternary amine. The lipids were compared for transfection of a β-galactosidase plasmid in COS.7 cells. Dioleyl lipids formulated with 50 mol% DOPE exhibited a rank order potency of hydroxyethyl 〉 hydroxypropyl 〉 hydroxybutyl 〉 hydroxypentyl 〉 DOTMA for transfection efficiency. These results suggest that packing of the polar head group may influence transfection efficiency of the liposomes, as lipids with larger polar groups were less effective. The hydroxyethyl derivative was synthesized with different alkyl chain substitutions to determine the effect on transfection efficiency. The order of transfection efficiency in COS.7 cells was dimyristyl 〉 dioleyl 〉 dipalmityl 〉 distearyl, demonstrating that changes in the aliphatic region can markedly affect activity.

The same series of cationic lipids formulated with 50 mol% DOPE described by Felgner et al.[12] for DNA transfection were also tested for their ability to enhance the biological activity of an ICAM-1 antisense oligonucleotide in a human lung carcinoma cell line, A549.[31] Comparison of lipids synthesized with either alkyl or acyl aliphatic groups using ICAM-1 antisense oligonucleotides suggest that alkyl lipids were slightly more effective than acyl lipids in A549 cells. In this study the differences in activity of the different cationic lipids was not as marked as in the DNA transfection assays. In A549 cells and human umbilical vein endothelial cells, the hydroxypropyl derivative 1,2-dioleyloxypropyl-3-dimethyl-hydroxypropyl ammonium bromide was consistently more effective than the other hydroxyalkyl quaternary amines. However, the hydroxyethyl, hydroxybutyl and DOTMA also enhanced the pharmacological activity of the antisense oligonucleotide. The hydroxypentyl quaternary amine was the least effective in the series. The rank order potency for different aliphatic substituted N-hydroxyethyl cationic lipids was similar to that described for DNA transfection, i.e., dimyristyl = dioleyl ⟩ dipalmityl ⟩ distearyl. The gel-to-liquid-phase transition for the different dialkyl-substituted lipids was below 0°C for the dioleyl lipid, 25°C for the dimyristyl lipid, 42°C for the dipalmityl lipid, and 56°C for the distearyl lipid.[31] Thus, cationic lipids with a gel-to-liquid-phase transition above 37°C were not as effective as those lipids in which the gel-to-liquid-phase transition was below 37°C, suggesting that it was important for the lipid to be in the liquid crystalline state for efficient delivery of nucleic acid to the cell.

IV.   EFFECT OF CO-LIPIDS

With the exception of the lipopolyamine lipids,[9] all cationic lipid preparations used for delivery of plasmid DNA to cells contain the fusogenic phospholipid phosphatidylethanolamine, with most liposomes formulated with DOPE. Cationic liposomes formulated with phosphatidylcholine in place of phosphatidylethanolamine did not enhance DNA transfection compared to the cationic lipid alone.[6,12] Our experience with delivery of phosphorothioate oligonucleotides to cells is similar to that with large plasmid DNA fragments, in that DOPE increases the activity of cationic lipids. Optimal activity was observed at an equal molar ratio of DOPE and cationic lipid.[31] Liposomes prepared with dioleoylphosphatidylcholine (DOPC) or cholesterol failed to enhance the activity of the antisense oligonucleotide. Methylated derivatives of DOPE were less effective enhancers of antisense oligonucleotide activity compared to DOPE. Cationic liposomes prepared from 1,2-dioleyloxypropyl-3-dimethyl-hydroxyethyl ammonium bromide and phosphatidylethanolamine derivatives containing different aliphatic groups were examined for their ability to enhance the activity of an ICAM-1 antisense oligonucleotide. Optimal activity was obtained with liposomes prepared with DOPE.[31] Activity decreased with saturated aliphatic chains and further decreased as the chain length decreased. The loss of activity with saturated aliphatic groups was probably a result of both an increase in gel-to-liquid-phase transition temperatures (bilayer stiffness) as well as possibly decreased lipid mixing in the vesicle, resulting in a heterogenous lipid vesicle.

Commercial preparations of a lipopolyamine molecule are supplied without a fusogenic lipid. Although they are reported to work well for DNA transfections we have been unable to demonstrate enhanced pharmacological activity for phosphorothioate oligonucleotides in A549 cells. The only cationic lipid which we have tested that did not require DOPE for enhanced delivery of phosphorothioate oligodeoxynucleotides to cells was 1,2-dimyristyloxypropyl-3-dimethyl-hydroxyethyl ammonium bromide (DMRIE). Liposomes prepared from 100% DMRIE were as effective as liposomes prepared from equal molar mixtures of DMRIE and DOPE.[31] Insights into the physical behavior of DMRIE could provide useful information on the mechanism(s) by which cationic liposomes facilitate delivery of oligonucleotides to cells.

V.   MECHANISM OF CATIONIC LIPOSOME-MEDIATED OLIGONUCLEOTIDE DELIVERY

The mechanism(s) by which cationic liposomes facilitate intracellular delivery of nucleic acids are not well understood. Cationic liposomes prepared from either DOTMA or DOTAP and DOPE aggregate and fuse in the presence of polyanions.[8,32] Cationic liposomes prepared with DOPC did not fuse, nor do they enhance delivery of nucleic acids to cells. Substitution of less fusogenic phosphatidylethanolamine (PE) derivatives resulted in decreased delivery of nucleic acids to the cells.[12,31] Thus, fusion of cationic liposome vesicles appears to be critical for their biological activity. Fusogenic lipids such as DOPE do not readily adopt a bilayer structure, but instead prefer to form a hexagonal array structure.[33,34] In addition to binding to nucleic acids, the cationic lipid component of the liposome may also help form stable lipid bilayers with fusogenic lipids.

Recently, Gershon et al.[35] studied the interaction of DNA with cationic liposomes using gel electrophoresis, metal shadowing electron microscopy, and fluorescence measurements of lipid mixing. Their data suggest that cationic liposomes bind to DNA polymers to form clusters of aggregated lipid vesicles on the DNA strand. At a critical concentration of lipid vesicles on the DNA strand, the vesicles fuse. In addition, the vesicles promote the DNA molecules to collapse, leading to condensed structures which are completely encapsulated within the fused lipid vesicles. The lipid fusion and DNA collapse appear to be interrelated. Preliminary data suggest that phosphorothioate and phosphodiester oligonucleotides also promote fusion of cationic lipid vesicles.[31] However, the initial interaction of oligonucleotides with cationic liposomes would be predicted to be different than the interaction with DNA plasmids as the size of the oligonucleotides (18 to 20 bases in length) would not physically allow the oligonucleotide to simultaneously interact with numerous lipid vesicles. It is possible that a single oligonucleotide could bind to two lipid vesicles and at higher oligonucleotide concentrations aggregates of lipid vesicles would occur. At oligonucleotide concentrations sufficient to neutralize the surface charge of the cationic liposomes the charge repulsion of the lipid vesicles would be eliminated, allowing the vesicles to fuse, entrapping the oligonucleotide in the process. The liposome-oligonucleotide complexes would then bind to cells.

Even more questions remain concerning the interaction of cationic liposomes with cellular membranes. For instance, do the liposomes initially interact with the negative charged carbohydrates on the cell surface, specific proteins on the cell surface, or the lipids in the cell membrane? Do the liposomes fuse with a cellular membrane and where does the fusion event take place? It is possible that the liposomes fuse directly with the plasma membrane, or, alternatively, the liposomes may be internalized and fuse with endosomal membranes or even caveoli membranes. Studies in endothelial cells suggest that delivery of the oligonucleotide to the cell was not inhibited by lysosomotropic amines such as chloroquine or NH_4Cl, suggesting that if the liposomes fuse with endosomal membranes it would probably occur in early endosomes.[5]

Although most studies suggest that cationic liposomes increase the pharmacological activity of phosphorothioate oligonucleotides by changing the cellular pharmacokinetics of the drug, it can not be ruled out that cationic lipids also enhance hybridization of the oligonucleotide to the target RNA in the cell. Pontius and Berg[36] report that cationic detergents increased renaturation of DNA strands by $\rangle$ 10^4-fold. The effect of the cationic detergent was to enhance the rate at which complementary strands encounter one another in solution. Studies are needed to determine the fate of the cationic lipid in the cell and to determine if a fraction of the cationic lipid remains associated with the oligonucleotide.

VI. *IN VIVO* ACTIVITY OF CATIONIC LIPOSOMES

Several papers have been published demonstrating that cationic liposome formulations enhance expression of plasmids in animals when administered locally or systemically.[37-43] Nabel et al.[38] used a balloon catheter to deliver a plasmid to specific arterial segments with no detectable systemic expression of the gene product. Stribling et al.[40] aerosolized a cationic liposome formulation (DOTMA/DOPE) in the presence of a chloramphenied acetyl transferase (CAT) expression plasmid, resulting in expression in the lung. Expression did not occur in the absence of cationic liposomes. CAT activity was expressed in lung tissue for at least 21 d following aerosolization and was not detected in other organs. Highest expression occurred in bronchiolar epithelium. Intraperitoneal injection of plasmid complexed with cationic liposomes resulted in expression of the plasmid DNA predominantly in spleen T-cells and in some bone marrow cells.[41] Following intravenous administration of cationic liposome-DNA complexes, the plasmids were widely distributed throughout the body with the highest level expression occurring in the lung, followed by the spleen.[37,43-45] Expression of the plasmid DNA was detected not only in the vascular compartment of the organs examined, but also in the extravascular compartment.

Only one study has been published in which an antisense oligonucleotide was administered in the presence of a cationic liposome formulation.[23] Intraperitoneal injection of a phosphorothioate oligonucleotide targeting the human nucleolar protein, p120, in the presence of DOTMA/DOPE liposomes inhibited intraperitoneal growth of a human melanoma in nude mice. The cationic liposome increased the activity of the oligonucleotide at least tenfold in this model. There are increasing numbers of reports demonstrating *in vivo* activity for phosphorothioate antisense oligonucleotides, all administered in the absence of cationic liposomes.[27,46-61] Results from these studies suggest that cationic liposome formulation of phosphorothioate are not obligatory to obtain pharmacological activity *in vivo*. Studies in our laboratory have demonstrated that phosphorothioate oligonucleotides selectively inhibit expression of the targeted gene product *in vivo* in the absence of cationic lipid formulations.[27,61] The oligonucleotides used for the *in vivo* studies were identified in a cell culture-based assay using cationic liposomes to enhance activity *in vitro*. In fact, the cationic liposome formulation appeared to be required for *in vitro* activity. These results suggest that the behavior of the oligonucleotide in cell culture-based experiments did not predict the pharmacokinetic behavior *in vivo*; however, pharmacological activity was predicted.

Experiments are currently in progress to determine the utility of cationic liposome formulations for *in vivo* applications of antisense oligonucleotides. Additional challenges present themselves for *in vivo* applications of liposome formulations such as generating liposomes of uniform size, stabilizing the oligonucleotide-liposome complexes in the presence of ions, and serum lipoproteins. However, preliminary data suggest that following intravenous administration cationic liposomes change the disposition of a phosphorothioate oligonucleotide.[61a] In the absence of cationic liposome, the major organs of disposition were liver, kidney, and bone marrow, with lung accumulating less than 0.5% of the total dose. In the presence of the cationic liposome formulation DMRIE:DOPE (50:50), the major organs for disposition of the oligonucleotide were liver, bone marrow, kidney, spleen, and lungs, with 50- to 100-fold more oligonucleotide accumulating in the lung. Thus, cationic lipid formulations would be expected to change the distribution of the oligonucleotide *in vivo* and may be used to selectively target certain organs with the oligonucleotide. It remains to be determined if the cationic liposomes will enhance the pharmacological activity of oligonucleotides for additional *in vivo* applications.

Preliminary data suggest that cationic liposomes were well tolerated for gene delivery studies.[38,40-43] These studies only examined effects after a single administration of the liposomes. It is unknown how they will be tolerated following repeated administration. This is of more concern for oligonucleotide-based therapeutics in that it would be expected that the liposomes would be administered more frequently than for gene therapy. Bottega and Epand[62] reported that some cationic lipid species were effective inhibitors of protein kinase C activity in an *in vitro* reaction. However, we have not observed changes in phorbol ester-induced phosphorylation patterns in cells treated with liposomes prepared from DOTMA/DOPE.[62a]

VII. CONCLUSIONS

Cationic liposome formulations have proven to be extremely useful for the identification of phosphorothioate antisense oligonucleotides which selectively inhibit gene expression in cell culture-based assays. The liposome preparations have been shown to reduce the amount of oligonucleotide required for cell culture-based assays by as much as 1000-fold. They have also facilitated obtaining reproducible and predictable inhibition of gene expression in cell culture-based assays, enabling investigations into the mechanism of action of oligonucleotides and to explore the cellular pharmacology of antisense oligonucleotides targeting various gene products. Application of cationic liposome for *in vivo* applications is just beginning to be explored. Preliminary data suggest that cationic liposome formulations change the pharmacokinetics of phosphorothioate oligonucleotides in mice, which could be used advantageously. Furthermore, studies using cationic liposomes to deliver DNA vectors *in vivo* demonstrate that they facilitate *in vivo* transfection of plasmid DNA, resulting in long-term expression in multiple organs which suggest that such formulations should also facilitate oligonucleotide delivery to cells in tissues. However, in contrast to DNA vectors, several studies have demonstrated that oligonucleotides are capable of inhibiting expression of targeted gene products in several tissues in the absence of cationic lipids. Additional studies are needed to adequately evaluate the utility of cationic liposomes for *in vivo* applications.

REFERENCES

1. Crooke, S. T., Therapeutic applications of oligonucleotides, *Annu. Rev. Pharmacol. Toxicol.*, 32, 329, 1992.
2. Stein, C. A. and Cheng, Y.-C., Antisense oligonucleotides as therapeutic agents — is the bullet really magical?, *Science*, 261, 1004, 1993.
3. Crooke, S. T. and Lebleu, B., *Antisense Research and Applications*, CRC Press, Boca Raton, FL, 1993.
4. Chiang, M.-Y., Chan, H., Zounes, M. A., Freier, S. M., Lima, W. F., and Bennett, C. F., Antisense oligonucleotides inhibit intercellular adhesion molecule 1 expression by two distinct mechanisms, *J. Biol. Chem.*, 266, 18162, 1991.
5. Bennett, C. F., Chiang, M.-Y., Chan, H., Shoemaker, J. E. E., and Mirabelli, C. K., Cationic lipids enhance cellular uptake and activity of phosphorothioate antisense oligonucleotides, *Mol. Pharmacol.*, 41, 1023, 1992.
6. Felgner, P. L., Gadek, T. R., Holm, M., Roman, R., Chan, H. W., Wenz, M., Northrop, J. P., Ringold, G. M., and Danielsen, M., Lipofection: a highly efficient, lipid-mediated DNA-transfection procedure, *Proc. Natl. Acad. Sci. U.S.A.*, 84, 7413, 1987.
7. Malone, R. W., Felgner, P. L., and Verma, I. M., Lipofectin mediated RNA transfection, *Proc. Natl. Acad. Sci. U.S.A.*, 86, 6077, 1989.
8. Stamatatos, L., Leventis, R., Zuckerman, M. J., and Silvius, J. R., Interactions of cationic lipid vesicles with negative charged phospholipid vesicles and biological membranes, *Biochemistry*, 27, 3917, 1988.
9. Behr, J.-P., Demeneix, B., Loeffler, J.-P., and Perez-Mutul, J., Efficient gene transfer into mammalian primary endocrine cells with lipopolyamine-coated DNA, *Proc. Natl. Acad. Sci. U.S.A.*, 86, 6982, 1989.
10. Pinnaduwage, P., Schmitt, L., and Huang, L., Use of a quaternary ammonium detergent in liposome mediated DNA transfection of mouse L-cells, *Biochim. Biophys. Acta*, 985, 33, 1989.
11. Gao, X. and Huang, L., A novel cationic liposome reagent for efficient transfection of mammalian cells, *Biochem. Biophys. Res. Commun.*, 179, 280, 1991.
12. Felgner, J. H., Kumar, R., Sridhar, C. N., Wheeler, C. J., Tsai, Y. J., Border, R., Ramsey, P., Martin, M., and Felgner, P. L., Enhanced gene delivery and mechanism studies with a novel series of cationic lipid formulations, *J. Biol. Chem.*, 269, 2550, 1994.

230

13. Guo, L. S. S., Radhakrishnan, R., Redemann, C. T., Brunette, E. N., and Debs, R. J., Cationic liposomes containing noncytotoxic phospholipid and cholesterol derivatives, *J. Liposome Res.*, 3, 51, 1993.
14. Walker, C., Selby, M., Erickson, A., Cataldo, D., and Valensi, J.-P., Cationic lipids direct a viral glycoprotein into the class I major histocompatibility complex antigen-presentation pathway, *Proc. Natl. Acad. Sci. U.S.A.*, 89, 7915, 1992.
15. Lin, M.-F., DaVolio, J., and Garcia, R., Cationic liposome-mediated incorporation of prostatic acid phosphatase protein into human prostate carcinoma cells, *Biochem. Biophys. Res. Commun.*, 192, 413, 1993.
16. Vickers, T., Baker, B. F., Cook, P. D., Zounes, M., Buckheit, R. W., Jr., Germany, J., and Ecker, D. J., Inhibition of HIV-LTR gene expression by oligonucleotides targeted to the TAR element, *Nucleic Acids Res.*, 19, 3359, 1991.
17. Yeoman, L. C., Daniels, Y. J., and Lynch, M. J., Lipofectin enhances cellular uptake of antisense DNA while inhibiting tumor cell growth, *Antisense Res. Dev.*, 2, 51, 1992.
18. Itoh, H., Mukoyama, M., Pratt, R. E., and Dzau, V. J., Specific blockade of basic fibroblast growth factor gene expression in endothelial cells by antisense oligonucleotide, *Biochem. Biophys. Res. Commun.*, 188, 1205, 1992.
19. Monia, B. P., Johnston, J. F., Ecker, D. J., Zounes, M., Lima, W. F., and Freier, S. M., Selective inhibition of mutant Ha-ras mRNA expression by antisense oligonucleotides, *J. Biol. Chem.*, 267, 19954, 1992.
20. Barbour, S. E. and Dennis, E. A., Antisense inhibition of group II phospholipase A_2 expression blocks the production of prostaglandin E_2 by $P388D_1$ cells, *J. Biol. Chem.*, 268, 21875, 1993.
21. Colige, A., Sokolov, B. P., Nugent, P., Baserga, R., and Prockop, D. J., Use of antisense oligonucleotide to inhibit expression of a mutated human procollagen gene (COL1A1) in transfected mouse 3T3 cells, *Biochemistry*, 32, 7, 1993.
22. Bennett, C. F., Chiang, M.-Y., Chan, H., and Grimm, S., Use of cationic lipids to enhance the biological activity of antisense oligonucleotides, *J. Liposome Res.*, 3, 85, 1993.
23. Perlaky, L., Saijo, Y., Busch, R. K., Bennett, C. F., Mirabelli, C. K., Crooke, S. T., and Busch, H., Growth inhibition of human tumor cell lines by antisense oligonucleotides designed to inhibit p120 expression, *Anti-Cancer Drug Design*, 8, 3, 1993.
24. Wagner, R. W., Matteucci, M. D., Lewis, J. G., Gutierrez, A. J., Moulds, C., and Froehler, B. C., Antisense gene inhibition by oligonucleotides containing C-5 propyne pyrimidines, *Science*, 260, 1510, 1993.
25. Bennett, C. F., Condon, T., Grimm, S., Chan, H., and Chiang, M.-Y., Inhibition of endothelial cell-leukocyte adhesion molecule expression with antisense oligonucleotides, *J. Immunol.*, 152, 3530, 1994.
26. Dean, N. M., McKay, R., Condon, T. P., and Bennett, C. F., Inhibition of protein kinase C-alpha expression in human A549 cells by antisense oligonucleotides inhibits induction of ICAM-1 mRNA, *J. Biol. Chem.*, 269, 16416, 1994.
27. Stepkowski, S. M., Tu, Y., Condon, T. P., and Bennett, C. F., Blocking of heart allograft rejection by ICAM-1 antisense oligonucleotide alone or in combination with other immunosuppressive modalities, *J. Immunol.*, in press.
28. Monia, B. P., Lesnik, E. A., Gonzalez, C., Lima, W. F., McGee, D., Guinosso, C., Kawasaki, A. M., Cook, P. D., and Freier, S. M., Evaluation of 2' modified oligonucleotides containing deoxy gaps as antisense inhibitors of gene expression, *J. Biol. Chem.*, 268, 14514, 1993.
28a. Monia, B. P., submitted.
29. Leventis, R. and Silvius, J. R., Interactions of mammalian cells with lipid dispersions containing novel metabolizable cationic amphiphiles, *Biochim. Biophys. Acta*, 1023, 124, 1990.

30. Yagi, K., Noda, H., Kurono, M. and Ohishi, N., Efficient gene transfer with less cytotoxicity by means of cationic multilamellar liposomes, *Biochem. Biophys. Res. Commun.*, 196, 1042, 1993.

31. Bennett, C. F., Mirejovsky, D. M., Tsai, Y. J., Felgner, J., Sridhar, C. N., Wheeler, C. J. and Felgner, P. L., Enhanced activity of antisense oligonucleotides with novel cationic liposome formulations, *submitted*.

32. Duzgunes, N., Goldstein, J. A., Friend, D. S. and Felgner, P. L., Fusion of liposomes containing a novel cationic lipid, N-[2,3-(dioeyloxy)propyl]-N,N,N-trimethyl-ammonium: induction by multivalent anions and asymmetric fusion with acidic phospholipid vesicles, *Biochemistry*, 28, 9178, 1989.

33. Cullis, P. R. and de Krujiff, B., The polymorphic phase behavior of phosphatidyletha-nolamines of natural and synthetic origins, *Biochim. Biophys. Acta*, 513, 31, 1978.

34. Gruner, S. M., Cullis, P. R., Hope, M. J. and Tilcock, C. P. S., Lipid polymorphism: the molecular basis of non-bilayer phases, *Annu. Rev. Biophys. Biophys. Chem.*, 14, 211, 1985.

35. Gershon, H., Ghirlando, R., Guttman, S. B. and Minsky, A., Mode of formation and structural features of DNA-cationic liposome complexes used for transfection, *Biochemistry*, 32, 7143, 1993.

36. Pontius, B. W. and Berg, P., Rapid renaturation of complementary DNA strands mediated by cationic detergents: a role for high-probability binding domains in enhancing the kinetics of molecular assembly processes, *Proc. Natl. Acad. Sci. U.S.A.*, 88, 8237, 1991.

37. Brigham, K., Meyrick, B., Christman, B., Conary, J., King, G., Berry, L. and Magnuson, M., Expression of human growth hormone fusion genes in cultured lung endothelial cells and in the lungs of mice, *Am. J. Respir. Cell Mol. Biol.*, 1, 95, 1989.

38. Nabel, E. G., Plautz, G. and Nabel, G. J., Site-specific gene expression *in vivo* by direct gene transfer into the arterial wall, *Science*, 249, 1285, 1990.

39. Felgner, P. L. and Rhodes, G., Gene therapeutics, *Nature*, 349, 351, 1991.

40. Stribling, R., Brunette, E., Liggitt, D., Gaensler, K. and Debs, R., Aerosol gene delivery in vivo, *Proc. Natl. Acad. Sci. U.S.A.*, 89, 11277, 1992.

41. Philip, R., Liggitt, D., Philip, M., Dazin, P. and Debs, R., *In vivo* gene delivery. Efficient transfection of T lymphocytes in adult mice, *J. Biol. Chem.*, 268, 16087, 1993.

42. Nabel, G. J., Nabel, E. G., Yang, Z.-Y., Fox, B. A., Plautz, G. E., Gao, X., Huang, L., Shu, S., Gordon, D. and Chang, A. E., Direct gene transfer with DNA-liposomes complexes in melanoma: expression, biological activity, and lack of toxicity in humans, *Proc. Natl. Acad. Sci. U.S.A.*, 90, 11307, 1993.

43. Zhu, N., Liggitt, D. and Debs, R., Systemic gene expression after intravenous DNA delivery into adult mice, *Science*, 261, 209, 1993.

44. Brigham, K., Meyrick, B., Christman, B., Magnuson, M., King, G. and Berry, L., *In vivo* transfection of murine lungs with a functioning prokaryotic gene using a lipo-some vehicle, *Am. J. Med. Sci.*, 298, 278, 1993.

45. Brigham, K. L. and Schreier, H., Cationic liposomes and DNA delivery, *J. Liposome Res.*, 3, 31, 1993.

46. Burch, R. M. and Mahan, L. C., Oligonucleotides antisense to the interleukin 1 receptor mRNA block the effects of interleukin 1 in cultured murine and human fibroblasts and in mice, *J. Clin. Invest.*, 88, 1190, 1991.

47. Whitsell, L., Rosolen, A. and Neckers, L. M., *In vivo* modulation of N-myc expression by continuous perfusion with an antisense oligonucleotide, *Antisense Res. Dev.*, 1, 343, 1991.

48. Chiasson, B. J., Hooper, M. L., Murphy, P. R. and Robertson, H. A., Antisense oligonucleotide eliminates *in vivo* expression of *c-fos* in mammalian brain, *Eur. J. Pharmacol.*, 227, 451, 1992.

49. Kitajima, I., Shinohara, T., Minor, T., Bibbs, L., Bilakovics, J., and Nerenberg, M., Human T-cell leukemia virus type I tax transformation is associated with increased uptake of oligodeoxynucleotides *in vitro* and *in vivo*, *J. Biol. Chem.*, 267, 25881, 1992.

50. Ratajczak, M. Z., Kant, J. A., Luger, S. M., Huiya, N., Zhang, J., Zon, G., and Gewirtz, A. M., *In vivo* treatment of human leukemia in a scid mouse model with *c-myb* antisense oligodeoxynucleotides, *Proc. Natl. Acad. Sci. U.S.A.*, 89, 11823, 1992.

51. Simons, M., Edelman, E. R., DeKeyser, J.-L., Langer, R., and Rosenberg, R. D., Antisense *c-myb* oligonucleotides inhibit arterial smooth muscle cell accumulation *in vivo*, *Nature*, 359, 67, 1992.

52. Offensperger, W.-B., Offensperger, S., Walter, E., Teubner, K., Igloi, G., Blum, H. E., and Gerok, W., *In vivo* inhibition of duck hepatitis B virus replication and gene expression by phosphorothioate modified antisense oligodeoxynucleotides, *EMBO J.*, 12, 1257, 1993.

53. Amaratunga, A., Morin, P. J., Kosik, K. S., and Fine, R. E., Inhibition of kinesin synthesis and rapid anterograde axonal transport *in vivo* by an antisense oligonucleotide, *J. Biol. Chem.*, 268, 17427, 1993.

54. Heilig, M., Engel, J. A., and Söderpalm, B., *C-fos* antisense in the nucleus accumbens blocks the locomotor stimulant action of cocaine, *Eur. J. Pharmacol.*, 236, 339, 1993.

55. Morishita, R., Gibbons, G. H., Ellison, K. E., Nakajima, M., Zhang, L., Kaneda, Y., Ogihara, T., and Dzau, V. J., Single intraluminal delivery of antisense cdc2 kinase and proliferating-cell nuclear antigen oligonucleotides results in chronic inhibition of neointimal hyperplasia, *Proc. Natl. Acad. Sci. U.S.A.*, 90, 8474, 1993.

56. Mojcik, C. F., Gourley, M. F., Klinman, D. M., Krieg, A. M., Gmelig-Meyling, F., and Steinberg, A. D., Administration of a phosphorothioate oligonucleotide antisense to murine endogenous retroviral MCF *env* causes immune effects *in vivo* in a sequence-specific manner, *Clin. Immunol. Immunopathol.*, 67, 130, 1993.

57. Osen-Sand, A., Catsicas, M., Staple, J. K., Jones, K. A., Ayala, G., Knowles, J., Grenningloh, G., and Catsicas, S., Inhibition of axonal growth by SNAP-25 antisense oligonucleotides *in vitro* and *in vivo*, *Nature*, 364, 445, 1993.

58. Pollio, G., Xue, P., Zanisi, M., Nicolin, A., and Maggi, A., Antisense oligonucleotide blocks progesterone-induced lordosis behavior in ovariectomized rats, *Mol. Brain Res.*, 19, 135, 1993.

59. Higgins, K. A., Perez, J. R., Coleman, T. A., Dorshkind, K., McComas, W. A., Sariento, U. M., Rosen, C. A., and Narayanan, R., Antisense inhibition of the p65 subunit of NF-kB blocks tumorigenicity and causes tumor regression, *Proc. Natl. Acad. Sci. U.S.A.*, 90, 9901, 1993.

60. Wahlestedt, C., Pich, E. M., Koob, G. F., Yee, F., and Heilig, M., Modulation of anxiety and neuropeptide Y-Y1 receptors by antisense oligodeoxynucleotides, *Science*, 259, 528, 1993.

61. Dean, N. M. and McKay, R., Inhibition of PKC-alpha expression in mice after systemic administration of phosphorothioate antisense oligonucleotides, *Proc. Natl. Acad. Sci. U.S.A.*, in press.

61a. Bennett, C. F. et al., unpublished data.

62. Bottega, R. and Epand, R. M., Inhibition of protein kinase C by cationic amphiphiles, *Biochemistry*, 31, 9025, 1992.

62a. Dean, N. M., unpublished observations.

Chapter 15

The Delivery of Oligonucleotides Using pH Sensitive Liposomes

Catherine Ropert, Patrick Couvreur, and Claude Malvy

CONTENTS

I. INTRODUCTION

Oligonucleotides with bases sequences complementary (antisense) to a specific RNA offer the exciting potential of selectively modulating the expression of an individual gene.[1,2] However, crucial problems such as the stability of oligonucleotides in relation to nuclease activity *in vitro* and *in vivo* and the low penetration into cells have to be solved. Since it is generally accepted that penetration of oligonucleotides into cells is an endocytotic phenomenon,[3,4] one may consider that the ultimate fate of oligonucleotides are the lysosomes where enzymatic degradation occurs. Thus, different chemical modifications have been performed in order to protect oligonucleotides and to improve their cellular uptake. Using another approach, two techniques have been proposed to increase oligonucleotide penetration into cells : electroporation[5] and microinjection,[6] both of which are supposed to introduce oligonucleotides directly into the cell cytoplasm, avoiding lyso somes. But these methods for cell incorporation have restricted applications in therapeutics trials. Thus, the development of carriers for the delivery of oligonucleotides may be considered as being more realistic for improving the *in vivo* efficacy of these molecules by protecting them against degradation and by increasing their delivery inside the cell. Liposomes are one of these particulate carriers which have the advantage of being already used in the clinic without any side effects.[7,8] Surprisingly, only a few attempts to use liposomes in the delivery of antisense oligonucleotides have been described.

In fact, liposomes are simply vesicles in which an aqueous volume is entirely enclosed by a membrane composed of lipid molecules. They form spontaneously when these phospholipids are dispersed in aqueous media.[9] They entrap quantities of materials within their aqueous compartment or within the membrane. Unfortunately, the majority of liposomes internalized by cells in culture enter through an endocytic pathway, so including a residence time in lysosomes.[10] Recently, pH-sensitive liposomes were developed to circumvent delivery to the lysosomes. They were designed based on the concept of

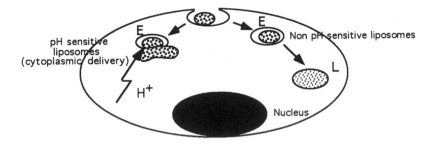

Figure 1 Schematic representation of the fate of pH-sensitive and pH-nonsensitive liposomes after cell endocytosis: release of pH-sensitive liposome content in cytoplasm from endosome; ending of pH-nonsensitive liposome content in lysomes with eventual degradation of this content.

viruses which contain proteins that can fuse with the endosomal membrane at pH $\cong$ 5 to 6, and deliver their genetic material before reaching the lysosomes.[11] Such liposomes become fusogenic when they are exposed to an acidic environment such as an endosome, a compartment that precedes the lysosomes.[12,13] The liposome might then transfer its contents into the cytoplasm before being conveyed to the lysosomes (Figure 1).

This chapter is a brief review on the usefulness of pH-sensitive liposomes for the delivery of oligonucleotides without residence time within lysosomes.

II. pH-SENSITIVE LIPOSOMES

Generally, the lipid used to design pH-sensitive liposomes is dioleyl phosphatidylethanolamine (DOPE), which has a phase transition temperature of $-16°C$. Phosphatidylethanolamine (PE) constitutes one such class of lipids which, when dispersed in a pure form, in aqueous medium, assembles into nonbilayer structures in H_{II} phase (cylindric phase)[14] (see review)[15] (Figure 2). But, in cellular membrane, PE is stabilized in lamellar phase in the surrounding mixed lipid/protein milieu. Thus, in the development of pH-sensitive liposomes, a series of PE stabilizers possessing titratable acidic headgroups have been utilized.[15] At neutral pH, all such stabilizing groups contain a negative charge which provides electrostatic repulsions able to block PE intermolecular interactions, thus preventing H_{II} formation.[15] Headgroup protonation caused by a pH decrease (like in the endosomes) induces fusion of the vesicles because PE reverts to the H_{II} phase (nonbilayer phase). The most common titratable acidic headgroups used are palmitoylhomocystein (PHC),[16] oleic acid (OA),[17] and cholesterol hemisuccinate (CHEMS).[18] In general, one can conceive of three possible interactions between the liposomes (modeled to understand fusion between endosomal and liposomal membranes): fusion, i.e., mixing of aqueous contents and mixing of bilayer components, which is the most interesting for the delivery of drugs to the cytoplasm; lysis, i.e., no mixing of aqueous contents, but leakage and mixing of bilayer contents; and mixing of bilayer components without leakage or mixing of aqueous contents. Obviously, the behavior of liposomes in acidic medium differs according to their phospholipid composition. Düzgünes et al.[17] have demonstrated irreversible changes in turbidity, as well as mixing of aqueous contents and mixing of lipids resulting from the acid-induced (pH <6.5) fusion of PE/OA vesicles. H_{II} phase transition of the PE component may be responsible for the fusion process as well as for the liposome content leakage. On the contrary, the effect of the acid pH on PE/CHEMS liposomes differs from PE/OA since no mixing of the aqueous content was observed. Apparently, in the case of the PE/CHEMS, protons induce a close apposition and destabilization followed by content leakage without a coalescence of the aqueous compartments.[19]

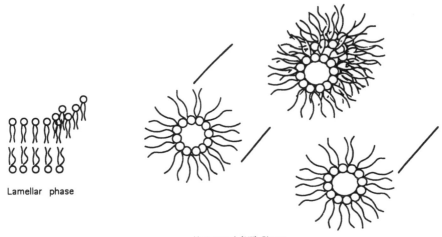

Lamellar phase

Hexagonal (HII) Phase

Figure 2 Schematic representations of the lamellar phase and hexagonal phase (H$_{II}$) of lipids in aqueous medium. (From Litzinger, D.C. and Huang, L., *Biochim. Biophys. Acta,* 1113, 201, 1992. With permission.)

The effect of divalent cations may be an important parameter to take into account in the design of pH-sensitive liposomes. Ca^{++} and Mg^{++} are able to induce the fusion of PE vesicles stabilized by the double amphiphilic chains at neutral pH.[20] In fact, these divalent cations act synergistically with protons and provoke liposome fusion at weakly acidic pH.

Therefore, the transposition into cells of the behavior of pH-sensitive liposomes in single reconstituted buffer acidic medium is very hazardous, since Mg^{++} and Ca^{++} components of cell compartments may have a great influence on the exact pH value which will induce delivery of the drug.

A few studies were reported on the efficiency of drug in pH-sensitive liposomes in cell culture and concerning oligonucleotides, a single study has only been reported. In fact, in cells the factors that control drug transfer by those vectors are still poorly understood; cytoplasmic delivery by these pH-sensitive liposomes appears incomplete.

III. pH-SENSITIVE LIPOSOMES AS DRUG DELIVERY SYSTEM

A. DELIVERY OF FLUORESCENT MARKER

pH-Sensitive liposomes composed of CHEMS/DOPE (1:2) were found to deliver entrapped calcein as well as fluorescein isothiocyanate dextran into the cytoplasm of various macrophage-like cells.[21] This was observed by fluorescence microscopy. On the contrary, in the presence of 20 mM NH$_4$Cl (which dissipates the pH gradients in acidic subcellular compartment) only punctate fluorescence was observed. Also, in addition to protons, divalent cations are required for cytoplasmic delivery by the DOPE/CHEMS liposomes as only vesicular fluorescence was revealed with the presence of ethylene diaminetrichloracetic (EDTA) in the medium. This strongly supports the idea that the transfer of calcein to the cytoplasm is a pH-dependent phenomenon. Of course, only vesicular fluorescence (endosomes and lysosomes) was revealed when cells were treated with the marker encapsulated into control pH nonsensitive liposomes. It is noteworthy that liposomes of CHEMS/DOPE are taken up by cells 5 to 10 times more efficiently than control pH nonsensitive liposomes composed of dioleyl-phosphatidyl-choline (DOPC). The same authors have related that the efficiency of calcein cytoplasmic delivery of liposomes composed of DOPE/CHEMS was more important than those composed of DOPE/OA evaluated by fluorescence microscopy.

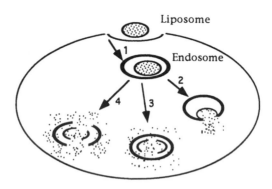

Figure 3 Proposed mechanisms of pH-sensitive liposome-mediated drug delivery from; (1) internalization of the liposomes into acidic endosomes; (2) fusion of the pH-sensitive liposome with the endosome membrane and release of contents into the cytoplasm; or (3) leakage of the liposome contents with subsequent escape into the cytoplasm; or (4) rupturing of the endosome membrane, causing release of liposome contents into the cytoplasm. (From Litzinger, D.C. and Huang, L., *Biochem. Biophys. Acta*, 1113, 201, 1992.)

It was surprising in regard to their delivery mechanism observed in acid buffer. Liposomes DOPE/OA induced aqueous content and lipid mixing, so the theoretical interaction with endosomal membrane, illustrated in step 2 (Figure 3), was fusion and expulsion of their content into the cytoplasm, the most favorable pathway for cytoplasmic delivery. In the case of CHEMS/DOPE liposomes, the decreased pH provoked destabilization and content leakage without vesicle fusion; thus, the drug released from liposomes would have to diffuse through endosomal membrane as illustrated in step 3 (Figure 3). However, it is unlikely that contents released from destabilized liposomes entered the cytoplasm from intact endosomes, since DNA and calcein, which are all unable to cross lipid membranes, were successfully delivered. Perhaps, as in step 4 (Figure 3), destabilization of liposomes caused rupturing of the endosome membrane to allow cytoplasmic delivery. In the light of these observations, it is obvious that further investigation of the delivery mechanism in cells is required.

Other authors[22] have used PHC/DOPE (8:2) liposomes covered with monoclonal antibodies in order to improve the cytoplasmic delivery, since receptor-mediated endocytosis is generally considered as more efficient than nonspecific adsorptive endocytosis. When such calcein-containing pH-sensitive immunoliposomes were incubated with target L929 cells, a diffuse fluorescence was observed throughout the cytoplasm. About half of the total cell-associated pH-sensitive immunoliposomes delivered their content into the target cell cytoplasm. This was considered as being greater than with DOPE/CHEMS liposomes. Incubation of L929 cells with control pH nonsensitive immunoliposomes (composed of phosphatidylcholine (PC)) resulted only in punctate fluorescence, indicating that these control liposomes were also bound and endocytosed by the target cells, but that release of calcein into the cytoplasm did not occur.

Although these studies gave useful qualitative information, data concerning the percent of fluorescence which was really released into the cytoplasm before reaching lysosomes were not available. Therefore, the efficacy of different pH-sensitive liposome compositions has been evaluated through the measure of the biological activity of encapsulated molecules.

B. DELIVERY OF DIPHTHERIA TOXIN: A FRAGMENT (DTA)

The fragment A of the diphtheria toxin inactivates eukaryotic protein synthesis and has been considered for killing tumor cells.[23] However, in its free form, DTA cannot diffuse through cell membranes and is, thus, nontoxic. When encapsulated into pH-nonsensitive immunoliposomes (DOPC/OA; 8:2), DTA failed to kill the target cells, contrary to the results obtained with DTA encapsulated into pH-sensitive immunoliposomes. In addition, DTA entrapped into pH-sensitive immunoliposomes was found nontoxic towards control cells, thus demonstrating target specificity.

Furthermore, cytoplasmic delivery of DTA to the macrophage-like P388D1 cell line by nontargeted liposomes composed of DOPE/CHEMS (2:1) was at least 100-fold more efficient than pH-insensitive DOPC/CHEMS liposomes.[21] However, the efficiency of cytoplasmic delivery of DTA by pH-sensitive composition is only 10%, the majority reaching lysosomes. But a single molecule of DTA is sufficient to kill a cell; this may explain the notable difference observed between the activity of pH-sensitive and pH-nonsensitive composition, although the real efficiency of cytoplasmic delivery by pH-sensitive liposome was low. This result has emphasized the heterogenous behavior of pH-sensitive liposomes in cells. Furthermore, for molecules such as oligonucleotides which require higher intracytoplasmic concentration to be efficient, this constitutes a major drawback.

DTA was used as a marker to compare the acid sensitivity of immunoliposomes with different phospholipid composition.[24] So, by combining DOPE with OA, PHC, or dipalmitoylsuccinylglycerol (DPSG), pH-sensitive liposomes (DOPE/amphiphile = 4:1) have been prepared. For these three compositions, the pH_{50} (pH at which 50% of lipid mixing occurs) was determined as being pH 6.9 for DOPE/PHC, pH 6 for DOPE/OA, and pH 5 for DOPE/DPSG. Furthermore, the $t_{1/2}$ (time required to induce 50% cell mortality) of delivery has been determined for the different immunoliposomes compositions. For DOPE/PHC, DOPE/OA, and DOPE/DPSG, the value of $t_{1/2}$ was 5 min, 15 min, and 25 min, respectively. The authors have suggested the following model: after membrane adsorption, the immunoliposomes are taken up by the cells and, within 5 min, encounter a luminal pH of approximately 6.2. At this stage, DOPE/PHC immunoliposomes release DTA into the cytoplasm. After approximately 15 min of internalization, the DOPE/OA immunoliposomes encounter a pH of approximately 6.0, allowing them to deliver DTA. As endocytosis proceeds, the pH encountered by the immunoliposomes still decreases. Approximately, 25 min after the cell internalization the DOPE/DPSG immunoliposomes encounter a pH around 5.0. This pH allows them to fuse and to deliver their DTA content into the cell cytoplasm. These values must, however, be taken with caution, because the observed cytotoxicity may also reflect the delivery by a small population of liposomes which have more rapidly reached a compartment with a sufficient acidity. This would lead to an underestimation of the average time required for endosome acidification.

C. DELIVERY OF DNA

Progress in gene therapy depends to a large degree on the development of delivery systems able to introduce DNA into the target cells. Virus transfection techniques use either a retrovirus[25] or an adenovirus,[26] but the risks of viral-based vectors have led to the development of liposomes for gene transfer. The encapsulation of DNA into liposomes is expected to increase its cellular uptake and protection against degradation. Many attempts to deliver DNA into cells have been successful with pH-sensitive liposomes. Wang and Huang[27] have used this type of carrier (DOPE/OA/CHOL [cholestrol]; 4:2:4) to deliver a gene coding for thymidine kinase into thymidine kinase-deficient cells. The rate of transfection with such a pH-sensitive composition was five- to tenfold higher than for pH-nonsensitive liposomes. We could expect to observe a more impressive difference of efficacy between the pH-sensitive and pH-nonsensitive liposomes as observed for DTA; even if the cytoplasmic delivery of pH-sensitive liposome was poor, a few DNA copies were necessary to obtain an effect. Surprisingly, this ratio of activity between the two compositions corresponded exactly to the ratio of the cell adsorbtion between DOPE and DOPC liposomes reported by Chu et al.[21] With these pH-sensitive liposomes bearing antibodies, 45% of the cells have been transfected at short term (a few hours after addition of liposomes) and 2% at long term (12 d: three cell divisions).

In order to improve the efficiency of transfection by liposomes, an easy method consisting of three cycles of freeze thawing a mixture of liposomes and DNA has been proposed.[28] This technique has been found to increase DNA entrapment (16 mg DNA/ μmol lipid). In the experiment reported, transfection efficiency has been determined by expression of a reporter plasmid pUCSV2CAT that carried the gene of *Escherichia coli* CAT (chloramphenicol acetyl transferase) under the control of SV40 early promoter activity. The acid sensitivity of the liposomes as determined by calcein release was directly related to the transfection activity. DOPE/DOSG (dioleylsuccinylglycerol) (1:1) liposomes induced a CAT activity of 5.48 mU/mg protein vs. 0.71 for pH-nonsensitive liposomes and only 0.07 for DOPC liposomes.

Legendre and Szoka et al.[29] have also compared the transfection efficiency of plasmid DNA encoding either for luciferase or for β-galactosidase when encapsulated into pH-sensitive liposomes (CHEMS/DOPE, 1:1) or pH-nonsensitive liposomes or when complexed with cationic liposomes (composed of dioleyloxypropyl-trimethylammonium/ DOPE [1:1] [Lipofectin®]). The cell-associated radiolabeled plasmid was also evaluated 2 h postincubation, using the various compositions mentioned above. It was confirmed that the level of cell association of radiolabeled plasmid encapsulated into pH-sensitive liposomes was 7 to 8 times more important than those of plasmids encapsulated in pH-nonsensitive ones. pH-sensitive liposomes mediated gene transfer with efficiency that were comprised between 1 and 30% of that obtained with cationic liposomes, while pH-nonsensitive liposome composition did not induce any detectable transfection.

D. DELIVERY OF OLIGONUCLEOTIDES

Few attemps to use liposomes in the delivery of antisense oligonucleotides have been reported, although their interest has been underlined.[30] Thierry and Dritschilo[31] have shown that liposomes increase the extravesicular localization of phosphodiester and phosphorothioate oligonucleotides in MOLT-3 cells.[31] Recently, Zephalti et al.[32] have shown that immunoliposomes increase at least 60-fold the efficiency of anti-*rev* phosphorothioate oligonucleotides in HIV chronically infected cells. Leonetti et al.[33] have used H-2K antibody-coated liposomes to deliver phosphodiester oligonucleotides. These oligonucleotides were directed against vesicular stomatitis virus N protein. The authors reported a significant inhibition of the virus titer (two log of inhibition at 0.8 μM). In this study, the effect was strictly dependent upon the use of an antibody specifically directed against a cell-surface protein on the target cells. In addition, this inhibition was influenced by the time elapsed between exposure of the cells to the liposome and the introduction of the virus. No protective effect was seen when infection immediately followed liposome binding. No hypothesis was suggested by the authors, but one may suppose that either the simultaneous presence of virus and liposomes on the cell membrane might inhibit liposome activity, or a delay might be necessary for encapsulated oligonucleotides to be present in the cell when the viral target is available. The authors have added in conclusion that, in these experiments, a large fraction of the endocytosed oligomers may be degraded in lysosomes.

Based on those considerations, we have used pH-sensitive liposomes (DOPE/OACHOL, 4:2:1) in order to avoid lysosomal degradation and to successfully achieve the transport of entrapped oligonucleotides into the cell cytoplasm.[34,35] It was observed in a preliminary experiment that DOPE liposomes with entrapped oligonucleotides remained pH sensitive *in vitro* despite their nonlamellar morphology. Indeed, 90% of ^{32}P-labeled oligonucleotides were released from DOPE liposomes after 90 min incubation at pH 5.5. On the contrary, only 15% of ^{32}P oligonucleotides were released at the same pH using

Table 1   Release of ^{32}P oligonucleotides from pH- and pH-nonsensitive liposomes at different acidic medium and at 37°C

Time in Hours	% of Oligo. Released from pH-Sensitive Liposomes			% of Oligo. Released from pH-Nonsensitive Liposomes		
	pH=4.5	pH=5.5	pH=7.5	pH=4.5	pH=5.5	pH=7.5
0	1	1	1	1	1	1
0.5	50	50	1	1	1	1
1	100	70	1	12	13	12
1.5	100	90	18	14	13	12

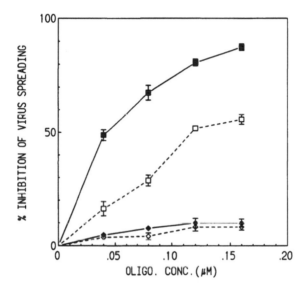

Figure 4 FIA-measured antiviral activity of encapsulated oligonucleotides in *de novo* infection. ■ Antisense oligonucleotide in pH-sensitive liposomes; □ antisense oligonucleotide in pH-nonsensitive liposomes; ◆ control oligonucleotide in pH-sensitive liposomes; ◊ control oligonucleotide in pH-nonsensitive liposomes. Results are the mean ±SD of three experiments.

pH-nonsensitive liposomes (Table 1). The efficiency of oligonucleotides was appreciated by the determination of the inhibition of the Friend murine retrovirus spreading quantified by a focus immunoassay (FIA). After 96 h, the FIA used for the immunologic detection revealed an IC$_{50}$ of about 0.04 μM with pH-sensitive liposomes, whereas this value was increased to 0.12 μM for pH-nonsensitive liposomes (Figure 4). Free oligonucleotides were unable to inhibit virus proliferation until the concentration of 100 μM. In fact, the activity of this oligonucleotide seems to be closely linked to its intracellular stability. Indeed, after 36 h incubation with cells, encapsulated oligonucleotides were completely protected in intracellular medium, whereas free oligonucleotides were completely degraded into cells within 6 h.

Retroviral inhibition was found to be specific since control oligonucleotides (reverse sense) encapsulated either in pH-sensitive or in pH-nonsensitive liposomes had no effect on viral proliferation. In the chronic infection model, known to be more resistant to drugs, 60% of inhibition was achieved for the pH-sensitive formulation, whereas only 30%

240

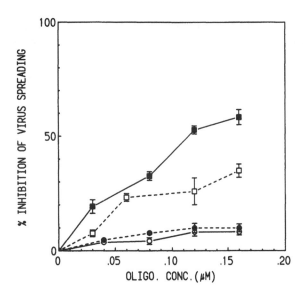

Figure 5 FIA-measured antiviral activity of encapsulated oligonucleotides in chronic infection. ■ Antisense oligonucleotide in pH-sensitive liposomes; □ antisense oligonucleotide in pH-nonsensitive liposomes; ● control oligonucleotide in pH-sensitive liposomes; ○ control oligonucleotide in pH-nonsensitive liposomes. Results are the mean ±SD of three experiments.

inhibition was obtained for pH-nonsensitive liposomes (oligonucleotide concentration of 0.16 μM) (Figure 5). pH-sensitive liposomes with antisense oligonucleotides were able to induce the inhibition of 100% of virus spreading, added at 0.16 μM twice with an interval of 24 h between each incubation. According to these results, pH-sensitive liposomes enabled a greater antiviral activity of oligonucleotides. In parallel, we have appreciated the cell association of ^{32}P oligonucleotide encapsulated in pH-sensitive and pH-nonsensitive liposomes. A twofold increase in cell association was observed when pH-sensitive liposomes were compared to pH-nonsensitive liposomes. So, the key question was whether or not pH-sensitive liposomes reacted as originally intended, because the viral inhibition obtained with such liposomes was only twice that of oligonucleotides entrapped in pH-nonsensitive liposomes. It was supposed that the higher association of DOPE containing liposomes with cells probably triggered a better delivery into the cytoplasm, which probably allowed the 100% viral inhibition. It is noteworthy that, with this model, no antibody was necessary to obtain the cell uptake of liposomes. In addition, the uptake of radiolabeled DOPE liposomes has been evaluated on infected and noninfected cells (Table 2). When associated with pH-sensitive liposomes, oligonucleotides were present in the cytoplasm of infected cells, whereas they remained only adsorbed on the membrane of noninfected cells. This suggested that the process of the virus exit facilitated probably the cellular uptake of liposomes, thus leading to a certain cell specificity even without antibodies.

IV. PLASMA STABILITY

The use of pH-sensitive liposomes in therapeutics would require that such vehicles remain stable in the circulation until binding and subsequent endocytosis by target cells. Large pH-sensitive immunoliposomes composed of DOPE/OA (4:1) as well as pH-nonsensitive immunoliposomes composed of DOPC/OA (4:1) and prepared by

Table 2 Distribution of oligonucleotide ^{14}C DOPE liposomes in Dunni cells according to time

	Time in Hours	% in Cytoplasm	% Associated with Membrane	% Noncell-associated Radioactivity
	6	1	1	98
	18	2	2	96
A	24	2	4	94
	36	3	20	77
	48	5	35	60
	6	1	1	98
	18	3	2	95
B	24	5	5	90
	36	8	12	80
	48	15	25	60
	6	2	1	97
	18	3	3	94
C	24	4	5	91
	36	20	5	75
	48	30	18	52

Note: A: Distribution of antisense or control oligonucleotide liposomes in noninfected Dunni cells; B: distribution of antisense oligonucleotides liposomes in infected cells; C: distribution of control oligonucleotide liposomes in infected cells.

reverse-phase evaporation revealed rapid leakage when incubated in mouse plasma at 37°C.[36] For the former composition, 70% of encapsulated calcein was released within 5 min incubation. Electron microscopy has revealed a high degree of aggregation of those DOPE/OA liposomes upon incubation in mouse plasma. On the contrary, small DOPE/OA liposomes obtained by sonication were found stable in plasma, but were unstable in phosphate buffer saline (PBS) at 37°C. These small liposomes reisolated after incubation in plasma were found stable in PBS at 37°C. This plasma stabilization may result from the insertion of plasma proteins into the liposomes bilayer, possibly with OA extraction.[37] This leads to a new composition of liposomes which are plasma stable but pH-nonsensitive, so unsuitable for cytoplasmic drug delivery.

In order to increase plasma stability of large pH-sensitive liposomes, cholesterol was included in the formulation. However, in DOPE/OA liposomes, the presence of cholesterol further increased OA transfer and the consequence was a decrease of the acid sensitivity of the liposomes. Inclusion of cholesterol at 40 mol% decreased the maximal calcein content release, in PBS at pH 5, from approximately 80 to 30%.[38] Moreover, this lipid exchange and, so, the loss of the pH sensitivity could explain, sometimes, the poor difference of the efficiency of pH-sensitive and pH-nonsensitive formulation.

This major drawback of liposomes designed with cholesterol has led to the use of cholesterol-free liposomes for *in vivo* delivery. To avoid lipid exchange, titratable double-chain amphiphiles have been incorporated into liposomes, but the problem of plasma stability remained unsolved. Another stabilizer has overcome the problem of the lipid exchange: DPSG.[39] After incubation with plasma, poor transfer of 1,2-DPSG included in DOPE liposomes has been observed and therefore DOPE/1,2-DPSG were found relatively stable in human plasma; this composition exhibited less than 20% release of entrapped calcein even after overnight incubation in human plasma at 37°C. Furthermore, DOPE/DPSG liposomes retained significant pH sensitivity.

V. POTENTIALITY OF USING pH-SENSITIVE LIPOSOMES FOR THE DELIVERY OF OLIGONUCLEOTIDES *IN VIVO* — PERSPECTIVES

The success of oligonucleotides *in vitro* has lead to widespread interest in extending the use of antisense compounds in the *in vivo* setting, thus bringing these agents into the therapeutic area. However, before it can be successfully accomplished a number of potential barriers to antisense therapeutics must be overcome. Obviously, as related above, stability and cellular permeability are two major drawbacks of oligonucleotides, avoided by chemical modifications. But other pharmaceutical problems, such as duration of action and target site, where desirable therapeutic effects will occur, remain. The carriers, like liposomes, may be used to deal with these requirements. pH-sensitive liposomes may have an important role in antisense pharmacology at the cellular distribution level; indeed, we have shown *in vitro* the superiority of the pH-sensitive formulation. But in regard to the body distribution, after i.v. administration like other carriers they were found to be rapidly cleared from the bloodstream and to concentrate in the liver.[40-42] Generally, when liposomes are injected into the bloodstream, their surface membrane rapidly becomes coated with various plasma proteins and they are taken up by the mononuclear phagocyte system (MPS). The function of cells of MPS located in several tissues (liver, spleen) is an important role in the host defense mechanism.

During the last few years, several new modifications of liposome technology have been created that may be of importance for delivery of oligonucleotides. One significant advance is the development of liposomes with very long circulation times. To avoid the accumulation in the liver and in the spleen, modifications of the surfaces of liposomes by the inclusion of certain negatively charged lipids such as ganglioside GM_1 have been performed. Inclusion of ganglioside GM_1 in the composition of some liposomes has been shown to prolong circulation half-life in mice by reducing uptake by the mononuclear phagocyte system.[43,44] It is supposed that GM_1 may alter the binding of plasma proteins and subsequently alter the interactions between liposomes and mononuclear phagocytes. Accordingly, GM_1 was inserted in plasma-stable, pH-sensitive liposomes composed of DOPE/DPSG.[38] Up to 5% (expressed in mol) GM_1 partial acid sensitivity was retained. This composition is promising for *in vivo* liposome-mediated drug delivery.

Another technology is the development of antibody-conjugated liposomes as a tool for drug delivery. Conjugating antibodies to the surface of liposomes could result in antigen-specific binding and cell uptake of the liposomes.[45,46] Successful drug delivery using immunoliposomes involves the use of pH-sensitive liposomes lipid composition to ensure the release of contents from the endosomal compartment. The goal would be the production of antigen-specific, pH-sensitive liposomes that combine access to target cells in tissues and the efficient delivery of liposome contents in the cytoplasm. In this view, only a single study *in vivo* has been reported until now. Wang and Huang[47] have entrapped a plasmid containing the E. coli gene under the control of mammalian cyclic adenosine monophosphate (cAMP)-regulated promoter in H-2K^K antibody-coated pH-sensitive liposomes (DOPE/OA/CHOL). The entrapped or free DNA was injected intraperitoneally into mice bearing ascites generated by lymphoma cells bearing the target antigene. About 20% of the injected immunoliposomes were taken up by target cells. The presence of the antibody onto liposomes also significantly decreased the nonspecific recognition of liposomes by the spleen. It is notable that liposomes have been injected in the same compartment as the target site, so limiting the interactions of liposomes and MPS. Significant CAT activity was detected in cells from mice treated with DNA entrapped in the pH-sensitive immunoliposomes. The requirement for acid sensitivity in the design was demonstrated when pH-insensitive immunoliposomes did not mediate efficient transfection (one fourth of pH-sensitive liposome efficiency).

In conclusion, there seem to be some reasonable possibilities for the application of liposome technology to the delivery of antisense oligonucleotides. Yet liposomes are carriers used in human clinical trials and could allow, in therapeutic trials, reduction of reinjection of oligonucleotides by protecting them from degradation. Furthermore, as shown in this review, pH-sensitive liposomes may prove valuable as a mean of releasing intact oligonucleotides (the major point) from the endosomal compartment in cytoplasmic cells, thus allowing them to reach their targets in the cytoplasm. However, the generalization of these types of carriers to various biological systems relies on a better understanding of the pathway followed by oligonucleotides encapsulated in pH-sensitive liposomes.

Currently, the challenge of a number of investigators is the production of long-lived, antigen-specific, and pH-sensitive liposomes to access target cells in tissues and deliver intact oligonucleotides to the cell cytoplasm.

REFERENCES

1. Hélène, C. and Toulmé, J.J., Specific regulation of gene expression by antisense, sense and antigene nucleic acids, *Biochem. Biophys. Acta*, 1049, 99, 1990.
2. Zon, G., Oligonucleotides as potential therapeutics agents, *Pharm. Res.*, 5, 539, 1988.
3. Loke, S.L., Stein, C.A., Zhang, X.H., Mori, K., Nakanishi, M., Subasinghe, C., Cohen, J.S., and Necker, L.M., Characterization of oligonucleotide transport into living cells, *Proc. Natl. Acad. Sci. U.S.A.*, 86, 3474–3478, 1989.
4. Yakubov, L.A., Deeva, E.A., Zarytova, V.F., Ivanova, E.M., Ryte, A.S., Yurchenko, L.V., and Vlassov, V.V., Mechanism of oligonucleotide uptake by cells: involvement of specific receptors?, *Proc. Natl. Acad. Sci. U.S.A.*, 86, 6454–6458, 1989.
5. Mir, L.M., Banou, H., and Paoletti, C., Introduction of definitive amounts of non permeant molecules into living cells after electropermeabilization: direct access to the cytosol, *Exp. Cell. Res.*, 175, 15, 1988.
6. Dash, P., Lotan, L., Knapp, M., Kandel, E. R., and Goelet, P., Selective elimination of mRNAs *in vivo* complementary oligonucleotide promote RNA degradation by an RnaseH like activity, *Proc. Natl. Acad. Sci. U.S.A.*, 84, 7896, 1987.
7. Ostro, M. J., Ed., *Liposomes: from Biophysics to Therapeutics,* Marcel Dekker, New York, 1987.
8. Poste, G., Liposomes targeting *in vivo*: problems and opportunities, *Biol. Cell,* 47, 19, 1983.
9. Bangham, A. D. and Horne, R. W., Negative staining of phospholipids and their structural modification by surface active agents as observed in the electron microscope, *J. Mol. Biol.*, 8, 660, 1964.
10. Straubinger, R.W., Hong, K., Friend, D.S., and Papahadjopoulos, D., Endocytosis of liposomes and intracellular fate of encapsulated molecules: encounter with a low pH compartment after internalization in coated vesicles, *Cell*, 32, 1069, 1983.
11. Marsh, M. and Helenius, A., Virus entry into animal cells, *Adv. Virus Res.,* 36, 107, 1989.
12. Yatvin, M.B., Kreutz, B.A., and Shinitzky, M., pH sensitive liposomes: possible clinical implications, *Science,* 210, 1253, 1980.
13. Cullen, J., Phillips, M.C., and Shipley, G.G., The effects of temperature on the composition and physical properties of the lipids of pseudomonas fluorescens, *Biochem. J.*, 125, 733, 1971.
14. Allen T.M., Hong, K., and Papahadjopoulos, D., Membrane contact, fusion, and hexagonal H_{II} transitions in phosphatidylethanolamine liposomes, *Biochemistry*, 29, 2976, 1990.
15. Litzinger, D. C. and Huang, L., Phosphatidylethanolamine liposomes: drug delivery, gene transfer and immunodiagnostic applications, *Biochim. Biophys. Acta*, 1113, 201, 1992.

16. Connor, J., Yatvin, M.B., and Huang, L., pH sensitive liposomes: acid-induced liposome fusion, *Proc. Natl. Acad. Sci. U.S.A.*, 81, 1715, 1984.

17. Düzgünes, N., Straubinger, R.M., Baldwin, P.A., Friend, D.S., and Papahadjopoulos, D., Proton induced fusion of oleic acid-phosphatidylethanolamine liposomes, *Biochemistry*, 24, 3091, 1985.

18. Bentz, J., Ellens, H., Lai, M.Z., and Szoka, F.C., On the correlation between HII phase and the contact induced destabilization, *Proc. Natl. Acad. Sci. U.S.A.*, 82, 5742, 1985.

19. Ellens, H., Bentz, J., and Szoka, F.C., pH induced destabilization of phosphatidylethanolamine containing liposomes: role of bilayer contact, *Biochemistry*, 24, 3099, 1985.

20. Ellens, H., Bentz, J., and Szoka, F.C., H^+ and Ca^{++} induced fusion and destabilization of liposomes, *Biochemistry*, 24, 3099, 1985.

21. Chu, C.J., Dijkstra, J., Lai, M.Z., Hong, K., and Szoka, F.C., Efficiency of cytoplasmic delivery by pH sensitive liposomes to cells in culture, *Pharm. Res.*, 7, 824, 1990.

22. Connor, J. and Huang, L., Efficient cytoplasmic delivery of a fluorescent dye by pH sensitive immunoliposomes, *J. Cell. Biol.*, 101, 582, 1985.

23. Collins, D. and Huang, L., Cytotoxicity of a Diphtheria toxin A fragment to toxin resistant murine cells delivered by pH sensitive immunoliposomes, *Cancer Res.*, 47, 735–739, 1987.

24. Collins, D., Maxfield, F., and Huang, L., Immunoliposomes with different acid sensitivities as probes for the cellular endocytic pathway, *Biochem. Biophys. Acta*, 987, 47, 1989.

25. Gilboa, E., Eglitis, M.A., Kantoff, P.W., and French Anderson, W., Transfer and expression of cloned genes using retroviral vectors, *Biotechniques*, 4, 504, 1986.

26. Rosenfeld, M.A., Siegfried, K., Yoshimura, K., Yoneyama, M., Fukayama, L.E., Stier, P.K., Paakko, P.K., Gilardi, P., Statford-Perricaudet, L.D., Perricaudet, M., Jallat, S., Pavirani, A., Lecocq, J.P., and Crystal, R.G., Adenovirus mediated transfer of a recombinant alpha-antitrypsin gene to the lung epithelium *in vivo*, *Science*, 252, 431, 1991.

27. Wang, C.Y. and Huang, L., Highly efficient DNA delivery mediated by pH sensitive immunoliposomes, *Biochemistry*, 28, 9508, 1989.

28. Zhou, X., Klibanov, A.L., and Huang, L., Improved encapsulation of DNA in pH sensitive liposomes for transfection, *J. Liposome Res.*, 2, 125, 1992.

29. Legendre, J.Y. and Svoka, F.C., Delivery of plasmid DNA into mammalian cell lines using pH sensitive liposomes: comparison with cationic liposomes, *Pharm. Res.*, 9, 1235, 1992.

30. Juliano, R.L. and Akhtar, S., Liposomes as a drug delivery system for antisense oligonucleotides, *Antisense Res. Dev.*, 2, 165, 1992.

31. Thierry, A.R. and Dritschilo, A., Intracellular availability unmodified, phosphorothioated and liposomally encapsulated oligodeoxyribonucleotides for antisense activity, *Nucleic Acid Res.*, 20, 5691, 1992.

32. Zephalti, O., Zon, G., and Leserman, L., Inhibition of HIV-1 in cultured cells with antisense oligonucleotides encapsulated in immunoliposomes, *Antisense Res. Dev.*, 3, 323, 1993.

33. Leonetti, J.P., Machy, P., Degols, G., Lebleu, B., and Leserman, L., Antibody targeted liposomes containing oligodeoxyribonucleotides complementary to viral RNA selectively inhibited viral replication, *Proc. Natl. Acad. Sci. U.S.A.*, 87, 2448, 1990.

34. Ropert, C., Lavignon, M., Dubernet, C., Couvreur, P., and Malvy, C., Oligonucleotides encapsulated in pH sensitive liposomes are efficient toward Friend retrovirus, *Biochem. Biophys. Res. Commun.*, 183, 879, 1992.

35. Ropert, C., Malvy, C., and Couvreur, P., Inhibition of the friend retrovirus by antisense oligonucleotides encapsulated in liposomes: mechanism of action, *Pharm. Res.*, 10, 1427, 1993.

36. Connor, J., Norley, N., and Huang, L., Biodistribution of pH sensitive immunoliposomes, *Biochim. Biophys. Acta*, 884, 474, 1986.
37. Liu, D., Zhou, F., and Huang, L., Characterization of plasma stabilized liposomes composed of dioleylphosphatidylethanolamine and oleic acid, *Biochem. Biophys. Res. Commun.*, 162, 326, 1989.
38. Liu, D. and Huang, L., Role of cholesterol in the stability of pH sensitive liposomes prepared by the detergent dialysis method, *Biochem. Biophys. Acta*, 981, 254, 1989.
39. Liu, D. and Huang, L., pH sensitive, plasma-stable liposomes with relatively prolonged residence in circulation, *Biochem. Biophys. Acta*, 1022, 348, 1990.
40. Allen, T.M., Interactions of liposomes and other drug carriers with the mononuclear phagocyte system, in *Liposomes as Drug Carriers,* Gregoriadis, G., Ed., John Wiley & Sons, Chichester, England, 1988, 37.
41. Turner, A.F., Presant, C.A., Proffitt, R.T., Williams, L.E., and Winsor, J.L., III, Liposomes: dosimetry and tumor depictions, *Radiology*, 166, 761, 1988.
42. Perez-Soler, R., Lopez-Berestein, G., Kasi, L.P., Cabanillas, F., Jahns, M., Glenn, H., Hersh, E.M., and Haynie, T., Distribution of technetium-99m-labeled multilamellar liposomes in patients with Hodgkin's disease, *J. Nucleic Med.*, 26, 743, 1985.
43. Allen, T.M., Hansen, C., and Rutledge, J., Liposomes with prolonged circulation times: factors affecting uptake by reticuloendothelial and other tissues, *Biochim. Biophys. Acta*, 981, 27, 1989.
44. Gabizon, A. and Papahadjopoulos, D., Liposome formulations with prolonged circulation time in blood and enhanced uptake by tumors, *Proc. Natl. Acad. Sci. U.S.A.*, 85, 6949, 1988.
45. Leserman, L., Machy, P., and Barbet, J., Cell specific drug transfer from liposomes bearing monoclonal antibodies, *Nature,* 293, 226, 1981.
46. Huang, A., Kennel, S.J., and Huang, L., Interactions of immunoliposomes with target cells, *J. Biol. Chem.*, 258, 14034, 1983.
47. Wang, C.Y. and Huang, L., pH sensitive immunoliposomes mediated target-cell-specific delivery and controlled expression of a foreign gene in mouse, *Biochemistry*, 84, 7851, 1987.

Chapter 16

Enhancing Endosomal Exit of Nucleic Acids Using pH-Sensitive Viral Fusion Peptides

Berndt Oberhauser, Christian Plank, and Ernst Wagner

CONTENTS

I. INTRODUCTION

Numerous low-molecular weight, pharmacologically active compounds are readily taken up into cells by passive diffusion processes through the cell membrane. Macromolecules and also small polar compounds such as peptides, proteins, nucleotides, or nucleic acids that can not penetrate through lipid membranes, require active cellular transport mechanisms for uptake into cells. Internalization may occur via specific transport channels, via receptor-mediated endocytosis or via general fluid-phase pinocytosis.

Receptor-mediated endocytosis is a specific process utilized by cells for the uptake of proteins or peptides, including low-density lipoprotein, transferrin, asialoglycoproteins, epidermal growth factor, insulin, and small vitamins such as folic acid. Many viruses and toxins also enter cells via receptor-mediated endocytosis. The first steps in this process include binding of the ligand to specific cell-surface receptors, followed by receptor clustering and internalization through coated vesicles into endosomal acidic compartments. The subsequent pathways are strongly dependent on the type of receptor/ligand pair; sorting processes may lead to degradative lysosomal compartments.

The efficient receptor-mediated endocytosis pathway can be subverted for drug delivery. Ligands such as transferrin have been successfully used for the import of low-molecular weight drugs, protein toxins, liposomes, or DNA molecules into cells by linkage of these agents to transferrin or to antireceptor antibodies.[1] However, a major

0-8493-4778-5/95/$0.00+$.50
© 1995 by CRC Press Inc.

roadblock to the delivery of nucleic acids[2-4] and other compounds into cells has been the exit of the delivered material from endosomes.[5-8] The material internalized by fluid-phase or receptor-mediated endocytosis is within the cell, but still separated from the cytoplasm by the vesicular membranes. Much of such internalization results in delivery to lysosomes followed by degradation therein, or recycling to the cell surface.

Viruses have acquired special mechanisms to release their genome from endosomes into the cytoplasm. In the case of membrane-free viruses such as adenoviruses,[9,10] the endosomal acidification process specifically activates viral coat protein domains that trigger disruption of the endosomal membrane; in the case of enveloped viruses such as influenza virus[11] the viral membrane fuses with endosomal membranes.

In this chapter we survey several naturally occurring membrane-destabilizing proteins and peptides which viruses or toxins utilize for delivery to the cytosol. We report on the capacity of synthetic peptides, derived from influenza virus sequences, to disrupt endosomes of living cells. The influenza peptides, when incorporated into DNA complexes, were found to greatly enhance (up to more than 5000-fold) receptor-mediated gene transfer. The concept that peptide-mediated endosome disruption is essential for an efficient delivery of nucleic acids may also be relevant for the delivery of other biologically active compounds such as antisense oligonucleotides, drugs, peptides, or proteins.

II. MATERIALS AND METHODS

A. SYNTHESIS OF PEPTIDES AND CONJUGATES

1. Peptide Synthesis

Peptides were assembled on an automatic synthesizer (Applied Biosystems 431A) by the solid phase method using *p*-alkoxybenzylalcohol resin (Bachem) as solid support and Fmoc-protected amino acids (Bachem). The carboxy terminal amino acid was coupled to the resin via the anhydride intermediate. Subsequent amino acids were coupled by the HBTU activation method.[12] The following side chain protecting groups were used: (Trt)Asn, (Trt)Cys (or (*t*-Bu)Cys in case of peptide GALA), (*t*-Bu)Glu, (Trt)His, (*t*-Bu)Ser. The peptides were cleaved from the resin and the side chain protecting groups were removed (except (*t*-Bu)Cys) by treatment of approximately 20 mg peptide-loaded resin with 1 ml of a mixture of phenol/ethanedithiol/thioanisol/water/TFA, 0.75:0.25:0.5:0.5:10, for 1.5 h at room temperature. The cleavage mixture was pipetted dropwise into 40 ml of ether while stirring. The mixture was left at 4°C for at least 1 h. Crude peptides were collected by centrifugation and the supernatants were discarded. Peptides were washed three times with 40 ml of ether and subsequently dried under a stream of argon followed by high vacuum.

Crude peptide INF3 (Cys-SH form) was dissolved in 1 ml of 20 mM ammonium bicarbonate, pH 8, containing 10 µl β-mercaptoethanol by stirring for 1 h at room temperature. After centrifugation the solution was purified by gel filtration (Sephadex® G–10, 10 x 300 mm, HBS, pH 7.3). The pooled product fractions were diluted twofold with water and loaded on a Mono Q® column (Pharmacia HR 5/5) in three portions (buffer A: 20 mM HEPES, pH 7.3; buffer B: A plus 3 M NaCl; gradient elution 0.5 ml/min, 0 to 50% B in 30 min; the peptide was eluted at 0.8 M NaCl). Peptides INF1 and INF2 were prepared in an analogous manner.

The S-*t*-Bu-protected peptide GALA was dissolved in a small volume 1 M triethylammonium bicarbonate (TEAB), pH 8, diluted to 100 mM TEAB and further purified by reverse-phase HPLC on a Nucleosil® 500-5C4 column (0.1% TFA-acetonitrile gradient). The peptide eluted at about 50% acetonitrile. A crude Cys-SH form of peptide GALA (about 5 mg, obtained by deprotecting the corresponding Trt-Cys peptide) was dissolved in 100 µl of 100 mM TEAB, pH 8, containing 1 µl β-mercaptoethanol and purified by gel filtration (Sephadex® G-10, 100 mM TEAB, 0.5 mM EDTA) and ion-exchange chromatography (Mono Q Pharmacia, buffer A: 20 mM HEPES, pH 7.3; buffer B: A plus 3 M NaCl; gradient elution 0 to 100% B; the peptide was eluted at 1.5 M NaCl). Peptides GALA-INF1, GALA-INF3, GLF, and GLF-delta were synthesized and purified in an analogous fashion.

When monomeric peptides were used in subsequent experiments, the C-terminal thiol group was blocked by reaction with a 1.3- to 5-fold molar excess of N-(hydroxyethyl)maleimide (1 h, room temperature [RT]). Excess maleimide was removed by gel filtration (Sephadex® G-25, 100 mM TEAB, pH 8) and the peptides were obtained as triethylammonium salt upon freeze drying.

Peptides were dimerized through disulfide bond formation at the C-terminal cysteines. The free thiol form of peptide INF3 (about 0.5 μmol/ml in 0.8 M NaCl, 20 mM HEPES, pH 7.3) was treated with half an equivalent of 2,2'-dithiobispyridine (10 mM in ethanol) and then concentrated to approximately half of the original volume by evaporation in a Speedvac® (Savant). Following reaction at room temperature overnight, the material was desalted (Sephadex® G-10, HBS) and finally purified by ion-exchange chromatography (Pharmacia Mono Q) as described above. Dimeric peptide INF3DI was eluted at 1.0 M NaCl. Peptide INF4DI was prepared in an analogous fashion.

The purity of the peptides was checked by analytical reverse-phase HPLC and their identities were confirmed by time-of-flight mass spectrometry[13] performed with a Finnigan MAT Lasermat® instrument or a Bio Ion® ^{252}Cf PD-MS instrument (Uppsala, Sweden).

2. Polylysine Conjugates and DNA Complexes

Conjugate synthesis — Human transferrin-poly(L)lysine (Tf-pL) conjugates with an average chain length of 290 lysines (Tf-pLys$_{290}$) were prepared as described.[14] The synthesis of influenza peptide-poly(L)lysine conjugates INF1-pLys and INF2-pLys was performed as described.[15] The plasmids pCMVL, containing the *Photinus pyralis* luciferase gene under control of the cytomegalovirus enhancer/promoter, and pCMV-β-gal, containing the β-galactosidase gene, have been described.[16,17]

DNA/Tf-pLys complexes — Electroneutral DNA complexes are obtained in an analogous fashion as described:[14,18] 6 μg of DNA in 250 μl HBS, and Tf-pLys$_{290}$ containing 8 μg transferrin conjugated to 4 μg pLys$_{290}$ (calculated as hydrobromide salt) in 250 μl HBS were mixed.

Complexes containing peptide-polylysine conjugates — DNA complexes were prepared as reported[15] by first mixing 6 μg of DNA in 150 μl HBS with 4 μg Tf-pLys$_{290}$ in 150 μl HBS, incubated for 30 min at room temperature, followed by mixing with 10 to 20 μg of influenza peptide-pLys$_{290}$ in 200 μl HBS. Complexes were incubated for another 30 min at room temperature.

Complexes containing peptides ionically bound to polylysine — DNA complexes were prepared as described[19] by first mixing 6 μg of DNA in 150 μl HBS with 4 μg Tf-pLys$_{290}$ in 150 μl HBS, incubation for 30 min, followed by mixing with 4 to 20 μg of pLys$_{290}$ in 100 μl HBS, and after a further 30 min addition of 1 to 30 μg of peptide in 100 μl HBS. Complexes were incubated for another 30 min.

B. DESTABILIZATION OF PHOSPHOLIPID MEMBRANES
1. Liposome Leakage Assay

The ability of synthetic peptides to disrupt liposomes was assayed by the release of a fluorescent dye from liposomes loaded with a self-quenching concentration of calcein. Liposomes were prepared from egg phosphatidylcholine (Avanti Polar Lipids) by reverse-phase evaporation[20] with an aqueous phase of 100 mM calcein (dissolved by addition of 3.75 equivalents of sodium hydroxide) and 50 mM NaCl, and extruded through a 100-nm polycarbonate filter to obtain a uniform size distribution.[21] The liposomes were separated from unincorporated material by gel filtration on Sephadex® G-25 with an iso-osmotic buffer (200 mM NaCl, 25 mM HEPES, pH 7.3). For the leakage assay at various pH values, the liposome stock solution was diluted (6 µl/ml) in 2× assay buffer (400 mM NaCl, 40 mM Na citrate of appropriate pH). An aliquot of 100 µl was added to 80 µl of a serial dilution of the peptide in water in a 96-well microtiter plate (final lipid concentration: 25 µM) and assayed for calcein fluorescence at 515 nm (excitation 495 nm) on a microtiter plate fluorescence photometer (Perkin Elmer) after 30 min of incubation at RT. The value for 100% leakage was obtained by addition of 1 µl of a 10% Triton® X-100 solution. The leakage units were calculated as reciprocal values of the peptide concentration, where 50% leakage was observed (i.e., the volume [µl] of liposome solution which is lysed to 50% per µg of peptide). Values below 20 units are extrapolated.

2. Erythrocyte Lysis Assay

Fresh human erythrocytes were washed with HBS several times and resuspended in 2× assay buffer of the appropriate pH (300 mM NaCl, 30 mM Na citrate) at a concentration of about 7×10^7/ml. An aliquot of 75 µl was added to 75 µl of a serial dilution of the peptide in water in a 96-well microtiter plate (cone type) and incubated for 1 h at 37°C with constant shaking. After removing unlysed erythrocytes by centrifugation (1000 rcf, 5 min) 100 µl of the supernatant were transferred to a new microtiter plate and hemoglobin absorption was determined at 450 nm (background correction at 750 nm). By adding 1 µl of a 10% Triton® X-100 solution prior to centrifugation, 100% lysis was determined. The hemolytic units were calculated as a reciprocal value of the peptide concentration, where 50% leakage was observed (i.e., the volume [µl] of erythrocyte solution which is lysed to 50% per µg of peptide). Values below 3 hemolytic units are extrapolated.

3. Release of Fluorescent Compounds from Endosomes

Fluorescein isothiocyanate dextran of average molecular weight 71 200 (Sigma FD-70S) was dissolved in DMEM (Gibco) and subjected to gel filtration (Sephadex® G-25 PD-10, Pharmacia; DMEM as eluent) to remove components of lower molecular weight.

A 200-mM solution of calcein (Sigma) containing 20 mM HEPES and 100 mM NaCl, adjusted to pH 7.3 with sodium hydroxide (690 mM final concentration), was diluted with medium (DMEM plus 10% FCS) to 10 mM prior to use in cell culture.

BNL Cl.2 cells or NIH3T3 cells were split into Lab-Tek chamber slides (Nunc 177402) at a density of 10,000 cells per well and grown overnight in DMEM plus 10% FCS. The medium was removed from the cells and 150 µl of DMEM containing 5 mg/ml FITC dextran or 10 mM calcein and 0.5 mg/ml of peptide INF3DI were added. Cells were incubated at 37°C for 15 min. The incubation medium was removed and the cells were incubated in fresh DMEM plus 10% FCS for another 15 min at 37°C. The cells were fixed in PBS containing 3% p-formaldehyde for 3 min. After removal of the culture chambers the cells were dried on air and mounted with coverslips in Mowiol.™ Fluorescence microscopy was performed with a Zeiss Axiophot® microscope using a CCD camera (Photometrics). Where indicated, cells were preincubated with 200 nM bafilomycin in DMEM plus 10% FCS for 30 min. Subsequent steps were as described above, with the exception that the medium always contained 200 nM bafilomycin.

C. DELIVERY OF DNA INTO CULTURED CELLS
1. Transfection of Cells

Adherent cell lines (BNL Cl.2 hepatocytes, NIH3T3 cells) were grown in 6-cm dishes for 1 d prior to transfection (DMEM medium with 10% FCS; 300,000 cells per dish). The medium was removed and 1.5 ml of DMEM plus 2% FCS and 500 µl of the DNA complexes were added. After 4 h of incubation the transfection medium was replaced by 4 ml of DMEM with 10% FCS. Harvesting of cells and luciferase assays were performed 24 h after transfection as previously described.[3] The light unit values shown in Figure 6a represent the total luciferase activity of the transfected cells. β-Galactosidase activity was assayed according to a published method.[22]

III. RESULTS AND DISCUSSION

A. MEMBRANE-DESTABILIZING PROTEINS AND PEPTIDES

Many biological functions are dependent on membrane fusion, disruption, or pore formation (see Tables 1 and 2 and references therein). These processes are involved in the entry of viruses or microorganisms into cells, in the cytolytic action of the complement system in higher organisms, and of toxins of insects, fish, and microorganisms, and in the action of antibiotic peptides of insects or frogs. These physically still not well-understood processes are also important in intracelluar vesicle fusion, in the fission and fusion of cells, and in sperm-egg fusion. In several cases the membrane-destabilizing element resides within a single protein or peptide and can be assigned to a short "fusion domain" of up to 30 amino acids (Table 3). Although these fusion domains are not directly related in sequence, they show homology in an alternating pattern of about 1 to 3 hydrophobic amino acids (Leu, Ile, Val, Phe, Tyr, or Met) followed by hydrophilic amino acids (Gly, Ser, Thr, Asn, Gln, Asp, Glu, Lys, or Arg). Under appropriate conditions the majority of these domains is expected to form an amphipathic α-helix that can interact with phospholipids. In the case of fusion/disruption peptide sequences that are activated by the low pH of endosomes, the acidification triggers a conformational change of the fusion domain from a random conformation to an amphipathic α-helix. Within the polar domains, acid-triggered sequences contain several acidic Glu or Asp residues, whereas pH-insensitive sequences have no acidic residues and/or a preference for positively charged Lys or Arg residues. Internal "helix-breaking" proline residues may result in kinks in the helix, as described for the structure of melittin.[42]

A well-studied fusogenic protein is the *influenza virus* hemagglutinin (HA), a homotrimer, each monomer of which consists of two polypeptides, HA-1 and HA-2, connected through a disulfide bond.[11,51] The fusion domain is located at the N-terminus of HA-2 and consists of a stretch of hydrophobic amino acids regularly interrupted by hydrophilic and acidic side chains. The intramolecular repulsion of the three acidic side chains (glutamic acid at positions 11 and 15, aspartic acid at position 19), which are negatively charged at neutral pH, inhibits the formation of an α-helix (Figure 1a). The fusion of viral and endosomal membrane is induced by an acid-induced conformational change of the protein, which exposes the N-terminal region of the HA-2 subunit, allowing it to interact with the endosomal membrane.[54] Another consequence of endosomal acidification is the protonation of the carboxyl groups in the N-terminal fusion peptide which then adopts an amphipathic α-helical conformation (Figure 1b,c). The amphipathic helices of the virus interact with the endosomal membrane, resulting in membrane fusion and the release of the viral genome into the cytoplasm. Synthetic peptides corresponding to the N-terminus of influenza HA-2 have been reported[55,56] to fuse artificial lipid membranes and also cause leakage of aqueous liposomal contents.

B. SYNTHETIC ENDOSOME-DISRUPTION PEPTIDES

Starting from the N-terminal peptide sequence of influenza HA subunit HA-2, several amphipathic peptides (see Table 4) have been synthesized. Peptide INF1 contains the sequence of the 20 amino-terminal amino acids of HA-2 and a C-terminal extension with a terminal cysteine. Peptide INF2 has a glycine to glutamic acid substitution at amino acid 4, and INF3 or INF4 both a glycine and an alanine to glutamic acid substitution at positions 4 and 7. According to the model presented in Figure 1, the glutamic acids at positions 4 and 7 (in addition to the acidic residues at positions 11, 15, and 19) should further destabilize an α-helix at neutral pH. This should result in a higher pH-specificity of membrane destabilization. An acidic amphipathic peptide, GALA, with pH specific

Table 1 **Membrane-active proteins**

Viral Fusion Proteins			pH Optimum	Ref.
N-Terminal fusion sequence				
Enveloped viruses				
Influenza virus		HA-2	pH 5	23,24
Vesicular stomahitis virus (VSV)		G	pH 5	25
Vaccinia virus		14 kDa	pH 5	26
Sendai virus		F1	pH 7	25,27
Measles virus		F	pH 7	24
Human immunodeficiency virus (HIV)		gp41	pH 7	28
Simian immunodeficiency virus (SIV)		gp41	pH 7	29,30
Unenveloped viruses				
Polio virus		vp1	pH 5	31
Coxackie virus		vp1	pH 5	31
Rhino virus		vp1	pH 5	
Internal fusion sequence				
Enveloped viruses				
Semliki Forest v.		E1	pH 5	23
Sindbis virus				23
Unenveloped viruses				
Rhesus rotavirus		vp5		32
Toxins of microorganisms				
Streptolysin O	*Streptococcus*	20 to 80 mer, 15-nm pores, binds cholesterol, sulfhydryl activated	pH 7	33
α-Toxin	*Staphylococcus*	Amphipathic, β-sheet structure, hexamer lesions, 2-nm pores	pH 7	
Hemolysin	*Escherichia coli*	8 amphipathic helixes	pH 7	34
Hemolysin	*Trypanosoma cruzi*	Related to perforins, C9	pH 5	35
Listeriolysin O	*Listeria monocytogenes*	Related to C9, streptolysin O, binds cholesterol, sulfhydryl activated	pH 5	36
Vertebrate immune system				
Perforin	Cytotoxic T-cells	Ca^{2+}-dependent insertion		37
Complement	C9 (MAC C5b-8,9$_{1-4}$)	10-nm channel, highly regulated		38,39
Sperm-egg fusion protein				
PH-30	α-Subunit, internal fusion sequence		pH 7	40

membrane fusion and disruption activity, had been designed by the group of Frank Szoka.[57-59] We synthesized an analogous peptide and several hybrid versions thereof that contain influenza peptide sequences at the N-terminus (see Table 4; peptides GLF, GLF-delta, GALA-INF1, and GALA-INF3). The rationale of including the influenza sequences was based on previous reports[55,60] that, apart from the amphipathic sequence, the free N-terminus (H_2N-Gly-Leu-Phe) is essential for membrane destabilization induced by influenza peptides.

Table 2 **Membrane-active peptides**

					Ref.
Defense toxins					
Melittin	Bee venom	26AA	amph. α-helix	Pro-kinked	41–43
Bombolitin	Bumblebee venom	17AA	amph. α-helix		44
Mastoparan	Wasp venom	14AA	amph. α-helix		44
Crabrolin	Hornet venom	13AA	amph. α-helix		44
Pardaxin	Moses sole fish (shark repellant)	33AA	amph. α-helix	Pro-kinked	45
Antibacterial peptides					
Sarcotoxin IA	Flesh fly (in hemolymph)				
Cecropins	Insects (humoral immune system, silk moth)	37AA	amph. α-helix	Pro-kinked	39
Maganin	Skin Xenopus laevis	23AA	amph. α-helix		46
Alameticin	Fungus (*Trichoderma viride*)	15–24AA	amph. α-helix	α-Amino-butyric acid	39
Bacterial toxins					
δ-Toxin	*Staphylococcus aureus*	26AA	amph. α-helix	Acid induced	47,48
Amoebapore	*Entamoeba histolytica*	25AA	amph. α-helix	Acid induced	49
Vertebrate immune system					
Defensins	Polynucleated neutrophils	29–34AA	SS-bridged β-sheet		50

Note: AA = amino acids.

We tested the membrane disruption activity of the peptides in liposome leakage (Figure 2) and erythrocyte lysis (Figure 3) assays. The peptides showed clear differences when their capacity of releasing calcein from liposomes was compared with the efficiency of releasing hemoglobin from erythrocytes, indicating that the peptides cause different sizes of membrane lesions. The pH dependence of leakage (see an example of pH titration in Figure 2b) is nicely correlated with the number and position of negatively charged amino acids in the amphipathic helix. Peptide INF1 showed a tenfold increased liposome leakage activity at pH 5 compared to pH 7 (see Figure 2). INF2 and INF3 are even more pH specific and have leakage activity only under acidic conditions. Dimerization of peptide INF3 by disulfide bond formation at the C-terminal cysteines (INF3DI) resulted in >20-fold enhanced liposome leakage activity. The rationally designed peptides GALA and GLF showed even higher pH-specific liposome leakage activity. Peptide GLF has the highest activity, showing that the three N-terminal amino acids of influenza HA-2 have a synergism with the GALA sequence.

In the erythrocyte lysis assay (Figure 3) the influenza peptides INF1, INF2, and INF3 showed no significant activity. Dimeric peptides INF3DI, INF4DI, and peptide GALA-INF3 (the corresponding influenza sequence elongated by a GALA sequence) showed the highest, pH-specific lysis of erythrocytes. GALA-INF1 showed similar lysis activity under acidic conditions, but also significant activity at pH 7. Peptides GLF and GALA showed about 100- and 500-fold lower erythrocyte lysis activity than the influenza peptide dimers. The contrasting behavior of GALA (and GLF) in liposome and erythrocyte leakage may be explained by the findings of Szoka and colleagues[59] that GALA forms small channels and does not mediate the release of larger molecules. Melittin, a

Table 3 Sequences of fusion peptides and proteins. Hydrophobic amino acids are outlined.

pH-Insensitive viral fusion proteins (N-terminus)

		I -I M R M
HIV-1	gp41	AVGVLGALFLGFLGAAGSTMGAASLTLTV–
HIV-2	gp41	GVFVLGFLGFLATAGSAMGAASLTLSA–
		L T R V
SIV	gp32	GVFVLGFLGFLATAG–
	P	GA
		(internal sequences)
Semliki Forest v.	E1	–PDYQCKVYTGVYPFMWGGAYCFCD–
Sindbis v.		–DYTCKVFGGVYPFMWGGA
Rh. rota v.	vp5	–GDYSFALPVGQWPVMTGGAVSLHS–
Sperm protein PH-30		–GKLICTGISSIPPIRALFAAIQIP–
Rubella v.	E2	–GRLICSTTAQYPPTR–

pH-Insensitive membrane-disruption peptides

Melittin	26 AA	GIGAVLKVLTTGLPALISWIKRKRQQ^CONH2
Bombolitin I	17 AA	IKITTMLAKLGKVLAHV^CONH2
Magainin 2	23 AA	GLGKFLHSAKKFGKAFVGEIMNS^CONH2
Pardaxin	33 AA	
		GFFALIPKIISSPLFKTLLSAVGSALSSSGGQE
Mastoparan	14 AA	INLKALAALAKKIL^CONH2
Crabrolin	13 AA	FLPLILRKIVTAL^CONH2
Cecropin	37 AA	
		KWKLFKKIEKVGQNIRDGIIKAGPAVAVVGQATQIAK^CONH2

Acid-triggered viral fusion proteins (N-terminus)

		T
		F K S N
		I S M G Q LVA
Influenza A,B	HA2	GLFGAIAGFIENGWEGMIDGWYG–
Influenza C	HA2	IFGIDDLIIGLLFVAIVEAGIGG
Polio 3	vp1	GIEDLISEVAQGALTLVP–
Polio 1	vp1	GLGQMLESMIDNTVREVGGAATS–
Rhino v. HRV-14	vp1	GLGDELEEVIVEKTKQTVASISSGP–
Coxackie v.	vp1	GPVEDAITAAIGRVADTVGT–
Vaccinia v.	14 kDa	MDGTLFPGDDDLAIPATEFFSTKAAKKP–
Rhino v. HRV-2	vp1	NPVENYIDEVLNEVLVVPNINSSNPTT–
Rhino v. HRV-89	vp1	S QPSTSVS

Acid-triggered membrane disruption peptides

Entamoeba		GEILxNLxTGLINTLENLLTxKGAD^COOH
Staphylococcus		MAQDIISTIGDLVKWIIDTVNKFTKK^COOH
δ-Toxin		A VEF L AE E I

Note: HRV = human rhinovirus.

pore-forming peptide of known high, but not pH-*specific erythrocyte lysis activity, was* included in the assays as standard.

We next asked whether or not peptides, when added to the medium of cultured cells, will be internalized and disrupt acidic vesicles instead of being degraded in lysosomes.

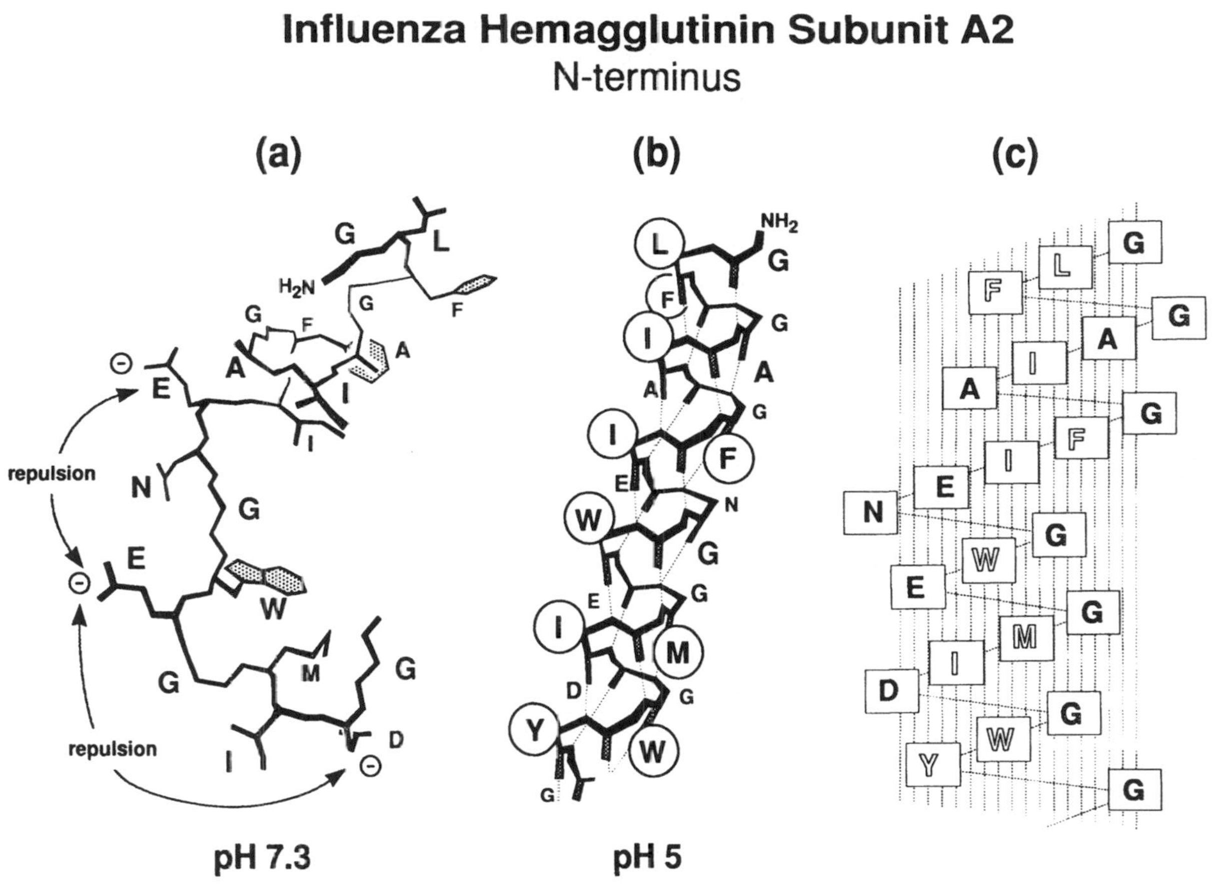

Influenza Hemagglutinin Subunit A2
N-terminus

Figure 1 N-Terminal fusogenic peptide of influenza hemagglutinin subunit HA-2. (a) Conformation at neutral pH, as determined by X-ray structure analysis.[52,53] (b), (c) Putative α-helical conformation at low pH, consistent with CD measurements.[24,56]

Table 4 **Sequences of synthetic peptides**

Inf HA-2	GLFGAIAGFI	ENGWEGMIDG	– – – – –	
INF1	GLFGAIAGFI	ENGWEGMIDG	GGC	
INF2	GLFEAIAGFI	ENGWEGMIDG	GGC	
INF3	GLFEAIEGFI	ENGWEGMIDG	GGC	
INF4	GLFEAIEGFI	ENGWEGnIDG	CA	
INF3DI	GLFEAIEGFI	ENGWEGMIDG	GGC	
	GLFEAIEGFI	ENGWEGMIDG	GGC	
GALA	WEAALAEALA	EALAEHLAEA	LAEALEALAA	GGSC
GLF	GLFGALAEALA	EALAEHLAEA	LAEALEALAA	GGSC
GLF-delta	GLFELAEALA	EALAEALAEA	LAEALEALAA	GGSC
GALA-INF1	GLFGAIAGFI	ENGWEGLAEA	LAEALEALAA	GGSC
GALA-INF3	GLFEAIEGFI	ENGWEGLAEA	LAEALEALAA	GGSC

Note: For Comparison Inf HA2, the N-terminal sequence of Influenza virus X-31 (H3N2) hemagglutinin subnit HA-2 is shown; INF1 and INF2 as described;[15] GALA, analogus to peptide sequence described;[57-59] C]C, cystine disulfide bond; n, norleucine.

For this purpose BNL Cl.2 hepatocytes or NIH3T3 cells were incubated with medium containing calcein or high-molecular weight (70-kDa) FITC-dextran with or without peptide (INF3DI or INF4DI) for 15 min at 37°C, followed by a 15-min incubation with normal medium. FITC-dextran, when taken up into cells by fluid-phase endocytosis, accumulates in internal vesicles that appear as the bright spots (Figure 4a,g) in fluorescence microscopy (see Materials and Methods). When the synthetic peptide at an appropriate concentration was included, no bright vesicles were found in most areas of the cell culture slide and the FITC-dextran had been released into the cytoplasm (Figure 4b,c,h). A similar peptide-mediated release of internalized calcein was observed (see Figure 4e,f). The presence of bafilomycin, which specifically blocks the endosomal proton pump, also blocked the peptide-mediated release of FITC-dextran from internal vesicles (see Figure 4d). The efficiency of endosome disruption is strongly dependent on a high concentration of membrane-active peptide when peptides are delivered by general fluid-phase endocytosis. At a concentration of about 100 µM peptide the majority of endocytic vesicles was disrupted within 15 to 30 min, as shown in Figure 4; at threefold lower concentration no significant leakage was detectable by fluorescence microscopy. However, the high concentrations of peptide may be avoided by measures that result in a colocalization of the peptide and the material to be delivered (see next chapter).

C. PEPTIDE-AUGMENTED GENE TRANSFER

We and others have investigated the concept of exploiting natural endocytosis pathways of protein ligands for the delivery of DNA macromolecules. Asialoglycoproteins[2,8,61,62] or synthetic galactose-containing peptides[16,63,64] (for targeting of hepatocytes via the asialoglycoprotein receptor), transferrins,[14,18,65] antibodies against CD3 (for targeting of T-cells), epidermal growth factor (our unpublished results), insulin,[66,67] antithrombomodulin antibody[68,69] for lung targeting, lectins such as WGA,[70] and other ligands have been modified with DNA-binding compounds such as the polycation polylysine.

One limiting step of gene transfer by receptor-mediated endocytosis is the exit of DNA from endosomes before being degraded in lysosomes. Inactivated adenovirus particles, when included in the transfection medium, provide an endosome-disruption activity and enhance the levels of receptor-mediated gene expression up to more than 2000-fold.[5,6] The highest gene expression levels were obtained when endosome-disruptive adenovirus particles were directly linked to the DNA complex.[7] Nevertheless, virus-free versions of the system with endosome-disruptive peptides replacing the virus particle might become attractive tools in gene transfer.

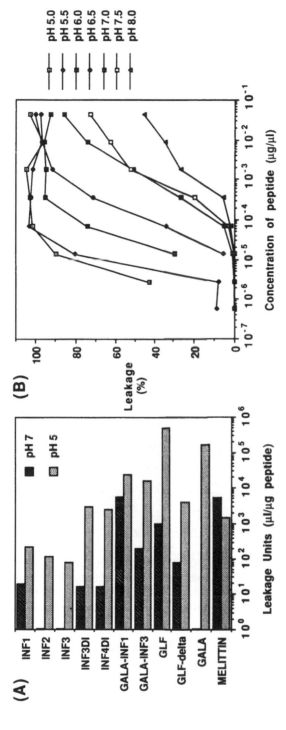

Figure 2 Liposome leakage assay. (A) Leakage units of various peptides at pH 7 and pH 5. (B) pH-dependent leakage using dilutions of peptide GLF indicated. Assays were performed as described in Materials and Methods. Leakage units of peptides given in (A) represent the reciprocal values of the concentration for 50% leakage and were calculated from analogous data as obtained for GLF in (B).

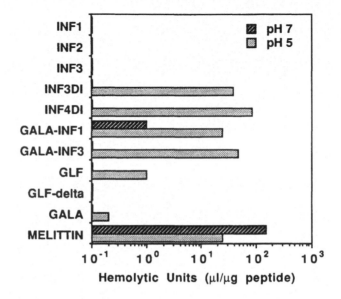

Figure 3 Erythrocyte lysis assay. Assays at pH 7 and pH 5 were performed using serial dilutions of indicated peptides, and hemolytic units were calculated as described in Materials and Methods.

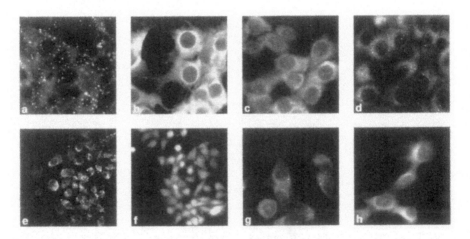

Figure 4 Peptide-mediated release of 70 kDa FITC-dextran or calcein from endosomes of BNL Cl.2 cells (a–f) or NIH3T3 cells (g–h). Fluorescence of internalized FITC-dextran without peptide (a,g); with peptide INF3DI (b,h); with peptide INF4DI (c); with INF4DI and bafilomycin (d); fluorescence of internalized calcein without (e) or with (f) INF3DI.

Our first synthetic, virus-like gene-transfer complexes (see Figure 5) consisted of plasmid DNA, a polylysine-conjugated ligand (transferrin, or a synthetic ligand binding to the asialoglycoprotein receptor), and polylysine-conjugated influenza peptides INF1 or INF2.[15,16] Modification of peptides with polylysine was found to substantially enhance their membrane-disruption activity.[15] The influenza peptide-polylysine conjugates augmented receptor-mediated gene transfer up to 500-fold in a series of cell lines. Supplying unconjugated peptides INF1, INF2, or the more potent peptides INF3DI or GALA-INF3,

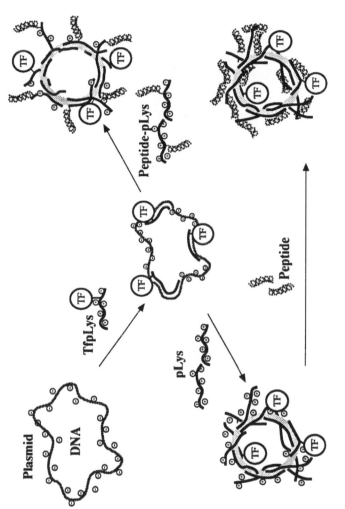

Figure 5 Synthetic virus-like gene-transfer complexes. Complexes consisting of DNA, transferrin-polylysine conjugates (Tf-pLys), and membrane-disruptive peptide either covalently (Peptide-pLys) or ionically (pLys, Peptide) bound to polylysine.

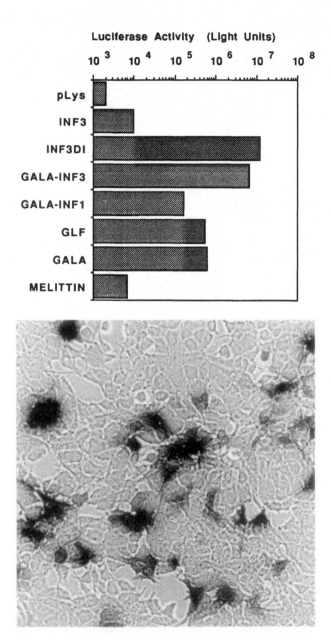

Figure 6 Transfection of BNL Cl.2 hepatocytes. (a) DNA complexes were prepared by mixing 6 μg of pCMVL-DNA in 250 μl HBS with 4 μg Tf-pLys$_{290}$ in 250 μl HBS, followed by mixing with 20 μg of poly(L)lysine$_{290}$ in 750 μl HBS and optimized amounts of peptides (30 μg INF3, 30 μg INF3DI, 20 μg GALA-INF3, 20 μg GALA, 10 μg GALA-INF1, 4 μg GLF, or 3 μg melittin) in 250 μl HBS. Mixing steps were performed at 30-min intervals. After an incubation for another 30 min, complexes were mixed with 0.5 ml DMEM plus 6% FCS and added to 300,000 cells. Harvesting of cells and luciferase assays were performed as described in Materials and Methods. pLys, cationic complex without peptide was used. (b) *In situ* demonstration of β-galactosidase expression in BNL Cl.2 cells. Transfections using pCMV-β-gal-DNA complexes were carried out as described. After 48 h, staining reactions were performed for 2 h. Approximately 15% of cells showed a strongly positive reaction for β-galactosidase.

to the transfection medium does not considerably augment transferrin-polylysine-mediated gene transfer (our published[15] and unpublished data).

In the first version of influenza peptide-linked complexes the numbers of polylysine-conjugated peptides per DNA complex were low because of technical reasons. The transfection efficiency, however, was found to correlate with an increasing number of peptides linked to polylysine. For this reason a new, more efficient version of gene-transfer complexes was generated that contains larger amounts of negatively charged, membrane-disruptive peptides ionically bound to polylysine (see Figure 5). As a first step in assembling the complex, half of the negative charges of the plasmid DNA encoding a reporter gene (luciferase or β-galactosidase) were saturated with transferrin-polylysine to provide a ligand for receptor-mediated endocytosis. The remaining nucleic acid charges were utilized for binding an excess of polylysine. To the resulting positively charged complex the negatively charged peptide was added to provide the endosome disrupting activity. The complexes were used for transfection of cell lines such as BNL Cl.2 hepatocytes or NIH3T3 fibroblasts.

A series of membrane-disruption peptides (Table 4) was tested, and a strong correlation of gene transfer efficiency (see Figure 6a) with the capacity of peptides to lyse erythrocytes in an acidic environment was found. Using the dimeric influenza peptide INF3DI, the highest luciferase expression levels (more than 5000-fold higher than in the absence of peptide) were obtained. Peptides GALA-INF3 and GALA also mediate gene transfer, with about 2-fold or 20-fold lower efficiency. Monomeric influenza peptide INF3 does not significantly enhance gene transfer. Peptides that possess a high hemolytic potential, but do not display the strong specificity for low pH (GALA-INF1 and melittin) or lack the negative charges necessary for binding to the DNA-complex (melittin), show only moderate augmentation of gene expression. When the endosomal acidification was inhibited with bafilomycin, a specific blocker of the endosomal proton pump, peptide-mediated gene expression was also considerably reduced (our unpublished results). Transfections of other cell types (NIH3T3 fibroblasts, B16 melanoma cells, or HeLa cells) with optimized complexes resulted in similar gene expression levels. In β-galactosidase gene-transfer experiments, depending on the cell type, 10 to 30% of cells showed a strong expression of the gene (BNL Cl.2 cells; see Figure 6b).

We consider our synthetic complexes as a first step toward highly efficient virus-like gene-transfer systems. Recently, other groups have also demonstrated the usefulness of membrane-active peptides to enhance DNA delivery. Midoux and colleagues[64] confirmed independently from our studies that synthetic influenza peptides are able to enhance receptor-mediated gene delivery. Two interesting reports from the laboratory of Legendre and Szoka[71] describe the use of membrane-active peptides in gene transfer: the cationic, amphipathic peptide gramicidin S when included into a DOPE lipid/DNA composition, strongly facilitates DNA delivery. Coupling the membrane-active GALA peptide to 50-nm, positively charged polyamidoamine cascade polymer particles allows DNA delivery with charge neutral polymer/DNA complexes.[72] We are currently extending the concept of synthetic viral gene-transfer systems *(i)* to the use of other membrane-destabilizing peptides, e.g., derived from picorna viruses; and *(ii)* to other ways of displaying these peptides in an ordered fashion thought to be required for optimal entry.

IV. CONCLUSIONS

Synthetic peptides derived from viral fusion sequences can mimic the process by which viruses destabilize endosomal membranes. At appropriate concentration the peptides are able to disrupt liposomes, erythrocytes, or endosomes of cultured cells in an acidification-dependent manner. This process can be exploited for the delivery of small compounds (e.g., calcein), medium-size molecules (e.g., FITC-dextran), or larger biopolymers (e.g., plasmid DNA) into the cytoplasm of cells. The peptides show a highly concentration-dependent, cooperative behavior in membrane disruption. Therefore, strategies have to be developed that specifically concentrate membrane-destabilizing peptides in sufficient quantities and active form into vesicles that contain the material (drug, protein, nucleic acid) to be delivered.

ACKNOWLEDGMENTS

We thank Helen Kirlappos, Karl Mechtler, and Christian Koch for their assistance. We appreciate the critical reading of this manuscript by Dr. Michael Buschle and Ciaran Morrison. We also thank Prof. Dr. Max L. Birnstiel, Dr. Matt Cotten, and Prof. Dr. Frank C. Szoka for helpful discussions. The research was done with the financial support of the Austrian Industrial Research Promotion Fund.

REFERENCES

1. Wagner, E., Curiel, D., and Cotten, M., Delivery of drugs, proteins and genes into cells using transferrin as a ligand for receptor-mediated endocytosis, *Adv. Drug Res. Rev.*, 14, 113, 1994.
2. Wu, G. Y. and Wu, C. H., Receptor-mediated *in vitro* gene transformation by a soluble DNA carrier system, *J. Biol. Chem.*, 262, 4429, 1987.
3. Cotten, M., Wagner, E., and Birnstiel, M. L., Receptor mediated transport of DNA into eukariotic cells, *Methods Enzymol.*, 217, 618, 1993.
4. Citro, G., Perrotti, D., Cucco, C., D'Agnano, I., Sacchi, A., Zupi, G., and Calabretta, B., Inhibition of leukemia cell proliferation by receptor-mediated uptake of c-myb antisense oligodeoxynucleotides, *Proc. Natl. Acad. Sci. U.S.A.*, 89, 7031, 1992.
5. Cotten, M., Wagner, E., Zatloukal, K., Phillips, S., Curiel, D. T., and Birnstiel, M. L., High-efficiency receptor-mediated delivery of small and large (48kb) gene constructs using the endosome disruption activity of defective or chemically inactivated adenovirus particles, *Proc. Natl. Acad. Sci. U.S.A.*, 89, 6094, 1992.
6. Curiel, D. T., Agarwal, S., Wagner, E., and Cotten, M., Adenovirus enhancement of transferrin-polylysine-mediated gene delivery, *Proc. Natl. Acad. Sci. U.S.A.*, 88, 8850, 1991.
7. Wagner, E., Zatloukal, K., Cotten, M., Kirlappos, H., Mechtler, K., Curiel, D. T., and Birnstiel, M. L., Coupling of adenovirus to transferrin-polylysine/DNA complexes greatly enhances receptor-mediated gene delivery and expression of transfected genes, *Proc. Natl. Acad. Sci. U.S.A.*, 89, 6099, 1992.
8. Cristiano, R. J., Smith, L. C., and Woo, S. L. C., Hepatic gene therapy: adenovirus enhancement of receptor-mediated gene delivery and expression in primary hepatocytes, *Proc. Natl. Acad. Sci. U.S.A.*, 90, 2122, 1993.
9. FitzGerald, D. J. P., Padmanabhan, R., Pastan, I., and Willingham, M. C., Adenovirus-induced release of epidermal growth factor and Pseudomonas toxin into the cytosol of KB cells during receptor-mediated endocytosis, *Cell*, 32, 607, 1983.
10. Seth, P., FitzGerald, D., Ginsberg, H., Willingham, M., and Pastan, I., Evidence that the penton base is involved in potentiation of toxicity of pseudomonas exotoxin conjugated to epidermal growth factor, *Mol. Cell. Biol.*, 4, 1528, 1984.
11. Wiley, D. C. and Skehel, J. J., The structure and function of the hemagglutinin membrane glycoprotein of influenza virus, *Annu. Rev. Biochem.*, 56, 365, 1987.
12. Knorr, R., Trzeciak, A., Bannwarth, W., and Gillessen, D., *Tetrahedron Lett.*, 30, 1927, 1989.
13. Macfarlane, R. D., Vemura, D., Veda, K., and Hirata, Y., *J. Am. Chem. Soc.*, 102, 875, 1980.

14. Wagner, E., Cotten, M., Mechtler, K., Kirlappos, H., and Birnstiel, M. L., DNA-binding transferrin conjugates as functional gene-delivery agents: synthesis by linkage of polylysine or ethidium homodimer to the transferrin carbohydrate moiety, *Bioconjugate Chem.*, 2, 226, 1991.

15. Wagner, E., Plank, C., Zatloukal, K., Cotten, M., and Birnstiel, M. L., Influenza virus hemagglutinin HA-2 N-terminal fusogenic peptides augment gene transfer by transferrin-polylysine/DNA complexes: towards a synthetic virus-like gene transfer vehicle, *Proc. Natl. Acad. Sci. U.S.A.*, 89, 7934, 1992.

16. Plank, C., Zatloukal, K., Cotten, M., Mechtler, K., and Wagner, E., Gene transfer into hepatocytes using asialoglycoprotein receptor mediated endocytosis of DNA complexed with an artificial tetra-antennary galactose ligand, *Bioconjugate Chem.*, 3, 533, 1992.

17. MacGregor, G. R. and Caskey, C. T., Construction of plasmids that express *E. coli* β-galactosidase in mammalian cells, *Nucleic Acids Res.*, 17, 2365, 1989.

18. Wagner, E., Cotten, M., Foisner, R., and Birnstiel, M. L., Transferrin-polycation-DNA complexes: the effect of polycations on the structure of the complex and DNA delivery to cells, *Proc. Natl. Acad. Sci. U.S.A.*, 88, 4255, 1991.

19. Plank, C., Oberhauser, B., Mechtler, K., Koch, C., and Wagner, E., Synthetic virus-like gene transfer systems: the influence of endosome-disruptive peptides on gene transfer, *J. Biol. Chem.*, 269, 12918, 1994.

20. Szoka, F. C. and Papahadjopoulos, D., Procedure for preparation of liposomes with large internal aqueous space and high capture by reverse-phase evaporation, *Proc. Natl. Acad. Sci. U.S.A.*, 75, 4194, 1978.

21. MacDonald, R. C., MacDonald, R. I., Menco, B. Ph. M., Takeshita K., Subbarao, N. K., and Hu, L., Small-volume extrusion apparatus for preparation of large, unilamellar vesicles, *Biochim. Biophys. Acta*, 1061, 297, 1991.

22. Lim, K. and Chae, C.-B., A simple assay for DNA transfection by incubation of the cells in culture dishes with substrates for β-galactosidase, *BioTechniques*, 7, 576, 1989.

23. White, J. M., Viral and cellular membrane fusion proteins, *Annu. Rev. Physiol.*, 52, 675, 1990.

24. Takahashi, S., Conformation of membrane fusion active 20-residue peptides with or without lipid bilayers. Implication of α-helix formation for membrane fusion, *Biochemistry*, 29, 6257, 1990.

25. Hoekstra, D., Membrane fusion of enveloped viruses: especially a matter of proteins, *J. Bioenerg. Biomembr.*, 22, 121, 1990.

26. Gong S., Lai, C., and Esteban, M., Vaccinia virus induces cell fusion at acidic pH and this activity is mediated by N-terminus of the 14-kDa virus envelope protein, *Virology*, 178, 81, 1990.

27. Gething, M. J., White, J. M., and Waterfield, M. D., Purification of the fusion protein of sendai virus: analysis of the NH2-terminal sequence generated during precursor activation, *Proc. Natl. Acad. Sci. U.S.A.*, 75, 2737, 1978.

28. Slepushkin, V. A., Andreev, S. M., Siderova, M. V., Melikyan, G. B., Grigoriev, V. B., Chumakov, V. M., Grinfeldt, A. E., Manukyan, R. A., and Karamov, E. V., Investigation on human immunodeficiency virus fusion peptides. Analysis of interrelations between their structure and function, *AIDS Res. Hum. Retroviruses*, 8, 9, 1992.

29. Martin, I., Defrise-Quertain, F., Mandieau, V., Nielsen, N. M., Saermark, T., Burny, A., Brasseur, R., Ruysschaert, J.-M., and Vandenbranden, M., Fusogenic activity of SIV (simian immunodeficiency virus) peptides located in the gp32 N-terminal domain, *Biochem. Biophys. Res. Commun.*, 175, 872, 1991.

30. Franchini, G., Identification of the fusion peptide of primate immunodeficiency viruses, *Science*, 244, 694, 1989.

31. Fricks, C. E. and Hogle, J. M., Cell-induced conformational change in poliovirus: externalization of the amino terminus of VP1 is responsible for liposome binding, *J. Virol.*, 64, 1934, 1990.

32. Mackow, E. R., Shaw, R. D., Matsui, S. M., Vo, P. T., Dang, M. N., and Greenberg, H. B., The rhesus rotavirus gene encoding protein VP3: location of amino acids involved in homologous and heterologous rotavirus neutralization and identification of a putative fusion region, *Proc. Natl. Acad. Sci. U.S.A.*, 85, 645, 1988.

33. Kehoe, M. A., Miller, L., Walker, J. A., and Boulnois, G. J., Streptolysin O precursor, sequence, *Infect. Immun.*, 55, 3228, 1987.

34. Oropeza-Werkerle, R. L., Muller, S., Briand, J. P., Benz, R., Schmid, A., and Goebel, W., Hemolysin-derived synthetic peptides with pore-forming and hemolytic activity, *Mol. Microbiol.*, 6, 115, 1992.

35. Andrews, N. W., Abrams, C. K., Slatin, S. L., and Griffiths, G., A T. cruzi-secreted protein immunologically related to the complement component C9: evidence for membrane pore-forming activity at low pH, *Cell*, 61, 1277, 1990.

36. Geoffroy, C., Gaillard, J.-L., Alouf, J. E., and Berche, P., Purification, characterization, and toxicity of the sulfhydryl-activated hemolysin lysteriolysin O from Listeria monocytogenes, *Infect. Immun.*, 55, 1641, 1987.

37. Ojcius, D. M. and Young, J. D., Cytolytic pore-forming proteins and peptides: is there a common structural motif?, *TIBS*, 16, 225, 1991.

38. Bhakdi, S. and Tranum-Jensen, J., Complement lysis: a hole is a hole, *Immunol. Today*, 12, 318, 1991.

39. Esser, A. F., Big MAC attack: complement proteins cause leaky patches, *Immunol. Today*, 12, 316, 1991.

40. Blobel, C. P., Wolfsberg, T. G., Turuck, C. W., Myles, D. G., Primakoff, P., and White, J. M., A potential fusion peptide and an integrin domain in a protein active in sperm-egg fusion, *Nature*, 356, 248, 1992.

41. Blondelle, S. E. and Houghten, R. A., Hemolytic and antimicrobial activities of the twenty-four individual omission analogues of melittin, *Biochemistry*, 30, 4671, 1991.

42. Dempsey, C. E., Bazzo, R., Harvey, T. S., Syperek, I., Boheim, G., and Campbell, I. D., Contribution of proline-14 to the structure and actions of melittin, *FEBS Lett.*, 281, 240, 1991.

43. Ikura, T., Go, N., and Inagaki, F., Refined structure of melittin bound to perdeuterated dodecylphosphocholine micelles and studied by 2D-NMR and distance geometry calculation, *Proteins*, 9, 81, 1991.

44. Argiolas, A. and Pisano, J. J., Bombolitins, a new class of mast cell degranulating peptides from the venom of the bumblebee Megabombus pennsylvanicus, *J. Biol. Chem.*, 260, 1437, 1985.

45. Shai, Y., Bach, D., and Yanovsky, A., Channel formation properties of synthetic pardaxin and analogues, *J. Biol. Chem.*, 265, 20202, 1990.

46. Marion, D., Zasloff, M., and Bax, A., A two-dimensional NMR study of the antimicrobial peptide maganin 2, *FEB*, 227, 21, 1988.

47. Thiaudiere, E., Siffert, O., Talbot, J.-C., Bolard, J., Alouf, J. E., and Dufourcq, J., The amphiphilic alpha-helix concept. Consequences on the structure of staphylococcal delta-toxin in solution and bound to lipids, *Eur. J. Biochem.*, 195, 203, 1991.

48. Alouf, J. E., Dufourcq, J., Siffert, O., Thiaudiere, E., and Geoffroy, Ch., Interaction of staphylococcal delta-toxin and synthetic analogues with erythrocytes and phospholipid vesicles, *Eur. J. Biochem.*, 183, 381, 1989.

49. Leippe, M., Ebel, S., Schoenberger, O. L., Horstmann, R. D., and Müller-Eberhard, H. J., Pore-forming peptide of pathogenic Entamoeba histolytica, *Proc. Natl. Acad. Sci. U.S.A.*, 88, 7659, 1991.

50. Lehrer, R. I., Ganz, T., and Selsted, M. E., Defensins: endogenous antibiotic peptides of animal cells. *Cell*, 64, 229, 1991.

51. Wiley, D. C., Skehel, J. J., and Waterfield, M. D., *Virology*, 79, 446, 1977.
52. Wilson, I. A., Skehel, J. J., and Wiley, D. C., Structure of the haemagglutinin membrane glycoprotein of influenza virus at 3 A resolution, *Nature*, 289, 366, 1981.
53. Wiley, D. C., Wilson, I. A., and Skehel, J. J., Structural identification of the antibody-binding sites of Hong Kong influenza haemagglutinin and their involvement in antigenic variation, *Nature*, 289, 373, 1981.
54. Carr, C. M. and Kim, P. S., A spring-loaded mechanism for the conformational change of influenza hemagglutinin, *Cell*, 73, 823, 1993.
55. Wharton, S. A., Martin, S. R., Ruigrok, R. W. H., Skehel, J. J., and Wiley, D. C., Membrane fusion by peptide analogues of influenza virus haemagglutinin, *J. Gen. Virol.*, 69, 1847, 1988.
56. Lear, J. D. and Grado, W. F., Membrane binding and conformational properties of peptides representing the NH2 terminus of influenza HA-2, *J. Biol. Chem.*, 262, 6500, 1987.
57. Subbarao, N. K., Parente, R. A., Szoka, F. C., Nadasdi, L., and Pongracz, K., pH-Dependent bilayer destabilization by an amphipathic peptide, *Biochemistry*, 26, 2964, 1987.
58. Parente, R. A., Nir, S., and Szoka, F. C., pH-Dependent fusion of phosphatidylcholine small vesicles. Induction by a synthetic amphipathic peptide, *J. Biol. Chem.*, 263, 4724, 1988.
59. Parente, R. A., Nir, S., and Szoka, F. C., Mechanism of leakage of phospholipid vesicle contents induced by the peptide GALA, *Biochemistry*, 29, 8720, 1990.
60. Murata, M., Sugahara, Y., Takahashi, S., and Ohnishi, S.-I., pH-Dependent membrane fusion activity of a synthetic twenty amino acid peptide with the same sequence as that of the hydrophobic segment of influenza virus hemagglutinin, *J. Biochem.*, 102, 957, 1987.
61. Wu, G. Y. and Wu, C. H., Evidence for targeted gene delivery to HepG2 hepatoma cells *in vitro*, *Biochemistry*, 27, 887, 1988.
62. Wilson, J. M., Grossman, M., Cabrerea, J. A., Wu, C. H., and Wu, G. Y., A novel mechanism for achieving transgene persistence *in vivo* after somatic gene transfer into hepatocytes, *J. Biol. Chem.*, 267, 11483, 1992.
63. Haensler, J. and Szoka, F. C., Synthesis and characterization of a trigalactosylated bisacridine compound to target DNA to hepatocytes, *Bioconjugate Chem.*, 4, 85, 1993.
64. Midoux, P., Mendes, C., Legrand, A., Raimond, J., Mayer, R., Monsigny, M., and Roche, A. C., Specific gene transfer mediated by lactosylated poly-L-lysine into hepatoma cells, *Nucleic Acids Res.*, 21, 871, 1993.
65. Wagner, E., Zenke, M., Cotten, M., Beug, H., and Birnstiel, M. L., Transferrin-polycation conjugates as carriers for DNA uptake into cells, *Proc. Natl. Acad. Sci. U.S.A.*, 87, 3410, 1990.
66. Huckett, B., Ariatti, M., and Hawtrey, A. O., Evidence for targeted gene transfer by receptor-mediated endocytosis: stable expression following insulin-directed entry of *neo* into HepG2 cells, *Biochem. Pharmacol.*, 40, 253, 1990.
67. Rosenkranz, A. A., Yachmenev, S. V., Jans, D. A., Serebryakova, N. V., Murav'ev, V. I., Peters, R., and Sobolev, A. S., Receptor-mediated endocytosis and nuclear transport of a transfecting DNA construct, *Exp. Cell Res.*, 199, 323, 1992.
68. Trubetskoy, V. S., Torchilin, V. P., Kennel, S., and Huang, L., Use of N-terminal modified poly(L-lysine)-antibody conjugates as a carrier for targeted gene delivery in mouse lung endothelial cells, *Bioconjugate Chem.*, 3, 323, 1992.
69. Trubetskoy, V. S., Torchilin, V. P., Kennel, S., and Huang, L., Cationic liposomes enhance targeted delivery and expression of exogenous DNA mediated by N-terminal modified poly(L-lysine)-antibody conjugate in mouse lung endothelial cells, *Biochim. Biophys. Acta*, 1131, 311, 1992.

70. Cotten, M., Wagner, E., Zatloukal, K., and Birnstiel, M. L., Chicken adenovirus (CELO virus) particles augment receptor-mediated DNA delivery to mammalian cells and yield exceptional levels of stable transformants, *J. Virol.*, 67, 3777, 1993.
71. Legendre, J. Y. and Szoka, F. C., Cyclic amphipathic peptide-DNA complexes mediate high-efficiency transfection of adherent cells, *Proc. Natl. Acad. Sci. U.S.A.*, 90, 893, 1993.
72. Haensler, J. and Szoka, F. C., Polyamidoamine cascade polymers mediate efficient transfection of cells in culture, *Bioconjugate Chem.*, 4, 372, 1993.

Chapter 17

Targeted Delivery of Anti-Hepatitis B Antisense Oligonucleotides

Ellen P. Carmichael, Henry C. Chiou, Mark A. Findeis,
George L. Spitalny, and June Rae Merwin

CONTENTS

I. INTRODUCTION

Antisense oligonucleotides have been designed and used to inhibit the replication of many viruses.[1] They have been shown to inhibit viral replication by interfering with the interactions of nucleic acids with other nucleic acids and with proteins via hybridization to their target complementary sequences.[2-7] When directed against mRNA, antisense oligonucleotides hybridize by Watson-Crick base pairing to prevent translation of the mRNA. When targeted to pre-mRNA, they can block such processes as polyadenylation, splicing, or nucleocytoplasmic transport.[8,9] In addition, antisense oligonucleotides have been shown to activate RNase H against certain sequences,[10-12] thereby destroying the target molecules. More recently, oligonucleotides that bind double-stranded DNA via Hoogsteen base pairing to form triple-helix structures have been studied. These triplex-forming oligomers appear to act by blocking transcription.[13-17] In theory, antisense technologies are alluring because they offer the possibility of exquisite specificity. This potentially eliminates the problems associated with most nucleoside analog antiviral agents, which are characterized by low therapeutic indices reflecting the less specific mechanism of action of these drugs.

In practice, however, there are many potential problems that must be overcome in order for oligonucleotides to become effective therapeutic agents. Issues of bioavailability, pharmacodynamics, and pharmacokinetics appear to be more complex for oligonucleotides than for simple nucleoside analogs. The stability *in vivo* of standard phosphodiester oligonucleotides is a major concern. Serum nucleases exist which will degrade oligonucleotides that are introduced into the bloodstream.[18-20] Degradation to monomers and dimers would certainly render the antisense oligonucleotide inactive. Conversely, degradation which results in larger fragments would destroy the specificity of the oligomer, but perhaps allow medium-size fragments to hybridize to alternative cellular sequences, resulting in unforeseen adverse effects.

0-8493-4778-5/95/$0.00+$.50
© 1995 by CRC Press Inc.

Oligonucleotides can be chemically modified to make them much more resistant to nuclease attack. Thus the potential therapeutic use of oligomers with modified backbone structures has received considerable attention. Phosphorothioates, which contain a sulfur for oxygen substitution, and methylphosphonates, which contain methyl-group substituents, have reduced susceptibility to nucleases without disruption of hybridization ability.[6,18,21] Recently, C-5 propyne-substituted phosphorothioate oligonucleotides have even been shown to have enhanced binding affinities for their complementary sequences.[22] However, administration of untargeted nuclease-resistant oligonucleotides *in vivo* via intravenous injection poses a number of potential problems. A large fraction of antisense oligonucleotides injected into mice is rapidly excreted into the urine.[6] In addition, much of the remaining oligonucleotide will be distributed to nontarget tissue. Thus, it may be necessary to introduce large amounts of untargeted oligomers into patients in order to achieve sufficient concentrations at the site of infection. This may render the cost of therapy prohibitive. Nuclease-resistant oligomers may also lead to long-term accumulation in a variety of tissues to which they were not intended, possibly resulting in the development of nonsequence-specific toxicities. The problem of adverse reactions resulting from the accumulation of partial degradation products that no longer have unique gene specificities may also be exacerbated with nuclease-resistant oligomers. In addition, even when degraded to monomers, such as in the slow degradation of phosphorothioate oligomers by S1 and P1 nucleases, the resulting release of modified mononucleotides could be deleterious because of incorporation into and mutagenization of cellular DNA.[7]

Targeting of oligonucleotides to the site of infection can be a means of limiting the problems discussed above. It can dramatically confine the uptake of the oligomer to the intended cell or tissue. Thus, an efficient and specific delivery system will restrict the accumulation of the oligomer and its degradation products to just the target tissue, possibly resulting in lower overall toxicity. Cells in culture and cells in the body take up charged oligonucleotides indiscriminately;[8,23] however, a delivery vehicle can greatly facilitate the cellular entry of these oligomers. Overall, targeted delivery can potentially increase the efficacy and the therapeutic index of intravenously administered oligonucleotides.

A variety of methods to help deliver oligonucleotides have been described. For example, liposomes have been used to deliver DNA to the liver *in vivo*.[24-26] However, cell-specific binding is not easily achieved and delivery is limited by rapid sequestration in the reticuloendothelial system. Both cationic lipids[27] and encapsulating liposomes[28,29] have been shown to enhance cellular uptake and the observed activity of antisense oligonucleotides *in vitro*. Polymeric nanoparticles bind oligonucleotides and can protect them from degradation. Thus, they may possess certain advantages to help improve delivery.[30] Approaches using liposomes and microparticles could become effective for *in vivo* delivery if the pharmacology and biodistribution of these particles could be controlled by their composition or by the use of a targeting ligand.

Polylysine (PL) is another agent that appears to enhance cellular uptake and nuclease resistance of bound oligomers.[24,31] Covalent conjugation of antisense oligonucleotides with poly-L-lysine potentiates their biological activity.[32,33] Coupling to this cationic polymer makes the oligonucleotide more hydrophobic, or lipophilic, and improves its penetration through the cell membrane. In addition, poly-L-lysine conjugation may reduce if not completely neutralize the negative charge of the oligomer, thus preventing it from interacting with nonspecific cellular binding proteins. The PL may also help to destabilize endosomal membranes and thus permit easier transit of oligonucleotides into the cytoplasm or nucleus.[7] This type of conjugation, as well as other modifications,[34-37] can increase the net intracellular internalization or retention of oligonucleotides.[38] Cholesteryl derivatives have been shown to enhance activity, but the pharmacology of these materials is complicated, including partial uptake by association with low-density lipoprotein particles.[39] In the absence of a targeting ligand, these approaches have limited promise for *in vivo* applications.

II. TARGETED OLIGONUCLEOTIDE DELIVERY TO HEPATOCYTES

We describe here a technology which circumvents some of the problems described above that are inherent in untargeted oligonucleotide-based antiviral therapy. The principle of this method is to covalently couple a polycation to a protein ligand that binds to specific cell-surface receptors. Anionic oligonucleotide DNA (ODN) is then bound electrostatically to the polycation. The ligand functions to selectively deliver the DNA-protein complex to cells displaying the appropriate receptor.

Liver cells provide ideal targets for *in vivo* delivery of DNA molecules by this strategy. Parenchymal hepatocytes express on their sinusoidal (blood-facing) surface a receptor that recognizes glycoproteins containing galactose-terminal residues.[40,41] The hepatic asialoglycoprotein receptor (ASGPr) is found only on hepatocytes, with 100,000 to 500,000 binding sites present per cell. Bound asialoglycoproteins (ASGP) are internalized in clathrin-coated pits by receptor-mediated endocytosis. Following internalization, the receptors are rapidly recycled, or replenished from substantial intracellular reservoirs. A large number of ligands can therefore be rapidly bound and internalized without compromising the ability of the liver to take up more ligand shortly thereafter. Thus, the ASGPr offers a number of pharmacokinetic advantages for *in vivo* targeted oligonucleotide delivery.

Removal of sialic acid from the abundant plasma protein, orosomucoid (OR) (α-acid glycoprotein), generates galactose-terminal asialoorosomucoid (ASOR), a high-affinity ligand for the ASGPr. Covalent coupling of the cationic polymer, PL, is readily achieved. A schematic representation of ASOR-PL-ODN complex is shown in Figure 1. Targeted delivery of oligonucleotide to hepatocytes is accomplished by specific binding of ASOR to the ASGPr and internalization by receptor-mediated endocytosis. Thus, the beneficial effects of poly-L-lysine coupling described above can be maintained with the added advantage of cell-specific targeting.

The innovative studies of Wu and Wu[42-44] and their colleagues[45-47] were the first to successfully implement DNA transfer by receptor-mediated delivery both *in vitro* and *in vivo*. This technology is versatile enough to deliver both low-molecular mass molecules[48-50] and macromolecules such as double-stranded plasmid DNA. More recently it was demonstrated that the targetable DNA carrier could also be used to deliver single-stranded oligomers *in vitro*.[51]

We have observed that liver-specific delivery of ASOR-PL-ODN complexes can be achieved *in vivo*. ASOR-PL-ODN complexes containing ^{125}I-labeled ASOR are targeted to the liver when injected intravenously into mice. Previous studies with complexes containing plasmid DNA have shown that 85% of the injected material is taken up by the liver within 10 to 20 min.[43,52] Of the complex found in the liver, 80% was found to be in the parenchymal cell population. When we tested the localization of ASOR-PL-ODN complexes we achieved similar results. Of the total counts distributed to the five major organs, 85% are detected in the liver. In addition, similar experiments conducted with complexes containing ^{32}P-labeled ODNs confirmed liver-specific delivery. Compared to untargeted oligonucleotides, double the amount of complexed ODN is delivered to the liver. More importantly, examination by fluorescence microscopy and cryoautoradiography showed that the complex is found predominantly in the parenchymal cells (unpublished results). In contrast, uncomplexed ODN is predominantly detected in the Kupffer cells. Targeting to parenchymal cells is an essential feature for treating infectious diseases of the liver, since the infections occur principally within the hepatocytes. Roughly one third of most substances (such as untargeted oligonucleotides and liposomes) injected into the

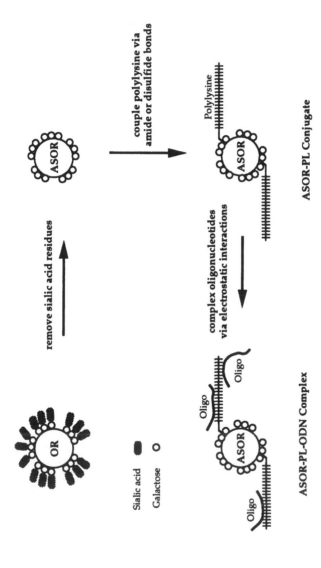

Figure 1 Schematic of the preparation of asialoorosomucoid-polylysine (ASOR-PL) conjugates and ASOR-PL-ODN complexes. Sialic acid is removed from orosomucoid (OR) by treatment with acid or neuraminidase to obtain asialoorosomucoid (ASOR). ASOR is crosslinked with PL using carbodiimide or thiol chemistry to obtain ASOR-PL conjugate in which the covalent coupling is via amide or disulfide bonds, respectively. ASOR-PL and antisense oligonucleotide (ODN) DNA to be targeted are combined to form an electrostatic complex. The ASOR on the surface of the complex will deliver the complexed ODN to asialoglycoprotein receptor-bearing cells.

bloodstream are internalized by the liver. However, these substances are taken up predominantly by the Kupffer cells of the reticuloendothelial system and not by the hepatocytes. Thus, the increase in liver uptake observed with ASOR-PL-ODN complexes only modestly indicates the advantage gained by hepatocyte-directed targeted delivery.

III. ACTIVITY OF TARGETED OLIGONUCLEOTIDES *IN VITRO* AND *IN VIVO*

The liver-specific human hepatitis B virus (HBV) is of major medical importance worldwide. Approximately 2 billion people have contracted HBV, and of those almost 350 million people remain chronically infected (World Health Organization Press Release, 1992). For comparison, that is roughly 140 times the number of people (14 million) that have been infected with human immunodeficiency virus. Despite the availability of safe and effective vaccines, HBV continues to be a major public health problem. It is a causative agent of chronic liver disease as well as of primary hepatocellular carcinoma in humans. An estimated 15 to 25% of individuals with chronic HBV infections will die prematurely of either cirrhosis or hepatocellular carcinoma.[53] The progression from acute to chronic infection most readily develops under circumstances where the host immune response fails to clear the hepatocytes that contain replicating HBV.[54] No specific or satisfactory treatment exists for HBV-infected patients.[55] Efforts to treat patients with nucleoside analogs that inhibit viral replication have had limited success due to associated toxicities and other deleterious side effects.[56-60] Currently, treatment of HBV with interferon (IFN) is the most successful therapy; however, more than half the patients in the chronic phase of infection do not respond to this treatment.[61]

Increased knowledge of the HBV replication cycle and the regulatory elements responsible for controlling viral transcription, genome replication, and DNA packaging has made it possible to more rationally design and use antisense oligonucleotides to interfere with these processes. The HBV genome consists of a small (approximately 3.2-kb) partially double-stranded DNA molecule whose sequence consists of four open reading frames (Figure 2). The life cycle of HBV[62,63] involves transcription of a greater-than-genome-length (3.5-kb) mRNA, which serves as a template both for the reverse transcription of the first strand of DNA and for protein synthesis in the cytoplasm. Because of the compact and overlapping organization of the HBV genome, it is possible to interfere with more than one step in the replication cycle with a single oligonucleotide.

We have expanded on the studies reported by Wu and Wu[51] that demonstrate that an antisense oligonucleotide (designed to bind to the polyadenylation consensus sequence of HBV) can be selectively internalized by hepatocytes and confer HBV antiviral activity *in vitro*. We have tested oligonucleotides directed against several other HBV functions and have shown that complexation of these oligonucleotides to ASOR-PL potentiates antiviral activity both *in vitro* and *in vivo*.

A. *IN VITRO* ACTIVITY

Cell culture systems for the evaluation of compounds which inhibit HBV replication have been developed by transformation of human liver cell lines with cloned dimeric HBV DNA.[64-66] These cell lines secrete hepatitis B surface antigen particles, nucleocapsids, and infectious virions. In the 2.2.15 cell line the intracellular nucleic acid profile is identical to that obtained by analyzing HBV-infected human liver tissue,[67] which makes it an attractive tissue culture model system for exploring the molecular biology of HBV and for evaluating potential antiviral compounds.

We have evaluated oligonucleotides designed to target some HBV functions that occur in the cytoplasm and some that occur in the nucleus. The choice of a target for antisense

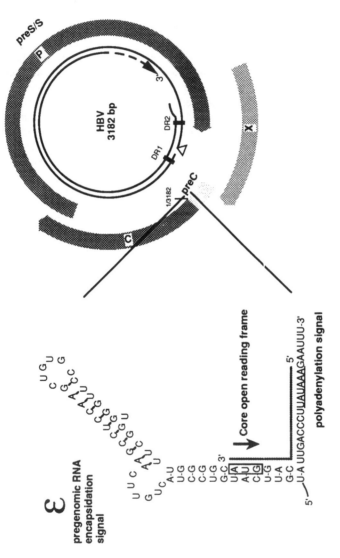

Figure 2 Translational map of the HBV genome. Positions of the four open reading frames are shown. C indicates core antigen, P indicates the viral DNA polymerase, preS/S indicate surface antigens, and X indicates the X antigen. The double-stranded genome is indicated by solid lines. The minus strand has a polymerase covalently linked to its 5' end (marked with a triangle). Dashes indicate the variability at the 3' end of the shorter plus strand. The relative positions of the repeats, DR1 and DR2, are shown. The left side of the panel shows the sequence and secondary structure, the pregenomic RNA encapsidation signal, and the position of the Poly A antisense oligonucleotide (solid line). Poly A encompasses the complementary strand of the AUG of core open reading frame and the HBV polyadenylation signal.

Table 1 **Effect of antisense oligonucleotides on HBV production in 2.2.15 cells**

	Protocol 1	Protocol 2
DRII(S)		
Free	>40 μM	2.5 μM
Complexed	2.5 μM	2.5 μM
Poly A		
Free	>40 μM	7.5 μM
Complexed	2.5 μM	3 μM

Note: Analysis of intracellular HBV DNA following two different protocols. Protocol 1 consisted of 1 addition of oligonucleotide or ASOR-PL-ODN on day 0 and analysis of HBV DNA intermediates on day 8. Protocol 2 consisted of 5 additions of oligonucleotide or ASOR-PS-ODN every 2 d for 10 d. Results are reported as the dose (μM) required to inhibit HBV DNA replication by 50%.

DNA oligonucleotide inhibition of HBV gene expression or viral replication is an important consideration. Interactions can occur in either the nucleus or the cytoplasm, and with mature mRNA, pre-mRNA, or with single-stranded DNA replication intermediates. While most efforts in the antisense field have been directed at translational inhibition of mRNA molecules, the actual mechanism of inhibition of gene expression by both oligos and vector-generated RNA remains controversial.[68] Antisense oligonucleotides that localize to the nucleus could act to block transcription initiation or elongation, RNA processing, or nucleocytoplasmic mRNA transport. Such duplexes might also induce an endogenous RNAse activity that results in the degradation of the target RNA molecules. Alternatively, cytoplasmic antisense oligonucleotides would be expected to function primarily by affecting the translation of mature mRNA molecules or single-stranded HBV DNA replication intermediates. In all cases, antisense oligonucleotides should work best when able to hybridize to single-stranded nucleic acids in regions containing little secondary structure, and should therefore be targeted to such regions, if possible. Some antisense sequences were designed to simultaneously interfere with viral processes in both compartments. For example, the 21-mer "Poly A" oligonucleotide is complementary to sequences involved in nuclear events (polyadenylation) and cytoplasmic events (the initiation codon for the viral core protein and a portion of the epsilon sequence that is required for packaging viral RNA) (Figure 2). A second oligonucleotide, named DRII(S), is targeted to the DR2 sequence of the HBV genome. It was designed to bind to the DR2 region and thus interfere with binding of the RNA primer to the site which initiates second (+)-strand DNA replication. Other oligonucleotides were directed against AUG initiation codons of various HBV-encoded proteins.

Using 2.2.15 cells,[65] we have examined the *in vitro* anti-HBV activity of these oligonucleotides. Our results indicate that antisense oligonucleotides that are complexed to ASOR-PL displayed considerable antiviral activity when used in a variety of treatment regimens, while free oligonucleotides were only effective in certain settings. For example, when oligonucleotide was added to the culture as a single dose during an 8-d incubation, the ODN complex displayed substantially more activity than free oligonucleotide (Table 1). When fresh oligonucleotides were added every other day for 10 d the activity of free and complexed ODN was virtually equivalent. These results suggest that the ODN complex possesses a high level of activity under a variety of culture conditions. Free oligonucleotide, on the other hand, required manipulation of test conditions to achieve significant activity. One of the possible explanations of these results is that the

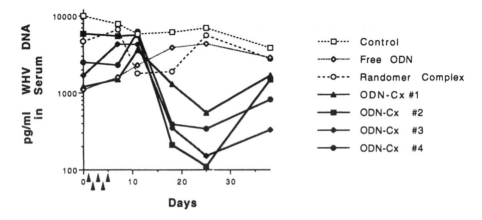

Figure 3 Levels of woodchuck hepatitis virus in serum. Animals were infected with virus several months prior to the start of the experiment. Animals were injected intravenously via an indwelling catheter, with five bolus injections administered daily for 5 d (solid triangles). Four animals received ASOR-PI-ODN complex; one animal received free ODN ; one animal was an untreated control.

complexed oligomer remained active within the cell for a much longer period of time than did uncomplexed oligonucleotide. The difference in activity of ODN alone vs. ASOR-PL-ODN would be predicted to be even more dramatic *in vivo,* where a much greater proportion of the complexed ODN should accumulate in the infected tissue. An *in vivo* study performed in woodchucks (see below, Figure 3) supports this assertion, though many more studies will have to be carried out to fully substantiate this claim. The results described above underscore both the stability and activity of antisense molecules delivered by the ASOR-PL technique.

B. *IN VIVO* ACTIVITY

To examine the efficacy of the ASOR-PL antisense delivery system *in vivo* we carried out a set of experiments using the woodchuck hepatitis virus (WHV) model. The WHV is a member of the hepadnaviridae family along with human HBV.[69,70] It is closely related to HBV, both in its genome structure and in its biology. Like human HBV carriers, WHV carrier animals are at a high risk of developing chronic hepatitis and hepatocellular carcinomas.[69] Studies of WHV have provided much of the basic information on the structure, replication, and gene products of HBV. Since the WHV system is one of the best-studied mammalian systems (other than chimpanzees) in which a hepadnavirus infection can be experimentally manipulated, it provides an important model in which to test the *in vivo* efficacy and toxicity of targeted antisense oligonucleotides.

Chronically infected woodchucks were treated on five successive days with intravenous bolus injections of ASOR-PL-ODN or free ODN. The antisense oligonucleotide contained the WHV sequence equivalent to the HBV Poly A oligomer. The cumulative amount of antisense DNA administered to each animal was 2.0 mg. Thus, each dose was equivalent to 133 μg/kg of body weight. At various times after treatment, blood samples were assayed for WHV DNA as an indicator of viral titer.

As shown in Figure 3, all four ASOR-PL-poly A-treated animals simultaneously exhibited a striking decrease in circulating viral titer by day 18 after initiation of treatment. The untreated control animal maintained a significantly higher viral titer. Two of the animals displayed a greater than tenfold decrease in the level of viral DNA detected in the circulation 12 d following cessation of treatment. The remaining two animals also exhibited decreased levels of virus, but the absolute reduction was less (between a five-and tenfold reduction). Though the effect of the targeted oligonucleotide was delayed, the reduced viral burden was maintained for more than 1 week and only slowly started to increase over time. The amount of viral DNA in the untreated control animal varied over time. This type of slow cycling of viral titer is typical of WHV infection. By contrast, the simultaneous rapid and sustained decrease in viral burden observed in the four treated woodchucks strongly suggests that administration of the oligonucleotide-protein complex produced a significant treatment effect. Treatment of one animal with uncomplexed oligonucleotide did not produce any evidence of viral inhibition. These data clearly suggest that complexation of the antisense oligonucleotide to ASOR-PL renders the oligonucleotide capable of inhibiting hepadnaviral replication, not only in tissue culture, but in animals.

IV. ADDITIONAL ADVANTAGES TO USING THE TARGETABLE DNA-CARRIER TECHNOLOGY

Rapid and specific targeting of oligonucleotides to hepatocytes *in vivo* by ASOR-PL conjugates should be advantageous in that a greater proportion of antisense oligomer will be delivered to and accumulate at the site of action compared to a general, nonspecific, systemic distribution strategy. ASOR-PL conjugates confer an additional advantage for DNA delivery in that binding of the conjugate appears to protect nucleic acids from degradation by nucleases. The protein conjugate may coat and condense DNA, thereby shielding it from nuclease attack. Greater than 75% of double-stranded plasmid DNA remained intact after 60 min of incubation in rat serum when the DNA was complexed to ASOR-PL. By contrast, uncomplexed plasmid was completely degraded in serum after only 5 to 15 min (unpublished data). Similar results, over a longer time course, were observed with oligonucleotide complexes. Oligomers were also shown to remain tightly bound to ASOR-PL conjugates even after 24 h of incubation in serum, suggesting that the complex will not dissociate during circulation to hepatocytes. Rapid targeted delivery and protection of antisense oligomers from degradation by serum nucleases should result in greatly enhanced accumulation of the active molecule in target cells compared to the use of naked oligomers.

Examination of the antiviral activities of complexed and uncomplexed phosphodiester oligonucleotides highlights the advantage conferred by ASOR-PL in preventing the degradation of complexed nucleic acids. Antisense phosphodiester oligonucleotides, containing the same sequence as the poly A phosphorothioate oligomer, were not active in inhibiting HBV DNA replication in cell culture (Figure 4), probably due to their high susceptibility to nuclease degradation. However, complexation of these phosphodiester oligonucleotides with ASOR-PL conjugates restored their antisense activity, presumably by protecting the oligonucleotide from degradation until it reached its site of action within the cell. This protection from degradation may make possible the use of nuclease-sensitive phosphodiester oligonucleotides in antisense therapies and thus circumvent some of the potential toxicity problems associated with the use of nuclease-resistant oligomers.

V. PRACTICAL ASPECTS OF OLIGONUCLEOTIDE COMPLEX PREPARATION

To develop a targetable oligonucleotide complex for use as a conventional injectable pharmaceutical, several practical requirements must be met. The targeting ligand, an appropriate polycation, and the oligonucleotide containing the desired base sequence must be available in sufficient quantity and at reasonable cost for commercial-scale

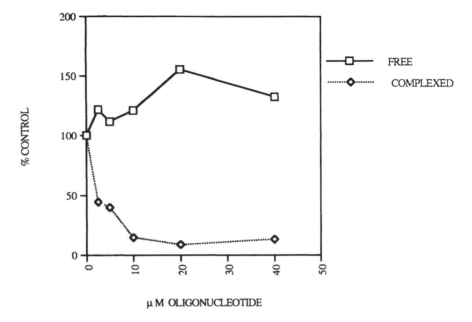

Figure 4 Dose-response curve for inhibition of HBV replication in HepG2 2.2.15 cells by phosphodiester oligonucleotide applied to the cell culture as either free or complexed to asialoarosomucoid-poly(L)lysine conjugate. The oligonucleotides were added to the cultures as described in Table 1 (protocol 2) and replication intermediates were quantitated by Southern analysis. Results have been normalized to the amount of HBV DNA in the untreated controls (100%).

production. Proteins such as OR and transferrin are readily available from human plasma; basic poly-amino acids such as PL are commercial reagents and are available in a variety of average molecular weights. If an analog of the natural phosphodiester backbone of DNA is used, it must be polyanionic (e.g., a phosphorothioate) to allow electrostatic complexation with the ligand-polycation conjugate. Economical industrial production of oligonucleotides (phosphodiesters, phosphorothioates, and other derivatives) of high quality on the gram scale or larger is a process that is currently being reduced to practice. Phosphorothioates are currently rather expensive, but as this technology matures these costs should come down.

A covalent bond between the ligand and the polycation ensures that an electrostatically bound oligonucleotide will be carried to the target cell. The chemistry of bioconjugate formation is diverse and well developed.[71] The chemistry used for crosslinking ligand and polycation is chosen to ensure that the receptor-binding ability of the ligand is not impaired. ASOR-PL conjugates are readily prepared using carbodiimide or thiol chemistry which does not interfere with carbohydrate groups that bind to the ASGP. Such conjugations between macromolecules invariably generate complex product mixtures. These are often sufficient for *in vitro* work; however, a more involved fractionation, in which unreacted starting materials are removed along with low-molecular weight byproducts, is preferred for *in vivo* applications.

In our experience, complexes containing oligonucleotides are easier to prepare in concentrated form than are complexes containing double-stranded plasmid DNA. Inadequate purification of ASOR-PL conjugates results in complexes with low solubilities and low recoveries of soluble complexed DNA or oligomer following filtration. However, in general, ASOR-PL-ODN complexes are more soluble than plasmid complexes.

ASOR-PL-plasmid complexes rarely can be prepared at greater than 1 mg of soluble complexed DNA per milliliter. Complexes containing oligonucleotides, on the other hand, can be prepared to greater than 20 mg ODN per milliliter, using ASOR-PL purified in a one-step reverse-phase chromatographic procedure.

VI. ASIALOOROSOMUCOID-POLYLYSINE CONJUGATES AND THEIR COMPLEXES

OR (also referred to as α_1-acid glycoprotein) is isolated from human plasma.[72-74] Removal of sialic acid generates galactose-terminal ASOR, a high-affinity ligand for the ASGPr. The stability of OR is sufficient to allow desialylation by acid hydrolysis at pH 1.5 for 1 h at 80°C.[75] This procedure is a more practical approach compared with the enzymatic hydrolysis of sialic acids using neuraminidase,[43] and the resulting ASOR is as effective a ligand. An added advantage of acid hydrolysis is that the conditions used should effectively neutralize any blood-borne pathogens, such as viruses, which may have persisted through plasma fractionation. OR is potentially available in large quantities as a byproduct of the plasma fractionation industry, which processes several million liters of plasma a year in the U.S. alone. After the isolation of albumin, OR is one of the main components in the supernatant of fraction V of the classical Cohn process.[72]

The polycation of choice in the synthesis of protein ligand-polycation conjugates has been poly-L-lysine. The two most favored strategies for crosslinking ligand and PL have been carbodiimide-mediated amide bond formation[51,76] and thiol chemistry-based crosslinking.[43,79] The resulting product mixtures are typically heterogeneous and require fractionation by one or more chromatographic techniques such as ion exchange[51] or gel filtration[42,77] to isolate the desired fraction.

An agarose gel retardation assay is used to determine the quantity of purified conjugate required to fully complex a selected oligonucleotide.[51] As cationic conjugate is added, the polyanionic phosphodiester or phosphorothioate diester backbone of the oligonucleotide is progressively neutralized and the electrophoretic mobility of the oligonucleotide is reduced. The ratio of conjugate to oligonucleotide at which the oligomer will not migrate into the gel is taken as full complexation and is used to prepare ODN complex for targeted delivery.

VII. FUTURE ISSUES

Cell specificity, protection against nuclease attack, and enhanced intracellular delivery have been achieved with the described soluble DNA carrier system. We have demonstrated that the oligonucleotides delivered to cells both *in vitro* and *in vivo* inhibit HBV DNA synthesis. However, the technology described will certainly be improved over time. It can be assumed that once the DNA is internalized into the endosome (the first intracellular destination of ASOR-PL ODN complexes), it is directed to the lysosomal compartment where a majority of it is destroyed. Despite this, enough DNA is clearly escaping lysosomal degradation so that activity is achieved. If such degradation could be averted or even attenuated, biological activity should be greatly enhanced.

The past several years have witnessed the use of adenovirus particles[78-82] and influenza proteins[83] in conjunction with soluble DNA carrier systems to dramatically increase the efficiency of protein[84] and gene delivery. This is likely due to the known activity of these viruses to allow their natural genomic DNA to escape lysosomal degradation. In their present state, although promising, these approaches are not practical because the viral constituents are immunogenic and will not allow repeat dosing for antiviral treatment. It is anticipated, however, that the next few years will witness major advances in establishing mechanisms similar to viral functions allowing endosomal escape which can be used in conjunction with ASOR-PL complexes to achieve even greater levels of targeted DNA delivery to cells.

Finally, we should emphasize that although the data presented here focus on delivery of HBV-specific oligonucleotides to the parenchymal cells of the liver, the principles may be applied more generally to other tissues and viruses.

REFERENCES

1. Agrawal, S., Antisense oligonucleotides as antiviral agents, *Trends Biotechnol.*, 10, 152, 1992.
2. Goodchild, J., Inhibition of gene expression by oligonucleotides, in *Oligodeoxynucleotides — Antisense Inhibitors of Gene Expression*, Cohen, J. S., Ed., CRC Press, Boca Raton, FL, 1989, 53.
3. Uhlmann, E. and Peyman, A., Antisense oligonucleotides: a new therapeutic principle, *Chem Rev.*, 90, 543, 1990.
4. Wickstrom, E., Ed., *Prospects for Antisense Nucleic Acid Therapy of Cancer and AIDS*, Wiley-Liss, New York, 1991.
5. Erickson, R. P. and Izant, J. G., *Gene Regulation: Biology of Antisense RNA and DNA*, Raven Press, New York, 1992.
6. Agrawal, S., Temsamani, J., and Tang, J. Y., Pharmacokinetics, biodistribution, and stability of oligodeoxynucleotide phosphorothioates in mice, *Proc. Natl. Acad. Sci. U.S.A.*, 88, 7595, 1991.
7. Stein, C. A. and Cheng, Y.-C., Antisense oligonucleotides as therapeutic agents — is the bullet really magical?, *Science*, 261, 1004, 1993.
8. Zamecnik, P. C. and Stephenson, M. L., Inhibition of Rous sarcoma virus replication and cell transformation by a specific oligodeoxynucleotide, *Proc. Natl. Acad. Sci. U.S.A.*, 75, 280, 1978.
9. Kim, S. K. and Wold, B. J., Stable reduction of thymidine kinase activity in cells expressing high levels of anti-sense RNA, *Cell*, 42, 129, 1985.
10. Walder, R. T. and Walder, J. A., Role of RNase-H in hybrid-arrested translation by antisense oligonucleotides, *Proc. Natl. Acad. Sci. U.S.A.*, 85, 5011, 1988.
11. Dash, P., Lotan, I., Knapp, M., Kandel, E. R., and Goelet, P., Selective elimination of mRNAs in vivo: complementary oligodeoxynucleotides promote RNA degradation by an RNaseH-like activity, *Proc. Natl. Acad. Sci. U.S.A.*, 84, 7896, 1987.
12. Shuttleworth, J. and Colman, A., Antisense oligonucleotide-directed cleavage of mRNA in Xenopus oocytes and eggs, *EMBO J.*, 7, 427, 1988.
13. Cooney, M., Czernuszewicz, G., Postel, E. H., Flint, S. J., and Hogan, M. E., Site specific oligonucleotide binding represses transcription of human c-myc gene *in vitro, Science*, 241, 465, 1988.
14. Postel, E. H., Flint, S. J., Kessler, D. J., and Hogan, M. E., Evidence that a triplex-forming oligodeoxyribonucleotide binds to the c-myc promoter in HeLa cells, thereby reducing c-myc mRNA levels, *Proc. Natl. Acad. Sci. U.S.A.*, 88, 8227, 1991.
15. Grigoriev, M., Praseuth, D., Robin, P., Hemar, A., Saison-Behmoaras, T., Dautry-Varsat, A., Thuong, N. T., Hélène, C., Harel-Bellan, A. A triple helix-forming oligonucleotide-intercalator conjugate acts as a transcriptional repressor via inhibition of NF kappa B binding to interleukin-2 receptor alpha-regulatory sequence, *J. Biol. Chem.*, 267, 3389, 1992.
16. Maher, L. J., III, Dervan, P. B., and Wold, B. J., Kinetic analysis of oligodeoxynucleotide-directed triple-helix formation on DNA, *Biochemistry*, 29, 8820, 1990.

17. Hélene, C., Thuong, N. T., and Harel-Bellan, A., Control of gene expression by triple helix-forming oligonucleotides: the antigene strategy, in *Antisense Strategies*, Baserga, R. and Denhardt, D., Eds., New York Academy of Sciences, New York, 1992, 660, 27–36.

18. Eckstein, R., Investigation of enzyme mechanisms with nucleoside phosphorothioates, *Annu. Rev. Biochem.*, 54, 367, 1985.

19. Wickstrom, E., Oligodeoxynucleotide stability in subcellular extracts and culture media, *J. Biochem. Biophys. Methods*, 13, 97, 1986.

20. Cazenave, C., Loreau, N., Thuong, N. T., Toulme, J. J., and Helene, C., Enzymatic amplification of translation inhibition of rabbit B-globin mRNA mediated by anti-message oligodeoxynucleotides covalently linked to intercalating agents, *Nucleic Acids Res.*, 15, 4717, 1987.

21. Miller, P. S., Agris, C., Aurelian, L., Blake, K., Murakami, A., Reddy, M., Spitz, S., and Ts'o, P. O. P., Control of ribonucleic acid function by oligonucleoside methylphosphonates, *Biochimie*, 67, 769, 1985.

22. Wagner, R. W., Matteucci, M. D., Lewis, J. G., Gutierrez, A. J., Moulds, C., and Froehler, B. C., Antisense gene inhibition by oligonucleotides containing C-5 propyne pyrimidines, *Science*, 260, 1510, 1993.

23. Loke, S. L., Stein, C. A., Zhang, X. H., Mori, K., Nakanishi, M., Subasinghe, C., Cohen, H. S., and Neckers, L. M., Characterization of oligonucleotide transport into living cells, *Proc. Natl. Acad. Sci. U.S.A.*, 86, 3474, 1989.

24. Clarenc, J. P., Degols, G., Leonetti, J. P., Milhaud, P., and Lebleu, B., Delivery of antisense oligonucleotides by poly(L-lysine) conjugation and liposome encapsulation, *Anti-Cancer Drug Design*, 8, 81, 1993.

25. de Smidt, P., Le Doan, T., de Falco, S., and van Berkel, T. J. C., Association of antisense oligonucleotides with lipoproteins prolongs the plasma half-life and modifies the tissue distribution, *Nucleic Acids Res.*, 19, 4695, 1991.

26. Ghosh, P. C. and Bachhawat, B. K., Targeting of liposomes to hepatocytes, in *Liver Diseases: Targeted Diagnosis and Therapy Using Specific Receptors and Ligands*, Wu, G. Y. and Wu, C. H., Eds., Marcel Dekker, New York, 1991, 87–103.

27. Bennett, C. F., Chiang, M.-Y., Chan, H., Shoemaker, J. E. E., and Mirabelli, C. K., Cationic lipids enhance cellular uptake and activity of phosphorothioate antisense oligonucleotides, *Mol. Pharm.*, 41, 1023, 1992.

28. Thierry, A. R. and Dritschilo, A., Liposomal delivery of antisense oligodeoxynucleotide, applications to the inhibition of the multidrug resistance in cancer cells, *Ann. N.Y. Acad. Sci.*, 660, 300, 1992.

29. Ropert, C., Lavignon, M., Duberner, C., Couvreur, P., and Malvy, C., Oligonucleotides encapsulated in pH sensitive liposome are efficient towards friend retrovirus, *Biochem. Biophys. Res. Commun.*, 183, 879, 1992.

30. Chavany, C., LeDoan, T., Couvreur, P., Puisieux, F., and Hélene, C., Polyalkylcyanoacrylate nanoparticles as polymeric carriers for antisense oligonucleotides, Pharm. Res., 9, 441, 1992.

31. Degols, G., Leonetti, J.-P., and Lebleu, B., Sequence-specific activity of antisense oligonucleotides conjugated to poly(L-Lysine) carriers, *Ann. N.Y. Acad. Sci.*, 660, 331, 1992.

32. Lemaitre, M., Bayard, B., and Lebleu, B., Specific antiviral activity of a poly(L-lysine)-conjugated oligodeoxyribonucleotide sequence complementary to vesicular stomatitis virus N protein mRNA initiation site, *Proc. Natl. Acad. Sci. U.S.A.*, 84, 648, 1987.

33. Leonetti, J. P., Degols, G., and Legleu, B., Biological activity of oligonucleotide-poly(L-lysine)-conjugates: mechansim of cell uptake, *Bioconjugate Chem.*, 1, 149, 1990.

34. Stein, C. A., Yakubov, L., Zhang, L.-M., and Tonkinson, J., Antisense oligonucleotides — promises and pitfalls, *Nucleic Acids Res. Symp. Ser.,* 24, 155–156, 1991.

35. Stevenson, M. and Iverson, P., Inhibition of human immunodeficiency virus type 1-mediated cytopathic effects by poly(L-lysine)-conjugated synthetic antisense oligodeoxyribonucleotides, *J. Gen. Virol.,* 70, 2673, 1989.

36. Citro, G., Perrotti, D., Cucco, C., D'Agnano, I., Sacchi, A., Zupi, G., and Calabretta, B., Inhibition of leukemia cell proliferation by receptor-mediated uptake of c-*myb* antisense oligodeoxynucleotides, *Proc. Natl. Acad. Sci. U.S.A.,* 89, 7031, 1992.

37. Letsinger, R. L., Zhang, G., Sun, D. K., Ikeuchi, T., and Sarin, P. S., Cholesterol-conjugated oligonucleotides: synthesis, properties, and activity as inhibitors of replication of human immunodeficiency virus in cell culture, *Proc. Natl. Acad. Sci. U.S.A.,* 86, 6553, 1989.

38. Boutourin, A. S., Guskova, L. V., Ivanova, E. M., Kobetz, N. D., Zarytova, V. F., Ryte, A. S., Yurchenko, L. V., and Vlassov, V. V., Synthesis of alkylating oligonucleotide derivatives containing cholesterol or phenazinium residues at their 3'-terminus and their interaction with DNA within mammalian cells, *FEBS Lett.,* 254, 129, 1989.

39. Krieg, A. M., Tonkinson, J., Matson, S., Zhao, Q., Saxon, M., Zhang, L.-M., Bhanja, U., Yakubov, L., and Stein, C. A., Modification of antisense phosphodiester oligonucleotides by a 5' cholesteryl moiety increases cellular association and improves efficacy, *Proc. Natl. Acad. Sci. U.S.A.,* 90, 1048, 1993.

40. Ashwell, G. and Morell, A. G., The role of surface carbohydrates in the hepatic recognition and transport of circulating glycoproteins, *Adv. Enzymol. Relat. Areas Mol. Biol.,* 44, 99, 1974.

41. Spiess, M., The asialoglycoprotein receptor: a model for endocytic transport receptors, *Biochemistry,* 29, 10009, 1990.

42. Wu, G. Y. and Wu, C. H., Receptor-mediated *in vitro* gene transformation by a soluble DNA carrier system, *J. Biol. Chem.,* 262, 4429, 1987.

43. Wu, G. Y. and Wu, C. H., Receptor-mediated gene delivery and expression *in vivo,* *J. Biol Chem.,* 263, 14261, 1988.

44. Wu, G. Y. and Wu, C. H., Evidence for targeted gene delivery to Hep G2 hepatoma cells *in vitro, Biochemistry,* 27, 887, 1988.

45. Wu, C. H., Wilson, J. M., and Wu, G. Y., Targeting genes: delivery and persistent expression of a foreign gene driven by mammalian regulatory elements *in vivo,* *J. Biol. Chem.,* 264, 16985, 1989.

46. Wu, G. Y., Tangco, M. V., and Wu, C. H., Targeted gene delivery: persistence of foreign gene expression achieved by pharmacological means, *Hepatology,* 12, 871, 1990.

47. Wilson, J. M., Grossman, M., Wu, C. H., Chowdhury, N. R., Wu, G. Y., and Chowdhury, J. R., Hepatocyte-directed gene transfer *in vivo* leads to transient improvement of hypercholesterolemia in low density lipoprotein receptor-deficient rabbits, *J. Biol. Chem.,* 267, 963, 1992.

48. Wu, G. Y., Wu, C. H., and Stockert, R. J., A model for the specific rescue of normal hepatocytes during methotrexate treatment of hepatic malignancy, *Proc. Natl. Acad. Sci. U.S.A.,* 80, 3078, 1983.

49. Wu, G. Y., Wu, C. H., and Rubin, M. I., Acetaminophen hepatotoxicity and targeted rescue: a model for specific chemotherapy for hepatocellular carcinoma, *Hepatology,* 5, 709, 1985.

50. Wu, G. Y., Keegan-Rogers, V., Franklin, S., Midford, S., and Wu, C. H., Targeted antagonism of galactosamine toxicity in normal rat hepatocytes *in vitro, J. Biol. Chem.,* 263, 4719, 1988.

51. Wu, G. Y. and Wu, C. H., Specific inhibition of hepatitis B viral gene expression *in vitro* by targeted antisense oligonucleotides, *J. Biol. Chem.,* 267, 12436, 1992.

52. Chowdhury, N. R., Wu, C. H., Wu, G. Y., Yerneni, P. C., Bommineni, V. R., and Chowdhury, J. R., Fate of DNA targeted to the liver by asialoglycoprotein receptor-mediated endocytosis *in vivo*, *J. Biol. Chem.*, 268, 11265, 1993.

53. Mast, E. E. and Alter, M. J., Epidemiology of viral hepatitis: an overview, *Seminars Virol.*, 4, 273–283, 1993.

54. Feitelson, M. A., Hepatitis B virus gene products as immunological targets in chronic infection, *Mol. Biol. Med.*, 6, 367, 1989.

55. Kurstak, E., Treatment of hepatitis B virus disease, in *Viral Hepatitis*, Springer-Verlag, New York, 1993, 119–127.

56. Hoofnagle, J. H., Minuk, G. Y., Dusheiko, G. M., Schafer, D. F., Johnson, R., Straus, S., and Jones, E. A., Adenine arabinoside 5′-monophosphate treatment of chronic type B hepatitis, *Hepatology*, 2, 784, 1982.

57. Dusheiko, G., DiBisceglie, A., Bowyer, S., Sachs, E., Ritchie, M., Schoub, B., and Kew, M., Recombinant leukocyte interferon treatment of chronic hepatitis B, *Hepatology*, 5, 556, 1985.

58. Lok, A. S. F., Novick, D. M., Karayiannis, P., Dunk, A. A., Sherlock, S., and Thomas, H. C., A randomized study of the effects of adenine arabinoside 5′-monophosphate (short or long courses) and lymphoblastoid interferon on hepatitis B virus replication, *Hepatology*, 5, 1132, 1985.

59. Davis, G. L. and Hoofnagle, J. H., Interferon in viral hepatitis: role in pathogenesis and treatment, *Hepatology*, 6, 1038, 1986.

60. Alexander, G. J. M., Fagan, E. A., Hegafty, J. E., Yeo, J., Eddeston, A. L., and Williams, R., Controlled clinical trial of acyclovir in chronic hepatitis B virus infection, *J. Med. Virol.*, 21, 81, 1987.

61. Korenman, J., Baker, B., Waggoner, J., Everhart, J. E., Di Bisceglie, A. M., and Hoofnagle, J. N., Long-term remission of chronic hepatitis B after alpha-interferon therapy, *Ann. Intern. Med.*, 114, 629, 1991.

62. Ganem, D. and Varmus, H. E., The molecular biology of the hepatitis B viruses, *Annu. Rev. Biochem.*, 56, 651, 1987.

63. Seeger, C., Summers, J., and Mason, W. S., Viral DNA synthesis, *Curr. Top. Microbiol. Immunol.*, 168, 41, 1991.

64. Sureau, C., Romet-Lemonne, J. L., Mullins, J. I., and Essex, M., Production of hepatitis B virus by a differentiated human hepatoma cell line after transfection with cloned circular HBV DNA, *Cell*, 47, 37, 1986.

65. Sells, M. A., Chen, M.-L., and Acs, G., Production of hepatitis B virus particles in Hep G2 cells transfected with cloned hepatitis B virus DNA, *Proc. Natl. Acad. Sci. U.S.A.*, 84, 1005, 1987.

66. Tsurimoto, T., Fujiyama, A., and Matsubara, K., Stable expression and replication of hepatitis B virus genome in an integrated state in a human hepatoblastoma cell line transfected with the cloned viral DNA, *Proc. Natl. Acad. Sci. U.S.A.*, 84, 444, 1987.

67. Sells, M. A., Zelent, A. Z., Shvartsman, M., and Acs, G., Stable expression and replication of hepatitis B virus genome in an integrated state in a human hepatoblastoma cell line transfected with the cloned viral DNA, *Proc. Natl. Acad. Sci. U.S.A.*, 84, 444, 1988.

68. Neckers, L., Whitesell, L., Rosolen, A., and Geselowitz, D., Antisense inhibition of oncogene expression, *Crit. Rev. Oncogenesis*, 3, 175, 1992.

69. Summers, J., Smolec, M. J., and Snyder, R., A virus similar to human hepatitis B virus associated with hepatitis and hepatoma in woodchucks, *Proc. Natl. Acad. Sci. U.S.A.*, 75, 4533, 1978.

70. Robinson, W. S., Genetic variation among hepatitis B and related viruses, *Ann. N.Y. Acad. Sci.*, 354, 371, 1980.

71. Wong, S. S., *Chemistry of Protein Conjugation and Cross-Linking*, CRC Press, Boca Raton, FL, 1993.

72. Schmid, K., *The Plasma Proteins, Structure, Function, and Genetic Control,* 2nd ed, Putnam, F. W., Ed., Academic Press, New York, 1975, chap. 4.
73. Whitehead, P. H. and Sammons, H. G., A simple technique for the isolation of orosomucoid from normal and pathological sera, *Biochim. Biophys. Acta,* 124, 209, 1966.
74. Succari, M., Foglietti M.-J., and Percheron, F., Two-step purification of human α_1-acid glycoprotein, *J. Chromatogr.,* 341, 457, 1985.
75. Schmid, K., Polis, A., Hunziker, K., Fricke, R., and Yayoshi, M., Partial characterization of the sialic acid-free forms of α_1-acid glycoprotein from human plasma, *Biochem. J.,* 104, 361, 1967.
76. Findeis, M. A., Wu, C. H., and Wu G. Y., Ligand-based carrier systems for delivery of DNA to hepatocytes, *Methods Enzymol.,* in press.
77. Wagner, E., Zenke, M., Cotten, M., Beug, H., and Birnstiel, M. L., Transferrin-polycation conjugates as carriers for DNA uptake into cells, *Proc. Natl. Acad. Sci. U.S.A.,* 87, 3410, 1990.
78. Cristiano, R. J., Smith, L. C., Kay, M. A., Brinkley, B. R., and Woo, S. L. C., Hepatic gene therapy: efficient gene delivery and expression in primary hepatocytes utilizing a conjugated adenovirus-DNA complex, *Proc. Natl. Acad. Sci. U.S.A.,* 90, 115148, 1993.
79. Michael, S. I., Huang, C., Romer, M. U., Wagner, E., Hu, P., and Curiel, D. T., Binding-incompetent adenovirus facilitates molecular conjugate-mediated gene transfer by the receptor-mediated endocytosis pathway, *J. Biol. Chem.,* 268, 6866, 1993.
80. Curiel, D. T., Wagner, E., Cotten, M., Birnstiel, M. L., Agarwal, S., Li, C., Loechel, S., and Hu, P., High-efficiency gene transfer mediated by adenovirus coupled to DNA-polylysine complexes, *Hum. Gene Ther.,* 3, 147, 1992.
81. Cotten, M., Wagner, E., Zatloukal, K., and Birnstiel, M. L., Chicken adenovirus (CELO Virus) particles augment receptor-mediated DNA delivery to mammalian cells and yield exceptional levels of stable t transformants, *J. Virol.,* 67, 3777, 1993.
82. Wagner, E., Zatloukal, K., Cotten, M., Kirlappos, H., Mechtler, K., Curiel, D. T., and Birnstiel, M. L., Coupling of adenovirus to transferrin-polylysine/DNA complexes greatly enhances receptor-mediated gene delivery and expression of transfected genes, *Proc. Natl. Acad. Sci. U.S.A.,* 89, 6099, 1992.
83. Wagner, E., Plank, C., Zatloukal, K., Cotten, M., and Birnstiel, M. L., Influenza virus hemagglutinin HA-2 N-terminal fusogenic peptides augment gene transfer by transferrin-polylysine-DNA complexes: toward a synthetic virus-like gene-transfer vehicle, *Proc. Natl. Acad. Sci. U.S.A.,* 89, 7934, 1992.
84. FitzGerald, D. J. P., Trowbridge, I. S., Pastan, I., and Willingham, M. C., Enhancement of toxicity of antitransferrin receptor antibody-*Pseudomonas* exotoxin conjugates by adenovirus, *Proc. Natl. Acad. Sci. U.S.A.,* 80, 4134, 1983.

Chapter 18

Design, Synthesis, and Cellular Delivery of Antibody-Antisense Oligonucleotide Conjugates for Cancer Therapy

Calvin S. R. Gooden and Agamemnon A. Epenetos

CONTENTS

I. INTRODUCTION

Antisense and antigene oligonucleotides provide an exciting avenue for the potential therapy of many significant diseases, including infection and cancer. There are now numerous examples in the literature where antisense oligonucleotides have been used to inhibit gene expression efficiently *in vitro* as well as a limited number of publications demonstrating specific *in vivo* activity of these reagents (for reviews see Carter and Lemoine[1] and Uhlmann and Peyman).[2] However, before the full therapeutic potential of these compounds can be realized it will be necessary to overcome many significant problems relating particularly to delivery mechanisms and inefficient uptake by cells. Perhaps the principal barrier to effective therapy with antisense compounds is for these polar compounds to cross the cell membrane. This results in inefficient localization of oligonucleotides into the appropriate compartment of the cell. Modified oligonucleotides containing phosphorothioate or methylphosphonate linkages have shown improved potential for use as therapeutic molecules in *in vitro* models, principally due to increased resistance to serum nucleases and better cell membrane permeation. Unfortunately, they have not become clinically practical owing to the expense of producing the large quantities necessary to achieve the desired effect and more importantly the lack of significant specificity for *in vivo* delivery to target tissue. Currently, there are no known methods to deliver large quantities of oligonucleotide specifically and efficiently to target cells *in vitro* or *in vivo*.

In the case of cancer chemotherapeutic agents which have a narrow therapeutic index, it has been a similar aim to increase the number of drug molecules delivered specifically to target cells. This is principally to reduce host toxicity and to increase antitumor efficacy. Attachment to antitumor antibodies is one strategy which is currently under investigation by a number of researchers to achieve this aim (for reviews see Byers and Baldwin[3] and Reisfeld *et al.*[4]). Likewise, there are many conjugates between monoclonal antibodies and plant or bacterial toxins (immunotoxins) which are under evaluation for possible clinical use.[5] A similar principle can be applied to antisense oligonucleotides.

0-8493-4778-5/95/$0.00+$.50
© 1995 by CRC Press Inc.

The site-specific delivery of antisense oligonucleotides by tumor-specific monoclonal antibody carriers should greatly improve the therapeutic index of these reagents. Antitumor monoclonal antibodies have been used to deliver many toxic molecules specifically to target cells (reviewed by Reisfeld *et al.*)[4] and in principle can be used similarly to deliver relatively nontoxic oligonucleotides. Also, with careful choice of target antigen, any improved internalization of the antibody-oligonucleotide conjugate (AbOL) and subsequent delivery of the oligonucleotide to the correct cellular compartment can be exploited. In summary, there is a strong case for exploring the potential of monoclonal antibodies to improve the delivery of antisense oligonucleotides to target cells.

In this chapter a brief but critical evaluation is made of some of the general problems inherent to this approach, and some ideas are advanced that may aid in optimization to make AbOL conjugates successful tools for cancer therapy.

II. CHOICE OF ANTIBODY AND ANTIGEN FOR OLIGONUCLEOTIDE DELIVERY SYSTEMS

The use of monoclonal antibodies for the delivery of antisense oligonucleotides is based on a clear-cut rationale: the achievement of site-specific delivery resulting in improved therapeutic index. Any form of antibody-guided therapy, however, will be dependent on efficient targeting of the tumor, and as such the interaction between antibody and antigen is the key for any subsequent events. Some of the ideal properties of monoclonal antibodies and antigen targets used in such studies are summarized in Table 1. These criteria would be required for the ideal immunoconjugate in an optimal situation. In reality, however most of the available monoclonal antibodies fall short of one or more of these ideals, resulting in limitations.

The best target antigens will be located only on the surface of tumor cells, will be accessible for antibody binding, and will not be shed into the circulation as circulating antigen will compete for binding to the antibody. Unfortunately, very few such tumor-specific antigens exist. Tumor cell surfaces resemble closely those of normal cells because a selective pressure exists (immune surveillance) for tumor cells to display a normal cell-surface phenotype. Hence, apart from a few notable exceptions, a quantitative rather than a qualitative difference in the expression of antigens on tumor cells prevails (tumor associated antigens), but in most cases this difference can be exploited. Apart from being operationally tumor specific, the antibody of choice should also possess high affinity for its target antigen as this will determine the amount and duration of antibody attachment to the tumor. Antibody cross reactivity with normal tissue and binding to circulating antigen can adversely affect tumor uptake of monoclonal antibody and is therefore also an important consideration.

Critically important for this approach is that bound AbOL is efficiently internalized and retained by tumor cells. Hence careful choice of an antibody/antigen system is vital to ensure adequate delivery of the oligonucleotide to its site of action. Immunotoxins which inhibit protein synthesis must be routed intracellularly, so that the toxic moiety can gain access to the cytosol. *In vitro* studies using ricin A-chain-antibody conjugates have revealed that only certain cell-surface receptor molecules can mediate the efficient translocation of A-chain across cell membranes. Furthermore, there is evidence that even if the target antigen is found on the cell surface and is efficiently internalized they may not make effective targets, because the immunotoxin is delivered to intracellular sites (e.g., lysosomes) that prevent release of active toxin into the cytoplasm.[6] These studies indicate that the route of internalization is paramount. As a result of this, cell-surface antigens that may serve as effective targets for AbOLs represent a very small proportion of the total cell-surface antigens. Following this the antigen-antibody combination must travel to post-endosomal compartments, avoiding lysosomal degradation.

Table 1 **Ideal characteristics of monoclonal antibody/antigen systems for targeted delivery of antisense oligonucleotides**

*	Target antigen well characterized biologically and chemically
*	Antigen found only on tumor cells (tumor specific)
*	Antigen exists on cell surface
*	Antigen not shed into the circulation
*	Antibody binds with high affinity
*	Antibody should be stable to linker chemistry
*	Antigen is rapidly and efficiently internalized

One antigenic target which has been rigorously explored as a target for immunotoxins is the transferrin receptor (TFR). The TFR is an integral membrane glycoprotein that is involved in the cell-surface binding of transferrin-iron complexes and their subsequent internalization into the cell. Internalization rapidly occurs via a coated pit pathway, although there is some evidence that anti-TFR antibodies are delivered to lysosomes upon internalization.[7,8] The TFR is found on both normal cells and tumor cells. The number of TFRs on a cell is proportional to the rate of proliferation of the cell,[9] and generally tumor cells express more TFR on their surface than normal cells do.[10] Despite the presence of TFR on many normal tissues, there have been many reports suggesting that anti-TFR immunotoxins may have a therapeutic role in certain settings where tumor cells are confined to a body cavity, and this may also be true for anti-TFR AbOLs. A major limitation of TFR is that it is expressed on some neurological tissue, and this could lead to severe toxicity.

Despite the requirement of efficient internalization, immunotoxins have been tested *in vitro* against oncofetal antigens such as carcinoembryonic antigen[11] and alpha-fetoprotein[12] for which no clear internalization pathway has been described. These immunotoxins may never be used *in vivo* as the targets exist at high concentrations in the serum, exhibit heterogeneity, and are widely distributed on normal tissues. A better candidate target antigen might be placental alkaline phosphatase (PLAP). PLAP is expressed by over 50% of ovarian tumors and seminomas, but only on normal placenta where its role is thought to be in aiding the transfer of maternal immunoglobulin to the fetus.[13] Antibodies bound to this antigen are reported to be efficiently internalized[14] into the endosome, and anti-PLAP immunotoxins have been effective in *in vitro* systems.[12]

III. BIODISTRIBUTION AND TARGETING

The aim of any drug delivery system should be to increase the therapeutic index of the drug (the ratio of the dose which causes a toxicity to that which causes a therapeutic effect). This may be attained by altering the distribution of drug in target sites vs. nontarget sites or by altering the kinetic profile of the drug.

Oligonucleotides are relatively nontoxic at high doses,[15] rapidly cleared from the circulation,[15,16] and internalized with low efficiency by mechanisms which are not completely understood. Modified oligonucleotides such as nonionic methylphosphonates have shown some improvement in membrane permeation, but are still far inferior to recognized specific internalization mechanisms.

As a consequence of the short *in vivo* half-life ($T_{1/2}\beta$ = 6 h in mice)[16,17] of the oligonucleotide, very large quantities would be required to maintain a therapeutic concentration. Studies with radiolabeled antibodies reveal them to have slow *in vivo* pharmacokinetics ($T_{1/2}\beta$ = 48 h in mice).[18] If AbOLs behave similarly *in vivo*, one of the potential advantages would be prolonged retention of the oligonucleotide in the circulation, leading to a sustained concentration during the course of treatment. We are currently measuring the pharmacokinetics and biodistribution of such conjugates.

IV. CONJUGATION TECHNIQUES AND CHOICE OF LINKER

Many diverse molecules have been conjugated to monoclonal antibodies with the ultimate aim of targeted therapy. These conjugates include a number of isotopes, cytotoxic agents, a range of plant, animal, or bacterial toxins, and more recently novel enzymes to activate prodrugs[19] or denature intracellular macromolecules such as DNA or RNA.[20]

Several classes of cytotoxic drugs have been investigated for the design of immunoconjugates including alkylating agents, antimetabolites, vinca alkaloids and their analogs, and, in particular, methotrexate.[4,21] Linkage to the antibody is obtained through endogenous reactive groups such as amino, hydroxyl, carboxyl, or sulfydryl groups which are not required for drug action.

Numerous chemical means have been described to couple toxins to antibodies.[22] A number of high-activity enzymes such as alkaline phosphatase and horse radish peroxidase have also been similarly conjugated for use in immunological assays.[23,24] The earliest attempts made use of homobifunctional reagents such as glutaraldehyde,[25] toluene diisocyanate,[26] or diethyl malonimidate[27] that cross linked free amino groups found on both proteins. This often resulted in a number of undesirable side products such as homopolymers and large protein aggregates.

The development of heterobifunctional cross linkers such as *m*-maleimidobenzoyl-*N*-hydroxysuccinimide ester (MBS), succinimidyl 4-(*N*-maleimidomethyl)cyclohexane-1-carboxylate (SMCC), and *N*-succinimidyl-3-(2-pyridyldithio)propionate (SPDP) has permitted the highly specific cross linking of many proteins by introducing different chemical functionalities onto each protein molecule. Most commercially available bifunctional cross linkers are insoluble in aqueous media owing to their nonpolar, organic nature, and have to be dissolved in water-miscible organic solvents for efficient cross linking. This may have undesirable effects on the protein. Water-soluble versions of these older cross linkers have been introduced, e.g., sulfo-MBS or sulfo-SMCC, where solubility is conferred by the presence of a sulfate group.

Specific cross-linking techniques generally involve two stages (Figure 1): firstly, a free thiol group is generated on one of the proteins, either by reductive cleavage of native cysteine residues or by chemical introduction of thiol groups. Reductive cleavage of disulfides is achieved under mild conditions with reagents such as dithiothreitol. The most commonly used reagents for the chemical introduction of sulfydryl groups are SPDP and 2-iminothiolane. The second stage involves modification of the second protein to introduce a chemical group which will react selectively with the free thiol groups on the first protein. Depending on the nature of this modification, the two proteins can then be coupled either via a cleavable disulfide (Figure 1) or a stable thioether linkage (Figure 2).

The same cross-linker technology has been applied to coupling of proteins with oligonucleotides, and there are now several publications detailing methods for creating conjugates which preserve the activity of each molecule.[28-33] Furthermore, the methodology for creating protein-oligonucleotide conjugates has been simplified to a kit form by Promega* (U.K.) using the heterobifunctional cross linker SPDP.[34,35] Generally, the oligonucleotide is derivatized with a reactive group, such as sulfydryl or amino group, at the 5′ end. To this is added the protein which has been derivatized by addition of the heterobifunctional cross linker. In our laboratory oligonucleotides synthesized with an amino group at the 5′ end and the heterobifunctional cross linker sulfo-MBS have been used to successfully create AbOLs (Figure 2). Depending on the reaction conditions, molar ratios of up to 10:1 (oligonucleotides:antibody) were achieved (unpublished results), although some loss of antibody affinity was noted at higher substitution ratios (unpublished results). Another novel technique for the preparation of AbOLs has been described by Kuijpers *et al.*[36] for a DNA-DNA hybridization approach to radioimmunotherapy.

---

*Promega, Southampton, U.K.

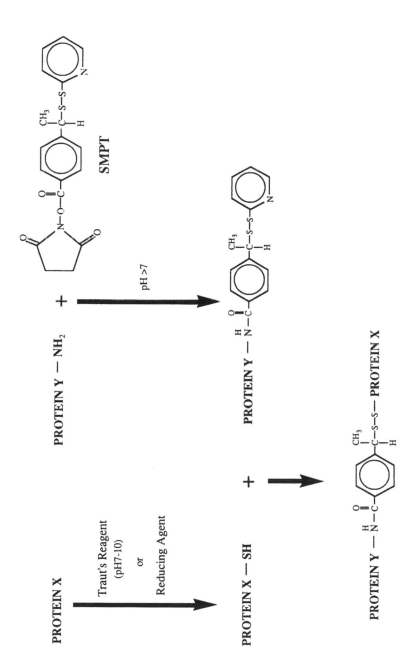

Figure 1 An example of protein-protein conjugation using 2-iminothiolane (Traut's reagent) and the heterobifunctional cross linker succinimidyloxycarbonyl-α-(2-pyridyldithio)-toluene (SMPT). A cleavable disulfide is introduced by use of this particular linker.

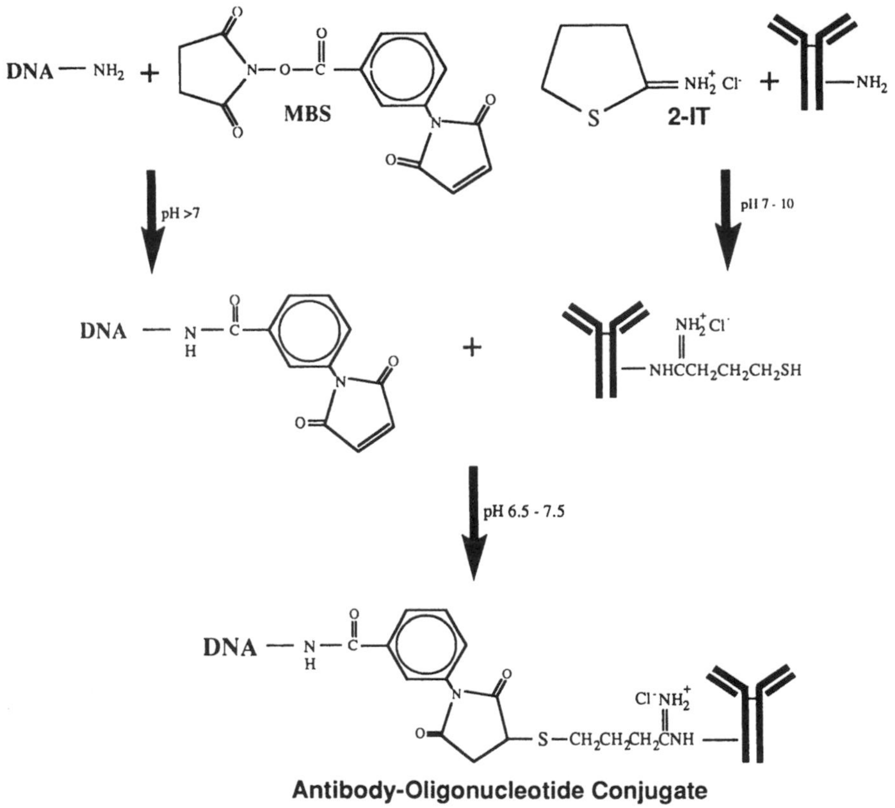

Antibody-Oligonucleotide Conjugate

Figure 2 Antibody-oligonucleotide conjugation using 2-iminothiolane and the heterobifunctional cross linker *m*-maleimidobenzoyl-*N*-hydroxysuccinimide ester (MBS). A stable thioether bond is created by the use of this linker.

The chosen antibody should be stable to the chemical reactions of modern linker chemistry, such that a number of oligonucleotides may be conjugated to a single antibody molecule without adversely affecting the integrity or binding capacity of any conjugate component. In the case of low M_r chemotherapeutic agents, extremely high substitution ratios have been achieved through the use of spacer molecules, such as albumin or various dextrans.[37] These carriers are firstly labeled with the cytotoxic agent before being conjugated to the antibody. Unfortunately, significantly reduced antibody affinity was reported with these carriers; therefore, they may not be useful for forming conjugates between monoclonal antibodies and higher M_r oligonucleotides. Site-specific linkage of the amino dextran spacer to carbohydrate groups on the Fc portion of the antibody molecule could result in very high substitution ratios without affecting the binding properties of the antibodies.[38,39]

A. ALTERNATIVE LINKERS

The formation of disulfide bridges through the use of reagents such as SPDP is a relatively slow process compared to thioether formation. Reagents such as MBS, however, form a strong thioether link between the protein and the oligonucleotide. These extremely stable links are not hydrolysed under *in vivo* conditions and hence will not release free oligonucleotide. Conceptually, the ideal linkers for these conjugates would release active oligonucleotide upon internalization, perhaps following a drop in pH in the endosomes.

The main drawback to the use of disulfides as cleavable cross links is that unless the accessible cysteine residue of the protein is utilized for the cross link,[40] cleavage of the cross link yields products which remain modified at one or more sites. A candidate, cleavable cross linker which does not contain disulfide bonds has recently been described.[41] This reagent, a maleimido derivative of 2-methylmaleic anhydride, is based on citaconic anhydride. It forms an acid-labile linkage and has so far been used to cross link proteins, but could also in theory be used to cross link oligonucleotides to antibody molecules. Above pH 7 the linkage is stable, but on mild acidification (pH 4 to 5) cleavage occurs to yield monomeric proteins, one of which is released in native form.

Another alternative type of cross-linking reagent which can be cleaved with low-energy UV light (365 nm) has been reported.[42] Such a linker could be useful for release of oligonucleotide at sites accessible to light.

V. STRATEGIES FOR THERAPY USING ANTIBODY DELIVERY SYSTEMS

Studies with radiolabeled monoclonal antibodies have revealed much about the *in vivo* behavior and pharmacokinetics of these molecules. When the radiolabeled antibody is injected a high plasma concentration is maintained due to its slow pharmacokinetics. This has the advantage of increased antibody binding to tumor but the disadvantage of significant irradiation of normal tissues and dose-limiting toxicity to the bone marrow. Conventional cytotoxic drugs such as doxorubicin, etc., have been conjugated to antitumor antibodies in the expectation that selective toxicity can be achieved. In these studies, specific *in vitro* toxicity was observed, but these data did not translate well to the *in vivo* situation where only slightly increased specificity was seen.[43] For tumors which are anatomically confined to a body region, intraregional injection has shown some success in antibody-based therapies and may have an important role in the treatment of locoregional diseases.[44] If AbOLs show similar behavior and hence have similar limitations, particular tumor models could allow alternative routes of delivery, e.g., intralymphatic or intraperitoneal.

Direct intratumoral administration may also be of value, since antibodies diffuse slowly through antigen-positive tissue, so that a high concentration of antibody can be provided for extended periods.[45] This particular route could be applicable to AbOLs, avoiding clearance by the reticuloendothelial system, receptors on hepatocytes, or other sites of metabolism.

VI. EVALUATION AND PERSPECTIVE

To be effective drugs, AbOLs will require a number of desirable features relating to the linker, the oligonucleotide, the antibody component, and the target antigen. In the development of AbOLs much can be learned from the experiences of investigators who have made similar conjugates of monoclonal antibodies with chemotherapeutic drugs and various toxins. AbOLs have a number of advantages over targeting these toxic agents in that oligonucleotides are relatively nontoxic and must be administered in high concentrations in order to produce an effect, a rationale supporting the use of immunoliposomes. The aim, therefore, is to increase toxicity by targeting to specific tissue. In the case of chemotherapeutic agents, the aim is often to reduce nonspecific toxicity by delivering the agent to the target tissue only.

Selection of target antigens and their corresponding monoclonal antibodies for development of effective AbOLs will be a multistep process. Experience with immunotoxins has taught us that simply because a monoclonal antibody binds to the surface of a cell does not mean it will make an effective immunotoxin. The importance of AbOL (or immunotoxin) internalization so that the oligonucleotide (or toxin) component can interact with its intracellular target has been discussed. Even then some immunotoxins that are efficiently internalized are ineffectual because they are directed to intracellular compartments from which the active toxin cannot reach the cytosol. Experience denotes that AbOLs can be expected to behave similarly.

The journey and potential barriers for an injected antibody required to bind tumor explain partially why such small amounts of the injected antibody (0.0007 to 0.01% of the injected dose per gram of tissue) targets to tumor.[46] After injection into the blood pool the antibody may bind to circulating antigen or with antigen present on blood cells that could evoke deleterious immune responses. The immunogenicity of murine monoclonal antibodies themselves is well documented and could preclude repeat administrations of the AbOL.[47] This is particularly important as oligonucleotides have transient effects. The linkage may also be disrupted, or the conjugate may have difficulty leaving the vasculature to enter the interstitial space. Antibody responses to haptens conjugated to monoclonal antibodies have also been reported.[48,49] In the case of AbOLs, an antioligonucleotide immune response could have serious consequences, and the potential for development of autoimmune responses must be fully investigated prior to clinical studies. Other host factors which will influence AbOL targeting include the location, size, antigen density, antigen accessibility, and cellular composition of the tumor. These drawbacks, coupled with factors such as tumor vascularization and vascular permeability, may lead one to consider combining AbOL treatments with therapies that do not require the targeting of all cells in the disease tissue.

Animal models will provide a comprehensive understanding of the pharmacokinetics and biodistribution of these immunoconjugates as well as information on the toxic side effects and optimal route of administration.

In the long term the problems of conjugation for large-scale production of these novel drugs as well as expense will be major factors in determining, if successful, whether or not this approach can be further developed. These problems of consistency in conjugation may be overcome somewhat if protein nucleic acids technology[50] continues to develop, when currently available protein cross linkers could be more easily incorporated.

The field of antisense and antigene oligonucleotides is still in its infancy, but the extraordinary potential for rational drug design is extremely attractive. Conjugation of the oligonucleotide to an antitumor monoclonal antibody is only one of a number of approaches which are currently under investigation for improving the therapeutic potential of these compounds.

REFERENCES

1. Carter, G. and Lemoine, N. R., Antisense technology for cancer therapy: does it make sense?, *Br. J. Cancer*, 67, 869, 1993.
2. Uhlmann, E. and Peyman, A., Antisense oligonucleotides: a new therapeutic principle, *Chem. Rev.*, 90, 543, 1990.
3. Byers, V. S. and Baldwin, R. W., Therapeutic strategies with monoclonal antibodies and immunoconjugates, *Immunology*, 65, 329, 1988.
4. Reisfeld, R. A., Yang, H. M., Muller, B., Wargalla, U. C., Schrappe, M., and Wrasidlo, W., Promises, problems, and prospects of monoclonal antibody-drug conjugates for cancer therapy, *Antibody Immunoconjugates Radiopharm.*, 2, 217, 1989.
5. Spooner, R. A. and Lord, J. M., Immunoroxins: status and prospects, *Trends Biotechnol.*, 8, 189, 1990.
6. Pirker, R., Fitzgerald, D. J. P., Willingham, M. C., and Pastan, I., Immunotoxins and endocytosis, *Lymphokines*, 14, 361, 1987.

7. Willingham, M. C. and Pastan, I., Ultrastructural immunocytochemical localization of the transferrin receptor using a monoclonal antibody in human KB cells, *J. Histochem. Cytochem.*, 33, 59, 1985.

8. Pastan, I., Willingham, M. C., and FitzGerald, D. J., Immunotoxins, *Cell*, 47, 641, 1986.

9. Sutherland, R., Delia, D., Schneider, C., Newman, R., Kemshead, J., and Greaves, M., Ubiquitous cell-surface glycoprotein on tumour cells is proliferation-associated receptor for transferrin, *Proc. Natl. Acad. Sci. U.S.A.*, 78, 4515, 1981.

10. Gatter, K., Brown, G., Trowbridge, I. S., Woolston, R., and Mason, D. Y., Transferrin receptors in human tissues: their distribution and possible clinical relevance, *J. Clin. Pathol.*, 36, 539, 1983.

11. Griffin, T. W., Haynes, L. R., and Demartino, J. A., Selective cytotoxicity of a ricin A-chain-anti-carcinoembryonic antigen antibody conjugate for a human adenocarcinoma, *J. Natl. Cancer Inst.*, 69, 799, 1982.

12. Tsukazazi, K., Hayman, E. G., and Ruoslahti, E., Effects of ricin A chain conjugates of monoclonal antibodies to human α-fetoprotein and placental alkaline phosphatase on antigen producing cells in culture, *Cancer Res.*, 45, 1834, 1985.

13. Makiya, R. and Stigbrand, T., Placental alkaline phosphatase has a binding site for the human immunoglobulin-G Fc portion, *Eur. J. Biochem.*, 205, 341, 1992.

14. Makiya, R. and Stigbrand, T., Placental alkaline phosphatase is related to human IgG internalization in HEp2 cells, *Biochem. Biophys. Res. Commun.*, 182, 624, 1992.

15. Iversen, P., *In vivo* studies with phosphorothioate oligonucleotides: pharmacokinetic prologue, *Anti-Cancer Drug Design*, 6, 531, 1991.

16. Agrawal, S., Temsamani, J., and Tang, J. Y., Pharmacokinetics, biodistribution and stability of oligodeoxynucleotide phosphorothioates in mice, *Proc. Natl. Acad. Sci. U.S.A.*, 88, 7595, 1992.

17. Emlen, W. and Mannik, M., Effect of size and strandedness on the *in vivo* clearance and organ localization of DNA, *Clin. Exp. Immunol.*, 56, 185, 1984.

18. Rowlinson, G., Paganelli, G., Snook, D., and Epenetos, A. A., Radiolocalisation of an anti-CEA monoclonal antibody (FO23C5) and its fragments in a colon carcinoma xenograft model, *Int. J. Biol. Markers*, 3, 259, 1988.

19. Senter, P. D., Wallace, P. M., Svensson, H. P., Vrudhula, V. M., Kerr, D. E., Hellström, I., and Hellström, K. E., Generation of cytotoxic agents by targeted enzymes, *Bioconjugate Chem.*, 4, 3, 1993.

20. Deonarain, M. P. and Epenetos, A. A., Targeting phosphodiesterases as a strategy for killing tumour cells, presented at The Tenth International Hammersmith Conference, Advances in the Applications of Monoclonal Antibodies in Clinical Oncology, Cyprus, May 3 to 5, 1993, 35.

21. Byers, V. S. and Baldwin, R. W., Therapeutic stratagies with monoclonal antibodies and immunoconjugates, *Immunology*, 65, 329, 1988.

22. Thorpe, P. E. and Ross, W. C. J., The preparation and cytotoxic properties of antibody-toxin conjugates, *Immunol. Rev.*, 62, 119, 1982.

23. Ishikawa, E., Imagawa, M., Hashida, S., Yoshitake, S., Hamaguchi, Y., and Ueno, T., Enzyme-labeling of antibodies and their fragments for enzyme immunoassay and immunohistochemical staining, *J. Immunoassay*, 4, 209, 1983.

24. Mahan, D. G., Morrison, L., Watson, L., and Haugneland, L. S., Phase change enzyme immunoassay, *Anal. Biochem.*, 162, 163, 1987.

25. Moolten, F. L., Capparell, N. J., Zajdel, S. H., and Cooperband, S. R., Antitumour effects of antibody-diphtheria toxin conjugates. II. Immunotherapy with conjugates directed against tumour antigens induced by simian virus 40, *J. Natl. Cancer Inst.*, 55, 473, 1975.

26. Moolten, F. L. and Cooperband, S. R., Selective destruction of target cells by diphtheria toxin conjugate to antibody directed against antigens on the cells, *Science*, 169, 68, 1970.

27. Philpott, G. W., Bower, R. J., and Parker, C. W., Improved selective cytotoxicity with an antibody-diphtheria toxin conjugate, *Surgery*, 73, 928, 1973.

28. Jablonski, E., Moomaw, E. W., Tullis, R. H., and Ruth, J. L., Preparation of oligodeoxynucleotide-alkaline phosphatase conjugates and their use as hybridization probes, *Nucleic Acids Res.*, 14, 6115, 1986.

29. Farmar, J. G. and Castaneda, M., An improved preparation and purification of oligonucleotide-alkaline phosphatase conjugates, *Biotechniques*, 11, 588, 1991.

30. Ghosh, S. S., Kao, P. M., and Kwoh, D. Y., Synthesis of 5'-oligonucleotide hydrazide derivatives and their use in preparation of enzyme-nucleic acid hybridization probes, *Anal. Biochem.*, 178, 43, 1989.

31. Ghosh, S. S., Kao, P. M., McCue, A. W., and Chappelle, H. L., Use of maleimide-thiol coupling chemistry for efficient syntheses of oligonucleotide-enzyme conjugate hybridization probes, *Bioconjugate Chem.*, 1, 71, 1990.

32. Li, P., Medon, P. P., Skingle, D. C., Lanser, J. A., and Symons, R. H., Enzyme-linked synthetic oligonucleotide probes: non-radioactive detection of enterotoxigenic *Escherichia coli* in faecal specimens, *Nucleic Acids Res.*, 16, 5275, 1987.

33. Urdea, M. S., Warner, B. D., Running, J. A., Stempien, M., Clyne, J., and Horn, T., A comparison of non-radioisotopic hybridization assay methods using fluorescent, chemiluminescent and enzyme labeled synthetic oligodeoxynucleotide probes, *Nucleic Acids Res.*, 16, 4937, 1988.

34. Cate, R. L., Ehrenfels, C. W., Wysk, M., Tizard, R., Voyta, J. C., Murphy, O. J., III, and Bronstein, I., Genomic southern analysis with alkaline-phosphatase-conjugated oligonucleotide probes and the chemiluminescent substrate AMPD, *Genet. Anal. Tech. Appl.*, 8, 102, 1991.

35. Schaap, A. P., Akhavan, H., and Romano, L. J., Chemi-luminescent substrates for alkaline-phosphatase — application to ultrasensitive enzyme-linked immunoassays and DNA probes, *Clin. Chem.*, 35, 1863, 1989.

36. Kuijpers, W. H. A., Bos, E. S., Kaspersen, F. M., Veeneman, G. H., and Van Boeckel, C. A. A., Specific recognition of antibody-oligonucleotide conjugates by radiolabeled antisense nucleotides: a novel approach for two-step radioimmunotherapy of cancer, *Bioconjugate Chem.*, 4, 94, 1993.

37. Baldwin, R. W. and Byers, V. S., Monoclonal antibodies and immunoconjugates for cancer treatment, *Cancer Chemotherapy and Biological Response Modifiers*, 409, 1987.

38. Shih, L. B., Xuan, H., and Sharkey, R. M., A fluorouridine-anti-CEA immunoconjugate is therapeutically effective in a human colonic cancer xenograft model, *Int. J. Cancer*, 46, 1101, 1990.

39. Shih, L. B., Goldenberg, D. M., Xuan, H., Lu, H., Sharkey, R. M., and Hall, T. C., Anthracycline immunoconjugates prepared by a site-specific linkage via an amino-dextran intermediate carrier, *Cancer Res.*, 51, 4192, 1991.

40. Masuho, Y., Kishida, K., Saito, M., Umemoto, N., and Hara, T., Importance of the antigen-binding valency and the nature of the cross-linking bond in ricin A-chain conjugates with antibody, *J. Biochem. (Tokyo)*, 91, 1583, 1982.

41. Blättler, W. A., Kuenzi, B. S., Lambert, J. M., and Senter, P. D., New heterobifunctional protein cross-linking reagent that forms an acid-labile link, *Biochemistry*, 24, 1517, 1985.

42. Senter, P. D., Tansey, M. J., Lambert, J. M., and Blättler, W. A., Novel photocleavable protein crosslinking reagents and their use in the preparation of antibody-toxin conjugates, *Photochem. Photobiol.*, 42, 231, 1985.

43. Greenfield, R. S., Daues, A., Edson, M. A., Gawlak, S., Fitzgerald-Kadow, K., Willner, D., and Braslawsky, G. R., Optimization of immunotherapy with adriamycin(hydrazone)-immunoconjugates in human B-lymphoma xenografts., *Antibody, Immunoconjugates, Radiopharm.*, 4, 107, 1991.
44. Hird, V., Maraveyas, A., Snook, D., Dhokia, B., Souter, W. P., Meares, C., Stewart, J. S. W., Mason, P., Lambert, H. E., and Epenetos, A. A., Adjuvant therapy of ovarian cancer with radioactive monoclonal antibody, *Br. J. Cancer*, 68, 403, 1993.
45. Riva, P., Arista, A., Sturiale, C., Moscatelli, G., Tison, V., Mariani, M., Seccamani, E., Lazzari, S., Fagioli, L., Franceschi, G., Sarti, G., Riva, N., Natali, P. G., Zardi, L., and Scassellati, G. A., Treatment of intracranial human glioblastoma by direct intratumoral administration of ^{131}I-labelled anti-tenascin monoclonal antibody BC-2, *Int. J. Cancer*, 51, 7, 1992.
46. Epenetos, A. A., Snook, D., Durbin, H., Johnson, P. M., and Taylor-Papadimitriou, J., Limitations of radiolabeled monoclonal antibodies for localization of human neoplasms., *Cancer Res.*, 40, 2979, 1986.
47. Courtenay-Luck, N. S., Epenetos, A. A., Moore, R., Larche, M., Pectasides, D., Dhokia, B., and Ritter, M. A., Development of primary and secondary immune responses to mouse monoclonal antibodies used in the diagnosis and therapy of malignant neoplasms, *Cancer Res.*, 46, 6489, 1986.
48. Reardan, D. T., Meares, C. F., Goodwin, D. A., McTigue, M., David, G. S., Stone, M. R., Leung, J. P., Bartholomew, R. M., and Frincke, J. M., Antibodies against metal chelates, *Nature*, 316, 265, 1985.
49. Kosmas, C., Snook, D., Gooden, C. S., Courtenay-Luck, N. S., McCall, M. J., Meares, C. F., and Epenetos, A. A., Development of humoral immune responses against a macrocyclic chelating agent (DOTA) in cancer patients receiving radioimmunoconjugates for imaging and therapy, *Cancer Res.*, 52, 904, 1992.
50. Nielsen, P. E., Egholm, M., Berg, R. H., and Buchardt, O., Peptide nucleic acids (PNAs): potential sense and anti-gene agents, *Anti-Cancer Drug Design*, 8, 53, 1993.

INDEX

Lightning Source UK Ltd.
Milton Keynes UK
UKHW022042220421
382475UK00003B/41

9 781138 558571